VOLUME ONE HUNDRED AND EIGHTEEN

CURRENT TOPICS IN DEVELOPMENTAL BIOLOGY

Hematopoiesis

CURRENT TOPICS IN DEVELOPMENTAL BIOLOGY

"A meeting-ground for critical review and discussion of developmental processes"
A.A. Moscona and Alberto Monroy (Volume 1, 1966)

VOLUME ONE HUNDRED AND EIGHTEEN

CURRENT TOPICS IN DEVELOPMENTAL BIOLOGY

Hematopoiesis

Edited by

EMERY H. BRESNICK
University of Wisconsin School of Medicine and Public Health, UW-Carbone Cancer Center; and UW-Madison Blood Research Program, Madison, WI, United States

ELSEVIER

AMSTERDAM • BOSTON • HEIDELBERG • LONDON
NEW YORK • OXFORD • PARIS • SAN DIEGO
SAN FRANCISCO • SINGAPORE • SYDNEY • TOKYO
Academic Press is an imprint of Elsevier

Academic Press is an imprint of Elsevier
50 Hampshire Street, 5th Floor, Cambridge, MA 02139, USA
525 B Street, Suite 1800, San Diego, CA 92101-4495, USA
The Boulevard, Langford Lane, Kidlington, Oxford OX5 1GB, UK
125 London Wall, London, EC2Y 5AS, UK

First edition 2016

Notices

Knowledge and best practice in this field are constantly changing. As new research and experience broaden our understanding, changes in research methods, professional practices, or medical treatment may become necessary.

Practitioners and researchers must always rely on their own experience and knowledge in evaluating and using any information, methods, compounds, or experiments described herein. In using such information or methods they should be mindful of their own safety and the safety of others, including parties for whom they have a professional responsibility.

To the fullest extent of the law, neither the Publisher nor the authors, contributors, or editors, assume any liability for any injury and/or damage to persons or property as a matter of products liability, negligence or otherwise, or from any use or operation of any methods, products, instructions, or ideas contained in the material herein.

ISBN: 978-0-12-803319-7
ISSN: 0070-2153

Publisher and Acquisition Editor: Zoe Kruze
Editorial Project Manager: Sarah Lay
Production Project Manager: Radhakrishnan Lakshmanan
Cover Designer: Mark Rogers

Typeset by SPi Global, India

CONTENTS

CONTRIBUTORS

E.H. Bresnick
UW-Carbone Cancer Center, University of Wisconsin School of Medicine and Public Health; UW-Madison Blood Research Program, Madison, WI, United States

K. Choi
Washington University, School of Medicine, St. Louis, MO, United States

E. de Pater
Erasmus MC, Rotterdam, The Netherlands

A.W. DeVilbiss
UW-Carbone Cancer Center, University of Wisconsin School of Medicine and Public Health; UW-Madison Blood Research Program, Madison, WI, United States

E. Dzierzak
Erasmus MC, Rotterdam, The Netherlands; MRC Centre for Inflammation Research and MRC Centre for Regenerative Medicine, University of Edinburgh, Edinburgh, United Kingdom

X. Gao
UW-Carbone Cancer Center, University of Wisconsin School of Medicine and Public Health; UW-Madison Blood Research Program, Madison, WI, United States

M.A. Goodell
Stem Cells and Regenerative Medicine Center, Baylor College of Medicine, Houston, TX, United States

K.J. Hewitt
UW-Carbone Cancer Center, University of Wisconsin School of Medicine and Public Health; UW-Madison Blood Research Program, Madison, WI, United States

T. Hoang
Laboratory of Hematopoiesis and Leukemia, Institute of Research in Immunology and Cancer (IRIC), University of Montreal, Montreal, QC, Canada

M. Jeong
Stem Cells and Regenerative Medicine Center, Baylor College of Medicine, Houston, TX, United States

K.D. Johnson
UW-Carbone Cancer Center, University of Wisconsin School of Medicine and Public Health; UW-Madison Blood Research Program, Madison, WI, United States

K.R. Katsumura
UW-Carbone Cancer Center, University of Wisconsin School of Medicine and Public Health; UW-Madison Blood Research Program, Madison, WI, United States

S. Keles
University of Wisconsin School of Medicine and Public Health, Madison, WI, United States

J.A. Lambert
Laboratory of Hematopoiesis and Leukemia, Institute of Research in Immunology and Cancer (IRIC), University of Montreal, Montreal, QC, Canada

M.W. Maijenburg
University of Pennsylvania, Philadelphia, PA, United States

R. Martin
Laboratory of Hematopoiesis and Leukemia, Institute of Research in Immunology and Cancer (IRIC), University of Montreal, Montreal, QC, Canada

S.C. McIver
UW-Carbone Cancer Center, University of Wisconsin School of Medicine and Public Health; UW-Madison Blood Research Program, Madison, WI, United States

D.T. Scadden
Massachusetts General Hospital, Boston; Harvard Stem Cell Institute; Harvard University, Cambridge, MA, United States

N.A. Speck
University of Pennsylvania, Philadelphia, PA, United States

S. Sumanas
Cincinnati Children's Hospital Medical Center, Cincinnati, OH, United States

N. Tanimura
UW-Carbone Cancer Center, University of Wisconsin School of Medicine and Public Health; UW-Madison Blood Research Program, Madison, WI, United States

J. Tober
University of Pennsylvania, Philadelphia, PA, United States

V.W.C. Yu
Massachusetts General Hospital, Boston; Harvard Stem Cell Institute; Harvard University, Cambridge, MA, United States

PREFACE

The process of developing the diverse blood cell repertoire from stem and progenitor cells termed hematopoiesis has been subject to considerable investigation. However, key steps in the complex process of hematopoiesis, including hematopoietic stem cell generation during embryogenesis, hematopoietic stem, and progenitor cell expansion to accommodate physiological and pathological requirements, and mechanisms that ensure hematopoietic stem and progenitor cell phenotypic integrity remain incompletely understood. Elucidating these mechanisms will yield concepts that can be extrapolated to diverse biological realms, and given the vital importance of the hematopoietic system, future efforts will invariably lead to medical breakthroughs that advance human health.

This issue of *Current Topics in Developmental Biology* reviews cutting-edge research on mechanistic and biological aspects of hematopoiesis. The initial chapters lay the groundwork for understanding the cellular origin of hematopoietic stem cells (Dzierzak and de Pater) and the pivotal role of the microenvironment or "niche" in non-cell-autonomously regulating hematopoietic stem cell genesis and function (Yu and Scadden). The next series of chapters probe into molecular mechanisms governing the development of hematopoietic stem and progenitor cells and for maintaining their unique phenotypes. Whereas a plethora of studies have used loss-of-function approaches to delineate proteins with activities essential for hematopoiesis, Hewitt et al. describe a new strategy aimed at discovering noncoding DNA sequences or *cis*-regulatory elements with nonredundant activities to control hematopoiesis. Editing specific *cis*-regulatory elements out of the mouse genome disrupts specific steps of hematopoiesis, yields instructive animal models, and provides unique systems for decoding the global *cis*-regulatory element network governing hematopoietic stem/progenitor cell transitions termed a "Hematopoietic Stem/Progenitor Cell Cistrome." Upstream of GATA-2, the generation and identity of endothelial cells with the unique capacity to form hematopoietic stem cells (hemogenic endothelial cells) requires an intriguing ETS transcription factor termed ETV2/ER71/Etsrp. Sumanas and Choi describe foundational studies on this critical protein, provide a unique perspective on how ETV2 functionally interfaces with a cohort of other ETS factors, and highlight the promise of forging ETV2-dependent regenerative medicine strategies. Having emerged from studies on mechanisms underlying leukemogenesis, the transcription factor Runx1, previously deemed Acute Myeloid Leukemia protein 1, has multiple crucial roles in the hematopoietic hierarchy. Tober et al. describe its activities to control hematopoietic stem and progenitor cell transitions and its relationship to a host of other components implicated in specific steps in hematopoiesis. A huge challenge is to assemble and decipher the complex protein networks governing hematopoiesis, and the basic helix-loop-helix protein Stem Cell Leukemia (SCL)/T-cell Acute Lymphocytic Leukemia Protein 1 (TAL1) is integral to such networks. SCL/TAL1 commonly colocalizes with GATA factors on chromatin and functions at multiple levels of the hematopoietic hierarchy. Hoang et al. provide a comprehensive perspective of how SCL/TAL1 controls normal hematopoiesis and is deregulated in leukemia. Moreover, they highlight how the mechanistic insights have illuminated new avenues to reprogram cellular phenotypes. Transcription factors commonly engage non-DNA binding coregulators that mediate activation or repression through enzymatic or nonenzymatic biochemical mechanisms. In certain cases, the same coregulator mediates activation and repression, often through elusive context-dependent mechanisms. DeVilbiss et al. describe concepts related to how GATA-1 negotiates coregulator ensembles to control the development and function of specific blood cell types. They highlight the concept of a "coregulator matrix," which, in effect, constitutes a code for how a given transcription factor functions in a specific environment—a code that exhibits fluidity dictated by the overall regulatory milieu. Progress on the role of noncoding regulatory RNAs in biological regulation has exploded in recent years, and this area is ripe for making discoveries to understand and modulate hematopoiesis. Jeong and Goodell provide an insightful overview of noncoding RNAs implicated in hematopoiesis and discuss prospective mechanisms. Noncoding RNA-dependent mechanisms are expected to function at all levels of the hematopoietic hierarchy and to converge upon the mechanisms involving ETV2, GATA-2, RUNX1, SCL/TAL1, and GATA-1 discussed in this volume. For the most part, integrating noncoding regulatory RNA mechanisms with the emerging factor-dependent pathways remains virgin territory.

In aggregate, this series of chapters highlights compelling mechanistic and biological advances, and the progress to date has emphasized the need to increase the depth of investigation into these problems—to get beyond limitations of dissecting complex mechanisms in cell populations, rather than in individual cells, to develop definitive insights as to whether murine mechanisms can be seamlessly extrapolated to humans, and to devise novel strategies for dissecting mechanisms within microenvironments impacted by diverse non-cell-autonomous regulatory inputs. It will be particularly instructive to determine how the respective mechanisms react to dynamic alterations in the microenvironment and systemic changes associated with pathologies and aging. The resulting knowledge will yield a veritable Pandora's box of scholarly insights that fill mechanistic/biological voids and transform existing concepts. Future mechanistic leveraging will yield desperately needed new strategies for treating nonmalignant (eg, anemia) and malignant hematopoietic disorders (eg, leukemia) and has potential to yield economically viable pipelines for engineering therapeutic blood products.

EMERY H. BRESNICK

CHAPTER ONE

Regulation of Blood Stem Cell Development

E. Dzierzak*[,†,1], E. de Pater*

*Erasmus MC, Rotterdam, The Netherlands

†MRC Centre for Inflammation Research and MRC Centre for Regenerative Medicine, University of Edinburgh, Edinburgh, United Kingdom

[1]Corresponding author: e-mail address: elaine.dzierzak@ed.ac.uk

Contents

Abstract

Understanding how the blood system is formed is an ongoing fundamental research challenge. Developmental biology has provided many insights into the molecules and processes that affect the formation of the blood tissues, both in health and disease. It is of particular interest for clinical transplantation therapies to understand how hematopoietic stem cells (HSCs)—the self-renewing purveyors of the adult blood system that produce over 10 different functionally specialized cell lineages and over 10^{11} cells daily—are generated during embryonic stages. Recent successes to reprogram the fate of adult differentiated cells to pluripotency and to other cell lineages now highlight the importance of identifying the cells and molecules that affect the *in vivo* developmental initiation of rare and robust transplantable HSCs. The close association of the developing hematopoietic and vascular system, hematopoietic cell mobility through the circulation, and the essential role of the embryonic hematopoietic system in adult hematopoietic cell development make this a formidable study. This chapter reviews the advances, controversies, and current state of our knowledge of the growing field of hematopoietic development, with a special focus on the regulation of the natural transdifferentiation of endothelial cells to HSCs within the developing embryo.

Current Topics in Developmental Biology, Volume 118
ISSN 0070-2153
http://dx.doi.org/10.1016/bs.ctdb.2016.01.001

1. INTRODUCTION

The dramatic and unexpected demonstration by Yamanaka and colleagues that fully differentiated skin cells could be reprogrammed to pluripotency using pivotal transcription factors changed a major paradigm in mammalian developmental biology—that of the fixed cell identity (Takahashi & Yamanaka, 2006). Since this major breakthrough, many researchers have been able to reprogram differentiated cells to lineage-specific stem or progenitor cells (reviewed by Ieda, 2013). Additional studies with induced pluripotent stem (iPS) cells or embryonic stem (ES) cells demonstrate controlled differentiation toward several adult lineages, such as neuronal, muscle, pancreatic cells, and even hematopoietic cells of the lymphoid, myeloid, and erythroid lineages (reviewed in Ohnuki & Takahashi, 2015). However, to program or reprogram cells toward a hematopoietic stem cell (HSC) has been extremely difficult and daunting challenge.

The gold-standard functional definition of a robust HSC is one that is based on its potential for long-lasting, high-level, multilineage hematopoietic repopulation in cell transplantation therapies. Clinical transplantation of HSC-containing adult bone marrow cells was the first successful "cell therapy" in humans. Beginning in the 1950s, this approach is currently part of the therapy used internationally for the treatment of leukemias and severe hematopoietic deficiencies (Takizawa, Schanz, et al., 2011). The potent adult HSC is the only cell type capable of sustained self-renewal and production of all types of hematopoietic cells (Yamamoto, Morita, et al., 2013).

To understand the biology of HSCs and the principles behind the growth of HSCs and repopulation following transplantation, the mouse has served as an excellent model. Inbred mouse strains, genetic markers, immunophenotyping, and cell sorting, as well as the ease of genetic manipulation make this the ideal model to study the biology and development of HSCs. Experimental transplantations of cells into adult murine recipients (irradiated so as to myeloablate the existing hematopoietic system) has broadened our knowledge of HSCs and what constitutes their supportive microenvironment (Copelan, 2006). In the adult, HSCs are maintained in specialized niches of the bone marrow and also are found at lower frequencies in the spleen and peripheral blood. Other vertebrate animal models such as chick, *Xenopus*, and zebrafish have also been instrumental in bringing us to our current knowledge of HSC development (Dzierzak & Speck, 2008; Orkin & Zon, 2008).

Already before the generation of the HSCs that reside in the bone marrow of the adult, the embryo requires blood cells for oxygenation, metabolite distribution, and tissue remodeling. This primitive embryonic blood is generated in a wave of hematopoietic cell commitment occurring in mesodermal precursors migrating to the early yolk sac (reviewed in Palis, Malik, et al., 2010). Cells with vascular endothelial and hematopoietic (and other) potential (generally referred to as hemangioblasts) contribute to the yolk sac blood islands and the production of primitive red blood cells, the first visible hematopoietic cells in the mouse and human embryos. This primitive wave of hematopoietic generation is not long-lived, although the primitive macrophages it produces form the tissue resident macrophages of the adult (reviewed in Davies & Taylor, 2015; Frame, McGrath, et al., 2013). More complex hematopoietic cells are produced in definitive waves. Erythroid-myeloid progenitors (EMP) are the first definitive multilineage progenitors and are produced in the yolk sac (McGrath, Frame, et al., 2015). Thereafter, more complex progenitors are generated in the yolk sac, aorta-gonad-mesonephros (AGM) region (Fig. 1A) and major vasculature of the embryo, the placenta, and the vascularized regions of the embryonic head. These cells have potential for the lymphoid lineages in addition to erythro-myeloid lineages (Cumano, Dieterlen-Lievre, et al., 1996). Finally, at midgestation in the mouse and at the first trimester of human development, the first transplantable HSCs that will sustain the life-long hematopoiesis in the adult, are generated (Bertrand, Giroux, et al., 2005; de Bruijn, Speck, et al., 2000; Gekas, Dieterlen-Lievre, et al., 2005; Ivanovs, Rybtsov, et al., 2011; Li, Lan, et al., 2012; Medvinsky & Dzierzak, 1996; Medvinsky, Taoudi, et al., 2008; Ottersbach & Dzierzak, 2005; Tavian, Robin, et al., 2001).

The *in vivo* waves of primitive and definitive progenitor generation that developmentally precede the HSC generating wave complicate *in vitro* studies, such as those that attempt to generate HSCs from ES/iPS cells and other cell types. With the exception of primitive erythrocytes, the majority of the differentiated blood cells produced in the early waves are similar to those in the adult. However, they do not originate from HSCs, and as such will not contribute in the long-term to the self-renewable adult blood system. Also greatly complicating studies to produce HSCs in culture is the lack of an exclusive HSC marker that enables recognition of these cells when they are produced, and the fact that to date, HSCs from any source cannot be cultured efficiently *in vitro*. The *de novo* generation of HSCs occurs in the embryo during a very restricted window of time from embryonic day E10.5 to E12.5 in the mouse (Zovein, Hofmann, et al., 2008) and gestational

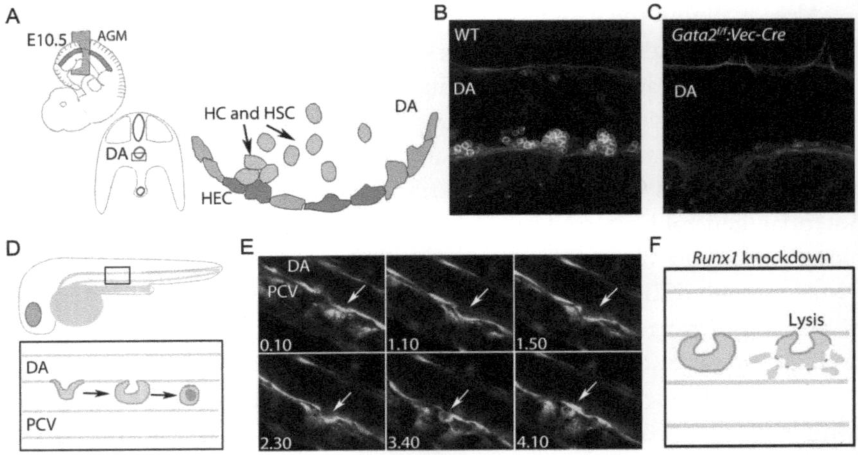

Fig. 1 The AGM, hematopoietic clusters and the process of endothelial-to-hematopoietic transition (EHT). (A) Graphical representation of EHT in the mouse. An E10.5 mouse embryo is depicted on the *left* where the *square* indicates the level of the cross-section shown in the *middle panel*. The *boxed* area indicates the ventral side of the dorsal aorta. The *right panel* shows where hemogenic endothelial cells in *purple* give rise to clusters of hematopoietic cells and HSCs indicated in *pink*. These shed into the lumen of the dorsal aorta and eventually populate the bone marrow. (B) and (C) *z* projection of a confocal image of (B) a wild-type (WT) mouse embryo and (C) a *Gata2*$^{f/f}$*:Vec-Cre* embryo. Endothelial cells are stained for CD31 (*red*) and hematopoietic cluster cells are stained for c-Kit (*green*). Note the size of WT clusters in (B) and the flat c-Kit positive cluster cells in (C), indicating that not only the number of clusters is severely reduced upon loss of Gata2 but also the process of EHT is affected. (D) Graphical representation of EHT in zebrafish. A schematic diagram of a Tg(*Fli:GFP*) zebrafish embryo (*top panel*) where the dorsal aorta (DA) and the posterior cardinal vein (PCV) are marked in *green* in a embryo. The *bottom panel* shows the *boxed area* where an endothelial cell bulges outward, rounds up to close the gap in the DA and sheds as a hematopoietic cell into the abluminal space between the DA and the PCV. (E) Single *z* scans of time-lapse confocal imaging of the process of EHT *in vivo* using a Tg(*Fli:GFP*) embryo. The DA and PCV are indicated in the first frame and the *arrow* indicates the cell that is undergoing EHT in each frame. The timing is indicated in the *bottom left corner* of each frame and the entire process of EHT is completed after 4 h. (F) Schematic representation of a cell undergoing EHT after knockdown of *Runx1* in the zebrafish embryo. EHT is greatly reduced, but the occasional cell undergoing EHT lyses halfway through the process. *E*, embryonic day; *DA*, dorsal aorta; *AGM*, aorta-gonad-mesonephros region; *HC*, hematopoietic cell; *HEC*, hemogenic endothelium; *HSC*, hematopoietic stem cell; *PCV*, posterior cardinal vein. (See the color plate.)

week 4–10 in the human (Ivanovs et al., 2011; Tavian et al., 2001). Therefore to understand which regulators are important for reprogramming and differentiating cells to HSCs, we first need to understand how HSCs are generated during embryonic development.

1.1 HSC Generation

The AGM is the first site of HSC generation in the mouse (Fig. 1A), identified through transplantations of dissected embryonic tissues into adult irradiated recipient mice (Medvinsky & Dzierzak, 1996; Muller, Medvinsky, et al., 1994). More refined dissections localized HSCs to the major vasculature—aorta, vitelline, and umbilical arteries at mouse embryonic day E10.5 (de Bruijn et al., 2000; North, Gu, et al., 1999), and more specifically to the ventral wall of the dorsal aorta (de Bruijn, Ma, et al., 2002; Taoudi & Medvinsky, 2007). Beginning at E11, HSCs are also found in other tissues—the yolk sac, placenta, and most recently the embryonic head (Gekas et al., 2005; Li et al., 2012; Medvinsky et al., 2008; Muller et al., 1994; Ottersbach & Dzierzak, 2005). Enrichment for HSCs in the AGM and these other tissues using antibodies (or other markers) that define the adult BM and/or fetal liver (FL) HSC in flow cytometric sorting, showed that long-term repopulating AGM HSCs are c-Kit^{+}CD34^{+} and Ly6a (Sca1)$^{+}$, but not CD150^{+} (de Bruijn et al., 2002; Ma, Robin, et al., 2002; McKinney-Freeman, Cahan, et al., 2012; Sanchez, Holmes, et al., 1996; Wood, May, et al., 1997). Further enrichment of AGM HSCs includes the use of hematopoietic cell markers CD41^{lo}, CD45, and Mac1 and markers typically defining endothelial cells, such as VE-cadherin, Flk1, CD31, and Tie2 (Li et al., 2012; Newman, Berndt, et al., 1990; Robin, Ottersbach, et al., 2011; Rybtsov, Sobiesiak, et al., 2011; Sanchez et al., 1996; Taoudi, Gonneau, et al., 2008; Yokomizo & Dzierzak, 2010). Knockin/knockout (North, de Bruijn, et al., 2002; North et al., 1999) and transgenic strategies (Ly6a (Sca1) GFP) (de Bruijn et al., 2002; Ma et al., 2002) identified other markers of HSCs in the AGM and in other tissues (Ottersbach & Dzierzak, 2005).

Immunostaining of embryo sections of the AGM region localized the expressing cells to the hematopoietic cluster cells and/or the endothelial cells lining the aorta (Fig. 1B). *In situ* hybridization studies examining hematopoietic transcription factor expression and transcription factor reporter knockin or transgenic reporter mice showed expression of *Runx1*, *SCL*, *Gata2*, *Gata3*, *Notch*, and other hematopoietic transcription factors in the cells of the AGM (de Bruijn et al., 2002; Fitch, Kimber, et al., 2012;

Marks-Bluth, Khanna, et al., 2015; North et al., 2002, 1999; Porcher, Swat, et al., 1996; Robert-Moreno, Espinosa, et al., 2005). Again, the expressing cells were localized in the hematopoietic cluster cells closely associated with aortic endothelium and/or within the ventral endothelial cell layer. Indeed, all amniote embryos examined to date have clusters of round hematopoietic-like cells associated with the aorta at the critical time in development when definitive hematopoiesis begins (reviewed in Jaffredo, Nottingham, et al., 2005). The developmental acquisition (c-Kit, CD45) and loss of other markers (Flk1), and the localization of the cells within the endothelium and the clusters gave further support for a fate change and differentiation to the hematopoietic lineage (Yokomizo & Dzierzak, 2010). These evidence add up to suggest that in the embryo, the vascular endothelial cells lining the major arteries possess hemogenic potential and generate hematopoietic progenitor and stem cells.

1.2 Tracking Hematopoietic Cell Generation from Embryonic Endothelium

The first experimental evidence linking hematopoietic cells in the clusters lining the dorsal aorta with aortic endothelial cells came from grafting and dye/retroviral marking studies in chick embryos (Jaffredo, Gautier, et al., 1998, 2000; Pardanaud & Dieterlen-Lievre, 1993). Until this time, microscopic observations over a period of about 100 years reported the close physical association of aortic hematopoietic cluster and endothelial cells (reviewed in Jaffredo et al., 2005). Jaffredo and colleagues were able to trace origins of the emergent hematopoietic cluster cells by the presence of markers that were used to exclusively label endothelial cells lining the aorta at a developmental time prior to hematopoietic cell appearance.

In the mouse, genetic lineage tracing studies with *Cre:Lox* recombination markers established the relationship between vascular endothelial cells (vascular endothelial cadherin (VEC)$^+$) with emerging hematopoietic cluster cells. Stable and temporally controlled recombination marking in *VEC-Cre:Rosa lox stop* reporter and *VEC-Cre ERT:Rosa lox stop* reporter mice, respectively, showed that the adult hematopoietic system and HSC arise from endothelial cells and that the cells with hemogenic potential are present in the embryonic vasculature from about E8 to E11 (Chen, Yokomizo, et al., 2009; Zovein et al., 2008). Moreover, when pivotal transcription factors such as *Runx1* or *Gata2* were deleted in VEC expressing embryonic cells (Fig. 1B), no vascular hematopoietic clusters and no HSCs were detected

(Chen et al., 2009; de Pater, Kaimakis, et al., 2013; Lim, Hosoya, et al., 2012; Ruiz-Herguido, Guiu, et al., 2012). Similar genetic marking studies, using another endothelial gene, *Tie2* as the *Cre* driver for reporter recombination, confirmed the precursor–progeny relationship of embryonic vascular endothelial cells with emerging hematopoietic cells (Liakhovitskaia, Gribi, et al., 2009; Schlaeger, Mikkola, et al., 2005; Tang, Harrington, et al., 2010). Additional support was provided from transgenic reporter mouse strategies. Enhancer elements that drive expression of transcription factors such as *Runx1 (+23)* in definitive HPSCs (hematopoietic stem and progenitor cells), facilitated not only the isolation of HPSC but also hemogenic endothelial cells that can be shown in OP9 cocultures to generate hematopoietic cells (Swiers, Baumann, et al., 2013).

Confocal microscopy and live-imaging facilitated the strongest proof of endothelial-to-hematopoietic transition (EHT). The physiologic evidence that EHT occurs *in vivo* in the major vasculature of the mouse embryo at the time of HSC generation was provided by Boisset and colleagues (Boisset, van Cappellen, et al., 2010). The time-lapse imaging, performed through thick sections of the E10.5 mouse AGM, visualized Ly6aGFP^{+}CD34^{+} aortic hemogenic endothelial cells transitioning to Ly6aGFP^{+}CD34^{+} hematopoietic cells that reside alone/in clusters on the luminal wall of the aorta. These cells coexpress additional markers, like c-Kit, that characterize functional AGM HSCs. Quantitation of EHT events revealed only two emerging hematopoietic cells per aorta, corresponding to the number of transplantable AGM HSCs at this developmental time point (Taylor, Taoudi, et al., 2010), thus indicating (along with/and verifying genetic data) that endothelial-to-HSC transition occurs *in vivo*. The process of EHT was also captured on video in zebrafish embryos (Bertrand, Chi, et al., 2010; Kissa & Herbomel, 2010). The zebrafish embryo shows round hematopoietic cells emerging from aortic endothelial cells in an abluminal direction, rather than in a luminal direction as in the mouse embryo (Fig. 1D and E). The emerged cells then migrate and enter into the circulation of the posterior cardinal vein. This was reported previously (Kissa, Murayama, et al., 2008) and counterintuitive at the time because in mouse and chick, clear luminal clusters were identified and it was assumed that the process would be similar in zebrafish. As in the mouse embryo, knockdown of *Runx1* severely affects the process of EHT (Fig. 1F). Taken together, these data confirm that the generation of hematopoietic progenitors and stem cells from endothelial cells is a normal *in vivo* physiologic process.

The latest refinement in understanding the birth of HSCs has been the identification of several populations of pre-HSCs and precursors. These studies showed that populations of endothelial cells harvested at E9.5 (before the generation of the first HSCs in mouse) are expressing VEC and $CD41^{lo}$, but not CD45 and CD43. Under specialized conditions in which AGM cells are reaggregated and cultured for several days, they transition to $CD43^{+}$ and $CD45^{+}$ HSCs that yield high-level hematopoietic chimerism following transplantation (Rybtsov, Batsivari, et al., 2014; Rybtsov et al., 2011; Taoudi et al., 2008). This *in vitro* system holds promise to reveal some of the relevant cell–cell interactions and molecular regulators of EHT.

1.3 Regulation of Hemogenic Endothelium

How are the specialized hemogenic endothelial cells that undergo EHT formed? What makes them different from other endothelial cells? Why does the ventral wall of the aorta in most species form hematopoietic clusters, while the dorsal wall does not? These issues are of great interest and several explanations are plausible, including differences in positional information that instructs the fate of the endothelial cells; differing mesodermal origins of the endothelial cells; and/or the effects of neighboring cells.

In chick embryos, both positional information and mesodermal origins play a role. Factors such as EGF and TGFa that emanate from dorsal tissues inhibit, whereas ventral tissue-emanating factors BMP4, FGF, VEGF, and TGFb1 promote hemogenic activity (Pardanaud & Dieterlen-Lievre, 1999). Chick embryo grafting experiments also reveal that different types of mesoderm (lateral plate and somitic) contribute to the ventral and dorsal walls, respectively (Pardanaud & Dieterlen-Lievre, 1993), and this was recently found in the zebrafish embryo (Pouget, Peterkin, et al., 2014). In addition, at the time of fusion of the paired dorsal aortae, the ventral endothelial cells express *Runx1* which marks the hemogenic endothelium. Upon hematopoietic differentiation, Notch signaling is downregulated and hematopoiesis is enhanced (Richard, Drevon, et al., 2013). As development proceeds, somitic cells begin to contribute to the ventral wall.

In the zebrafish embryo, the stage at which the divergence of endothelial and hemogenic endothelial cell fate occurs is as yet unknown. In the posterior lateral plate mesoderm, it is unclear whether endothelial- and hemogenic endothelial-cell fates are diverged yet, but several signaling molecules are implicated in the specification of hemogenic endothelial cells. In the lateral plate mesoderm, FGF signaling provided by the somite is required for the start of the hemogenic program in cells which can be later defined as hemogenic endothelial cells. Also Notch3 function in the somite is non-cell autonomously required for HSC specification (downstream of wnt16) (Kim, Melick, et al., 2014; Kobayashi, Kobayashi-Sun, et al., 2014; Lee, Manegold, et al., 2014). In *Xenopus*, hemogenic endothelial cells require short-range VEGFa, also provided by the somite (Leung, Ciau-Uitz, et al., 2013). From experiments in zebrafish (and chick) we know that hemogenic endothelial cells (and endothelial cells) migrate underneath the somite to the midline where they receive multiple signaling molecules. Hemogenic endothelial cells receive and require long-range VEGFa and Notch signaling. This is facilitated through junctional adhesion molecule 1a (Jam1a) in hemogenic endothelial cells interacting with Jam2a in the somite. This interaction facilitates the movement underneath the somite and allows the Notch signaling to take place (also received from the somite) (Kobayashi et al., 2014). Notch signaling is required to start the hemogenic program in endothelial cells. Once the dorsal aorta is formed, a ventral BMP signal further enhances the hemogenic program together with hedgehog (HH) signaling. All HPSCs require BMP and HH signaling at this time of their generation, and FGF signaling from the somite functions to restrict the BMP signal and the initiation of the hematopoietic program to the ventral cells of the dorsal aorta (Gering & Patient, 2005; Pouget et al., 2014; Wilkinson, Pouget, et al., 2009). The cells undergoing EHT are then shed into the area between the dorsal aorta and the posterior cardinal vein. For EHT to be accomplished, Notch signaling needs to be downregulated in hemogenic endothelial cells through activation of Gpr183 (Zhang, He, et al., 2015).

In the mouse embryo it is as yet unclear whether there are diverse mesodermal origins of aortic endothelial cells. However, it is thought that hemogenic fate is determined as early as E8 (Jaffredo et al., 2005; North et al., 2002). Not all endothelial cells of the dorsal aorta are capable of generating HSCs, and endothelial cells and hemogenic endothelial cells are intermixed in the dorsal aorta, as suggested by Ly6a-GFP expression at E10.5 (de Bruijn et al., 2002; Solaimani Kartalaei, Yamada-Inagawa, et al., 2015). Hemogenic endothelial cells are fully functioning endothelial cells and are defined by a different genetic program which sets them apart from endothelial cells (Solaimani Kartalaei et al., 2015; Swiers et al., 2013). Positional information is key to hemogenic and hematopoietic fate. Before the onset of HSC generation in the AGM, a coculture of this tissue with ventral gut tissue is capable of inducing HSC generation in the AGM region, while culture with dorsal tissue inhibits the generation of HSCs even after the onset of HSC generation (Peeters, Ottersbach, et al., 2009). BMP4 is a ventrally expressed factor and plays a role in HSC generation in the mouse AGM (Durand, Robin, et al., 2007). Only some endothelial cells and some hematopoietic cluster cells are BMP-activated (Crisan, Kartalaei, et al., 2015). Since all AGM HSCs are ventrally localized (Taoudi & Medvinsky, 2007), and all HSCs at the time of their *in vivo* generation are BMP-activated (Crisan et al., 2015), the ventral positional BMP4 signaling factor likely plays an important role in hemogenic endothelial cells and their specification toward HSCs. Notch1 and Jagged1 are also expressed in some ventral endothelial cells in the aorta (Guiu, Bergen, et al., 2014; Robert-Moreno et al., 2005; Robert-Moreno, Guiu, et al., 2008). Notch signaling is required for both endothelial and hematopoietic development, but only the Notch ligand Jagged1 is required for HSC generation. Also in mouse, Notch signaling is downregulated upon the formation of hematopoietic cells, but this is more transient (Richard et al., 2013) and Notch is again required for fetal hematopoiesis (Gerhardt, Pajcini, et al., 2014). Unlike BMP4, HH expression is not ventrally restricted. HH affects HSC generation in AGM explants most likely through its effects on niche cells (Peeters et al., 2009).

1.4 Hemogenic Transcriptional Program

Extrinsic molecular regulators (BMP, HH, and others) induce the hemogenic transcriptional program. The intrinsic hemogenic molecular program is hallmarked by the expression of pivotal transcription factors (Solaimani Kartalaei et al., 2015; Swiers et al., 2013), the "heptad" factors. The heptad transcription factor complex comprised of Gata2, Runx1, Scl/Tal1, Lmo2, Fli1, Erg, and Lyl1 (Wilson, Foster, et al., 2010) was identified by global screening of the regulatory elements of known hematopoietic genes in a hematopoietic progenitor cell line, HPC7. Most such hematopoietic gene enhancers contain consensus binding sites for, and were shown to bind, the heptad complex of factors. Hence, these factors are thought to work in harmony in hematopoietic progenitors and stem cells. Since all heptad factors are found to be upregulated during EHT, and the genetic deletion of each factor individually reveals pivotal roles in EHT and HSC generation, they are likely to act in combinatorial manner to promote EHT and hematopoietic progenitor/stem cell development.

Several of the heptad transcription factors belong to the *Ets* family of transcription factors (*Fli1*, *Scl/Tal1*, *Erg*, and *Lyl1*). The Ets family of factors have overlapping transcriptional targets, but due to differential expression they regulate distinct processes in HSC generation (Dooley, Davidson, et al., 2005; Patterson, Gering, et al., 2005; Robb, Drinkwater, et al., 1995; Shivdasani, Mayer, et al., 1995). Scl/Tal1 and Fli1 are important for endothelial formation (Bussmann, Bos, et al., 2010; Spyropoulos, Pharr, et al., 2000; Visvader, Fujiwara, et al., 1998), and Erg and Lyl1 play an important role in HSC maintenance (Capron, Lecluse, et al., 2006; Taoudi, Bee, et al., 2011). Scl and Fli1 start the hemogenic program and regulate expression of *Lmo2*, *Gata2*, and *Runx1*, where the program is further enhanced by Gata2 (Gao, Johnson, et al., 2013; Landry, Kinston, et al., 2008; Taoudi et al., 2011; Zhu, Traver, et al., 2005).

Several independent studies showing the transcriptional activation and complex formation between the individual members of the heptad support the combinatorial function of the heptad factors (Gering & Patient, 2005; Pimanda, Ottersbach, et al., 2007; Wilson, Foster, et al., 2010). The individual roles of each heptad factor in HSC generation and maintenance are beginning to be revealed through similar conditional knockout approaches. For example, Scl and Lyl1 are highly homologous Ets transcription factors with the same major downstream gene targets, but *Scl* and *Lyl1* knockouts show very different phenotypes that can only be partially explained by their expression patterns. In embryonic development, *Scl* expression starts earlier than *Lyl1* expression. However, the overexpression of *Lyl1* cannot rescue the hematopoietic differentiation defect seen in $Scl^{-/-}$ ESC (Chan, Follows, et al., 2007). In contrast, Scl and Lyl1 have overlapping functions in HSC maintenance. *Lyl1* can rescue loss of *Scl*, thus explaining the mild defect in $Scl^{-/-}$ HSCs (Souroullas, Salmon, et al., 2009). Also Erg and Fli1 have a dual role in the HSC and megakaryocytic lineage, since double heterozygous mice have a much more severe phenotype than the single heterozygous mutant mice (Kruse, Loughran, et al., 2009).

Gata2 and Runx1 are the most widely studied transcription factors in hematopoietic development. *Gata2* expression precedes *Runx1* expression in hematopoietic cell development. Mouse germline knockouts of *Gata2* die at E10.5, just before the onset of HSC generation (Tsai, Keller, et al., 1994). These $Gata2^{-/-}$ embryos die with FL anemia and $Gata2^{-/-}$ ES cells cannot contribute to definitive hematopoiesis in chimeric embryos/mice. Furthermore, in E10.5 $Gata2^{+/-}$ mouse embryos there is a 90% decrease in HSCs (Ling, Ottersbach, et al., 2004). This led to the conclusion that Gata2 is required for HSC generation. Since *Gata2* expression persists after generation of HSCs from hemogenic endothelium, it was postulated that it should function thereafter. A conditional knockout approach showed that

Gata2 functions not only in the generation of HSCs from hemogenic endothelium but also after the generation of HSC for their maintenance (de Pater et al., 2013; Gao et al., 2013). The importance of Gata2 was also shown in the zebrafish model, where *gata2b* plays an important role in EHT (Butko, Distel, et al., 2015).

Mouse germline knockouts of *Runx1* die at E12.5 with FL anemia and vascular hemorrhaging (Mukouyama, Chiba, et al., 2000; Okuda, van Deursen, et al., 1996). *Runx1* expression in the E10.5 embryo is localized to the dorsal aorta, and Runx1 expression is used as a marker for hemogenic endothelial cells and HSCs (North et al., 2002; Swiers et al., 2013). All AGM HSCs express Runx1 and *Runx1*$^{+/-}$ AGMs contain only 50% of the wild-type numbers of HSCs. Runx1 expression persists after the generation of HSCs (Chen et al., 2009). However, unlike Gata2, the Runx1 requirement is much more restricted to the generation of HSCs, although the exact timing of the Runx1 requirement is still under debate (Liakhovitskaia, Rybtsov, et al., 2014; Tober, Yzaguirre, et al., 2013). Also in zebrafish, *Runx1* is required for EHT. Upon *Runx1* knockdown, EHT is very rare and the cells which try to undergo EHT eventually lyse (Fig. 1F; Kissa & Herbomel, 2010).

The difference in the phenotypes in the conditional knockout studies of *Runx1*, *Gata2*, and the other factors shows that we do not fully understand the role of each of the heptad factors in HSC generation, collectively and individually, and leaves much to explore.

1.5 Reprogramming

The reprogramming of cells toward an HSC fate would open new doors to future regenerative therapy strategies and provide insight into the regulatory mechanisms of HSC generation and even leukemia. Findings in the field of developmental biology of HSC generation have been the guide to reprogram cells toward an HSC fate. Two main strategies have been used to obtain HSCs *in vitro*. The first uses extrinsic factors to differentiate ES cells toward a hematopoietic fate, while the second uses the overexpression of transcription factors to (de)differentiate cells to an HSC.

With the first strategy (ES cell differentiation; Fig. 2), the stepwise addition of a variety of factors induces stages of ES cell differentiation leading to primitive streak, mesoderm, hemangioblast/hemogenic endothelium, and then hematopoietic cells. Some of these factors include Nodal/Activin, Wnt, and BMP (Irion, Clarke, et al., 2010; Nostro, Cheng, et al., 2008; Pearson, Sroczynska, et al., 2008; Sturgeon, Ditadi, et al., 2014). BMP is

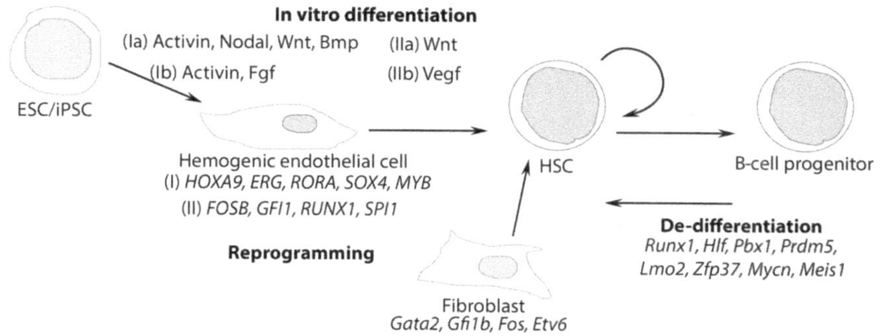

Fig. 2 Different reprogramming strategies in an effort to obtain HSCs. Two strategies are used. The first approach differentiates ES cells or iPS cells toward a hematopoietic progenitor and/or HSC fate via a hemogenic endothelial intermediate step through the use of a combination of growth and developmental factors. The second approach uses a reprogramming strategy by overexpressing combinations of transcription factors in starting nonhematopoietic cells such as fibroblasts or hematopoietic cells such as B-cell progenitors. (See the color plate.)

important for induction of Flk1$^+$ cells, which are believed to serve as hemangioblasts/hemogenic endothelial cells and is a necessary step in generating definitive hematopoietic cells *in vivo*. Under the influence of Wnt, a definitive hematopoietic progenitor is formed, whereas by blocking Wnt a primitive progenitor is formed (Sturgeon et al., 2014). Others have used Activin and FGF to induce hemangioblast formation and Vegf to induce committed hematopoietic precursors (Pearson et al., 2008). These studies were unsuccessful in generating (detectable) repopulating HSCs, suggesting that an additional factor is required for HSC generation.

With the second strategy, the reprogramming or de-differentiation approach utilizes transcription factor transduction/expression to induce an HSC fate in nonhematopoietic or differentiated hematopoietic cells, respectively (Fig. 2). Mouse fibroblasts, when transfected with multiple transcription factor genes—*Gata2*, *Gfi1b*, *Fos*, and *Etv6* (Pereira, Chang, et al., 2013)—and cultured for 35 d, were reprogrammed and found to express a hemogenic transcriptional program. Following a further culture of 20 d, definitive hematopoietic progenitors, but no HSCs, were found. A similar strategy with human EB-derived CD34$^+$CD45$^+$ progenitors transfected with *HOXA9*, *ERG*, *RORA*, *SOX4*, and *MYB* resulted in short-term repopulating progenitors (Doulatov, Vo, et al., 2013). Endothelial cells, purified from human umbilical cord blood and subsequently transfected with *FOSB*, *GFI1*, *RUNX1*, and *SPI1* (and cultured up to 40 d) resulted in transplantable multipotent progenitor cells (Sandler, Lis, et al., 2014). As mentioned earlier, these studies are possibly hampered by the inefficiency of culturing HSCs *in vitro*. This has been circumvented by the Rossi team, who used a B-cell progenitor population, harvested from mouse peripheral blood as a starting cell (Riddell, Gazit, et al., 2014). These B-cell progenitors were transfected with a mix of transcription factor genes—*Hlf, Pbx1, Prdm5, Lmo2, Zfp37* with *Mycn* and *Meis1* included for higher efficiency—and were immediately transplanted back into irradiated recipient mice. The transfected donor cells resulted in long-term multilineage hematopoietic repopulation of the recipients. By circumventing a culture period, the loss of HSC fate or growth (as can be expected in attempts of HSC culture) is possibly avoided (Riddell et al., 2014).

Why is it that these transcription factor cocktails differ so much? Most of these studies began with a list of candidate transcription factors, including the heptad factors. Are some factors important for HSC generation *in vitro*, but not *in vivo*? In fact, several of the genes used for these studies are known oncogenes such as *Hlf, Mycn, cMyb, Hoxa9, Meis1, Pbx1*, and *Etv6*. Most of these genes have not been implicated in HSC generation *in vivo* and are not even expressed at the HSC generation stage. Their clear function in oncogenic transformations of hematopoietic cells, could explain their importance for *ex vivo* iHSCs (induced HSC) strategies. During normal fetal stages *Myb, Mycn, Meis1*, and *Pbx1* play a role in maintaining HSCs (de Boer, Yeung, et al., 2011; DiMartino, Selleri, et al., 2001; Laurenti, Varnum-Finney, et al., 2008; Mucenski, McLain, et al., 1991). Also, some genes in these transfection cocktails could circumvent the use the heptad factors since they are direct targets of the heptads. For example, *Gfi1* and *Gfi1b* are targets of Runx1, Scl, Gata2, and Erg and are required for the repression of the endothelial program in hemogenic endothelial cells at early stages of EHT (Lancrin, Mazan, et al., 2012; Lichtinger, Ingram, et al., 2012; Wilson, Timms, et al., 2010). Surprisingly, some of the transcription factor genes used to induce a hematopoietic fate (*Sox4, Prdm5*, and *Zfp37*), have not been previously implicated in hematopoiesis.

The fact that cells overexpressing these transcription factors can be reprogrammed over a period of time to hematopoietic progenitor and/or HSC fate is proof that their overall genetic/epigenetic program is flexible. The current challenge is to understand the stepwise changes that occur. While the overexpression of oncogenes for iHSC production rules out their clinical use, these cell systems are valuable tools for identifying bioactive compounds and the sequence of the cues directing HSC production. This new interplay between the discoveries from the fields of developmental biology and stem cell reprogramming is complementary and synergistic, and currently is stimulating more intensive and exciting studies in developmental hematopoiesis.

REFERENCES

Bertrand, J. Y., Chi, N. C., et al. (2010). Haematopoietic stem cells derive directly from aortic endothelium during development. *Nature, 464*(7285), 108–111.

Bertrand, J. Y., Giroux, S., et al. (2005). Characterization of purified intraembryonic hematopoietic stem cells as a tool to define their site of origin. *Proceedings of the National Academy of Sciences of the United States of America, 102*(1), 134–139.

Boisset, J. C., van Cappellen, W., et al. (2010). In vivo imaging of haematopoietic cells emerging from the mouse aortic endothelium. *Nature, 464*(7285), 116–120.

Bussmann, J., Bos, F. L., et al. (2010). Arteries provide essential guidance cues for lymphatic endothelial cells in the zebrafish trunk. *Development, 137*(16), 2653–2657.

Butko, E., Distel, M., et al. (2015). Gata2b is a restricted early regulator of hemogenic endothelium in the zebrafish embryo. *Development, 142*(6), 1050–1061.

Capron, C., Lecluse, Y., et al. (2006). The SCL relative LYL-1 is required for fetal and adult hematopoietic stem cell function and B-cell differentiation. *Blood, 107*(12), 4678–4686.

Chan, W. Y., Follows, G. A., et al. (2007). The paralogous hematopoietic regulators Lyl1 and Scl are coregulated by Ets and GATA factors, but Lyl1 cannot rescue the early Scl-/- phenotype. *Blood, 109*(5), 1908–1916.

Chen, M. J., Yokomizo, T., et al. (2009). Runx1 is required for the endothelial to haematopoietic cell transition but not thereafter. *Nature, 457*(7231), 887–891.

Copelan, E. A. (2006). Hematopoietic stem-cell transplantation. *The New England Journal of Medicine, 354*(17), 1813–1826.

Crisan, M., Kartalaei, P. S., et al. (2015). BMP signalling differentially regulates distinct haematopoietic stem cell types. *Nature Communications, 6*, 8040.

Cumano, A., Dieterlen-Lievre, F., et al. (1996). Lymphoid potential, probed before circulation in mouse, is restricted to caudal intraembryonic splanchnopleura. *Cell, 86*(6), 907–916.

Davies, L. C., & Taylor, P. R. (2015). Tissue-resident macrophages: Then and now. *Immunology, 144*(4), 541–548.

de Boer, J., Yeung, J., et al. (2011). The E2A-HLF oncogenic fusion protein acts through Lmo2 and Bcl-2 to immortalize hematopoietic progenitors. *Leukemia, 25*(2), 321–330.

de Bruijn, M. F., Ma, X., et al. (2002). Hematopoietic stem cells localize to the endothelial cell layer in the midgestation mouse aorta. *Immunity, 16*(5), 673–683.

de Bruijn, M. F., Speck, N. A., et al. (2000). Definitive hematopoietic stem cells first develop within the major arterial regions of the mouse embryo. *The EMBO Journal, 19*(11), 2465–2474.

de Pater, E., Kaimakis, P., et al. (2013). Gata2 is required for HSC generation and survival. *The Journal of Experimental Medicine, 210*(13), 2843–2850.

DiMartino, J. F., Selleri, L., et al. (2001). The Hox cofactor and proto-oncogene Pbx1 is required for maintenance of definitive hematopoiesis in the fetal liver. *Blood, 98*(3), 618–626.

Dooley, K. A., Davidson, A. J., et al. (2005). Zebrafish scl functions independently in hematopoietic and endothelial development. *Developmental Biology, 277*(2), 522–536.

Doulatov, S., Vo, L. T., et al. (2013). Induction of multipotential hematopoietic progenitors from human pluripotent stem cells via respecification of lineage-restricted precursors. *Cell Stem Cell, 13*(4), 459–470.

Durand, C., Robin, C., et al. (2007). Embryonic stromal clones reveal developmental regulators of definitive hematopoietic stem cells. *Proceedings of the National Academy of Sciences of the United States of America, 104*(52), 20838–20843.

Dzierzak, E., & Speck, N. A. (2008). Of lineage and legacy: The development of mammalian hematopoietic stem cells. *Nature Immunology, 9*(2), 129–136.

Fitch, S. R., Kimber, G. M., et al. (2012). Signaling from the sympathetic nervous system regulates hematopoietic stem cell emergence during embryogenesis. *Cell Stem Cell, 11*(4), 554–566.

Frame, J. M., McGrath, K. E., et al. (2013). Erythro-myeloid progenitors: "Definitive" hematopoiesis in the conceptus prior to the emergence of hematopoietic stem cells. *Blood Cells, Molecules & Diseases, 51*(4), 220–225.

Gao, X., Johnson, K. D., et al. (2013). Gata2 cis-element is required for hematopoietic stem cell generation in the mammalian embryo. *The Journal of Experimental Medicine, 210*(13), 2833–2842.

Gekas, C., Dieterlen-Lievre, F., et al. (2005). The placenta is a niche for hematopoietic stem cells. *Developmental Cell, 8*(3), 365–375.

Gerhardt, D. M., Pajcini, K. V., et al. (2014). The Notch1 transcriptional activation domain is required for development and reveals a novel role for Notch1 signaling in fetal hematopoietic stem cells. *Genes & Development, 28*(6), 576–593.

Gering, M., & Patient, R. (2005). Hedgehog signaling is required for adult blood stem cell formation in zebrafish embryos. *Developmental Cell, 8*(3), 389–400.

Guiu, J., Bergen, D. J., et al. (2014). Identification of Cdca7 as a novel Notch transcriptional target involved in hematopoietic stem cell emergence. *The Journal of Experimental Medicine, 211*(12), 2411–2423.

Ieda, M. (2013). Direct reprogramming into desired cell types by defined factors. *The Keio Journal of Medicine, 62*(3), 74–82.

Irion, S., Clarke, R. L., et al. (2010). Temporal specification of blood progenitors from mouse embryonic stem cells and induced pluripotent stem cells. *Development, 137*(17), 2829–2839.

Ivanovs, A., Rybtsov, S., et al. (2011). Highly potent human hematopoietic stem cells first emerge in the intraembryonic aorta-gonad-mesonephros region. *The Journal of Experimental Medicine, 208*(12), 2417–2427.

Jaffredo, T., Gautier, R., et al. (1998). Intraaortic hemopoietic cells are derived from endothelial cells during ontogeny. *Development, 125*(22), 4575–4583.

Jaffredo, T., Gautier, R., et al. (2000). Tracing the progeny of the aortic hemangioblast in the avian embryo. *Developmental Biology, 224*(2), 204–214.

Jaffredo, T., Nottingham, W., et al. (2005). From hemangioblast to hematopoietic stem cell: An endothelial connection? *Experimental Hematology, 33*(9), 1029–1040.

Kim, A. D., Melick, C. H., et al. (2014). Discrete Notch signaling requirements in the specification of hematopoietic stem cells. *The EMBO Journal, 33*(20), 2363–2373.

Kissa, K., & Herbomel, P. (2010). Blood stem cells emerge from aortic endothelium by a novel type of cell transition. *Nature, 464*(7285), 112–115.

Kissa, K., Murayama, E., et al. (2008). Live imaging of emerging hematopoietic stem cells and early thymus colonization. *Blood, 111*(3), 1147–1156.

Kobayashi, I., Kobayashi-Sun, J., et al. (2014). Jam1a-Jam2a interactions regulate haematopoietic stem cell fate through Notch signalling. *Nature, 512*(7514), 319–323.

Kruse, E. A., Loughran, S. J., et al. (2009). Dual requirement for the ETS transcription factors Fli-1 and Erg in hematopoietic stem cells and the megakaryocyte lineage. *Proceedings of the National Academy of Sciences of the United States of America, 106*(33), 13814–13819.

Lancrin, C., Mazan, M., et al. (2012). GFI1 and GFI1B control the loss of endothelial identity of hemogenic endothelium during hematopoietic commitment. *Blood, 120*(2), 314–322.

Landry, J. R., Kinston, S., et al. (2008). Runx genes are direct targets of Scl/Tal1 in the yolk sac and fetal liver. *Blood, 111*(6), 3005–3014.

Laurenti, E., Varnum-Finney, B., et al. (2008). Hematopoietic stem cell function and survival depend on c-Myc and N-Myc activity. *Cell Stem Cell, 3*(6), 611–624.

Lee, Y., Manegold, J. E., et al. (2014). FGF signalling specifies haematopoietic stem cells through its regulation of somitic Notch signalling. *Nature Communications, 5*, 5583.

Leung, A., Ciau-Uitz, A., et al. (2013). Uncoupling VEGFA functions in arteriogenesis and hematopoietic stem cell specification. *Developmental Cell, 24*(2), 144–158.

Li, Z., Lan, Y., et al. (2012). Mouse embryonic head as a site for hematopoietic stem cell development. *Cell Stem Cell, 11*(5), 663–675.

Liakhovitskaia, A., Gribi, R., et al. (2009). Restoration of Runx1 expression in the Tie2 cell compartment rescues definitive hematopoietic stem cells and extends life of Runx1 knockout animals until birth. *Stem Cells, 27*(7), 1616–1624.

Liakhovitskaia, A., Rybtsov, S., et al. (2014). Runx1 is required for progression of CD41+ embryonic precursors into HSCs but not prior to this. *Development, 141*(17), 3319–3323.

Lichtinger, M., Ingram, R., et al. (2012). RUNX1 reshapes the epigenetic landscape at the onset of haematopoiesis. *The EMBO Journal, 31*(22), 4318–4333.

Lim, K. C., Hosoya, T., et al. (2012). Conditional Gata2 inactivation results in HSC loss and lymphatic mispatterning. *The Journal of Clinical Investigation, 122*(10), 3705–3717.

Ling, K. W., Ottersbach, K., et al. (2004). GATA-2 plays two functionally distinct roles during the ontogeny of hematopoietic stem cells. *The Journal of Experimental Medicine, 200*(7), 871–882.

Ma, X., Robin, C., et al. (2002). The Ly-6A (Sca-1) GFP transgene is expressed in all adult mouse hematopoietic stem cells. *Stem Cells, 20*(6), 514–521.

Marks-Bluth, J., Khanna, A., et al. (2015). SMAD1 and SMAD5 expression is coordinately regulated by FLI1 and GATA2 during endothelial development. *Molecular and Cellular Biology, 35*(12), 2165–2172.

McGrath, K. E., Frame, J. M., et al. (2015). Distinct sources of hematopoietic progenitors emerge before HSCs and provide functional blood cells in the mammalian embryo. *Cell Reports, 11*(12), 1892–1904.

McKinney-Freeman, S., Cahan, P., et al. (2012). The transcriptional landscape of hematopoietic stem cell ontogeny. *Cell Stem Cell, 11*(5), 701–714.

Medvinsky, A., & Dzierzak, E. (1996). Definitive hematopoiesis is autonomously initiated by the AGM region. *Cell, 86*(6), 897–906.

Medvinsky, A., Taoudi, S., et al. (2008). Analysis and manipulation of hematopoietic progenitor and stem cells from murine embryonic tissues. *Current Protocols in Stem Cell Biology*(Suppl. 4). Chapter 2: Unit 2A.6.1–2A.6.25.

Mucenski, M. L., McLain, K., et al. (1991). A functional c-myb gene is required for normal murine fetal hepatic hematopoiesis. *Cell, 65*(4), 677–689.

Mukouyama, Y., Chiba, N., et al. (2000). The AML1 transcription factor functions to develop and maintain hematogenic precursor cells in the embryonic aorta-gonad-mesonephros region. *Developmental Biology, 220*(1), 27–36.

Muller, A. M., Medvinsky, A., et al. (1994). Development of hematopoietic stem cell activity in the mouse embryo. *Immunity, 1*(4), 291–301.

Newman, P. J., Berndt, M. C., et al. (1990). PECAM-1 (CD31) cloning and relation to adhesion molecules of the immunoglobulin gene superfamily. *Science, 247*(4947), 1219–1222.

North, T. E., de Bruijn, M. F., et al. (2002). Runx1 expression marks long-term repopulating hematopoietic stem cells in the midgestation mouse embryo. *Immunity, 16*(5), 661–672.

North, T., Gu, T. L., et al. (1999). Cbfa2 is required for the formation of intra-aortic hematopoietic clusters. *Development, 126*(11), 2563–2575.

Nostro, M. C., Cheng, X., et al. (2008). Wnt, activin, and BMP signaling regulate distinct stages in the developmental pathway from embryonic stem cells to blood. *Cell Stem Cell, 2*(1), 60–71.
Ohnuki, M., & Takahashi, K. (2015). Present and future challenges of induced pluripotent stem cells. *Philosophical Transactions of the Royal Society of London. Series B, Biological Sciences, 370*(1680).
Okuda, T., van Deursen, J., et al. (1996). AML1, the target of multiple chromosomal translocations in human leukemia, is essential for normal fetal liver hematopoiesis. *Cell, 84*(2), 321–330.
Orkin, S. H., & Zon, L. I. (2008). Hematopoiesis: An evolving paradigm for stem cell biology. *Cell, 132*(4), 631–644.
Ottersbach, K., & Dzierzak, E. (2005). The murine placenta contains hematopoietic stem cells within the vascular labyrinth region. *Developmental Cell, 8*(3), 377–387.
Palis, J., Malik, J., et al. (2010). Primitive erythropoiesis in the mammalian embryo. *The International Journal of Developmental Biology, 54*(6-7), 1011–1018.
Pardanaud, L., & Dieterlen-Lievre, F. (1993). Emergence of endothelial and hemopoietic cells in the avian embryo. *Anatomy and Embryology (Berl), 187*(2), 107–114.
Pardanaud, L., & Dieterlen-Lievre, F. (1999). Manipulation of the angiopoietic/hemangiopoietic commitment in the avian embryo. *Development, 126*(4), 617–627.
Patterson, L. J., Gering, M., et al. (2005). Scl is required for dorsal aorta as well as blood formation in zebrafish embryos. *Blood, 105*(9), 3502–3511.
Pearson, S., Sroczynska, P., et al. (2008). The stepwise specification of embryonic stem cells to hematopoietic fate is driven by sequential exposure to Bmp4, activin A, bFGF and VEGF. *Development, 135*(8), 1525–1535.
Peeters, M., Ottersbach, K., et al. (2009). Ventral embryonic tissues and Hedgehog proteins induce early AGM hematopoietic stem cell development. *Development, 136*(15), 2613–2621.
Pereira, C. F., Chang, B., et al. (2013). Induction of a hemogenic program in mouse fibroblasts. *Cell Stem Cell, 13*(2), 205–218.
Pimanda, J. E., Ottersbach, K., et al. (2007). Gata2, Fli1, and Scl form a recursively wired gene-regulatory circuit during early hematopoietic development. *Proceedings of the National Academy of Sciences of the United States of America, 104*(45), 17692–17697.
Porcher, C., Swat, W., et al. (1996). The T cell leukemia oncoprotein SCL/tal-1 is essential for development of all hematopoietic lineages. *Cell, 86*(1), 47–57.
Pouget, C., Peterkin, T., et al. (2014). FGF signalling restricts haematopoietic stem cell specification via modulation of the BMP pathway. *Nature Communications, 5*, 5588.
Richard, C., Drevon, C., et al. (2013). Endothelio-mesenchymal interaction controls runx1 expression and modulates the notch pathway to initiate aortic hematopoiesis. *Developmental Cell, 24*(6), 600–611.
Riddell, J., Gazit, R., et al. (2014). Reprogramming committed murine blood cells to induced hematopoietic stem cells with defined factors. *Cell, 157*(3), 549–564.
Robb, L., Drinkwater, C. C., et al. (1995). Hematopoietic and lung abnormalities in mice with a null mutation of the common beta subunit of the receptors for granulocyte-macrophage colony-stimulating factor and interleukins 3 and 5. *Proceedings of the National Academy of Sciences of the United States of America, 92*(21), 9565–9569.
Robert-Moreno, A., Espinosa, L., et al. (2005). RBPjkappa-dependent Notch function regulates Gata2 and is essential for the formation of intra-embryonic hematopoietic cells. *Development, 132*(5), 1117–1126.
Robert-Moreno, A., Guiu, J., et al. (2008). Impaired embryonic haematopoiesis yet normal arterial development in the absence of the Notch ligand Jagged1. *The EMBO Journal, 27*(13), 1886–1895.
Robin, C., Ottersbach, K., et al. (2011). CD41 is developmentally regulated and differentially expressed on mouse hematopoietic stem cells. *Blood, 117*(19), 5088–5091.

Ruiz-Herguido, C., Guiu, J., et al. (2012). Hematopoietic stem cell development requires transient Wnt/beta-catenin activity. *The Journal of Experimental Medicine*, *209*(8), 1457–1468.

Rybtsov, S., Batsivari, A., et al. (2014). Tracing the origin of the HSC hierarchy reveals an SCF-dependent, IL-3-independent CD43(-) embryonic precursor. *Stem Cell Reports*, *3*(3), 489–501.

Rybtsov, S., Sobiesiak, M., et al. (2011). Hierarchical organization and early hematopoietic specification of the developing HSC lineage in the AGM region. *The Journal of Experimental Medicine*, *208*(6), 1305–1315.

Sanchez, M. J., Holmes, A., et al. (1996). Characterization of the first definitive hematopoietic stem cells in the AGM and liver of the mouse embryo. *Immunity*, *5*(6), 513–525.

Sandler, V. M., Lis, R., et al. (2014). Reprogramming human endothelial cells to haematopoietic cells requires vascular induction. *Nature*, *511*(7509), 312–318.

Schlaeger, T. M., Mikkola, H. K., et al. (2005). Tie2Cre-mediated gene ablation defines the stem-cell leukemia gene (SCL/tal1)-dependent window during hematopoietic stem-cell development. *Blood*, *105*(10), 3871–3874.

Shivdasani, R. A., Mayer, E. L., et al. (1995). Absence of blood formation in mice lacking the T-cell leukaemia oncoprotein tal-1/SCL. *Nature*, *373*(6513), 432–434.

Solaimani Kartalaei, P., Yamada-Inagawa, T., et al. (2015). Whole-transcriptome analysis of endothelial to hematopoietic stem cell transition reveals a requirement for Gpr56 in HSC generation. *The Journal of Experimental Medicine*, *212*(1), 93–106.

Souroullas, G. P., Salmon, J. M., et al. (2009). Adult hematopoietic stem and progenitor cells require either Lyl1 or Scl for survival. *Cell Stem Cell*, *4*(2), 180–186.

Spyropoulos, D. D., Pharr, P. N., et al. (2000). Hemorrhage, impaired hematopoiesis, and lethality in mouse embryos carrying a targeted disruption of the Fli1 transcription factor. *Molecular and Cellular Biology*, *20*(15), 5643–5652.

Sturgeon, C. M., Ditadi, A., et al. (2014). Wnt signaling controls the specification of definitive and primitive hematopoiesis from human pluripotent stem cells. *Nature Biotechnology*, *32*(6), 554–561.

Swiers, G., Baumann, C., et al. (2013). Early dynamic fate changes in haemogenic endothelium characterized at the single-cell level. *Nature Communications*, *4*, 2924.

Takahashi, K., & Yamanaka, S. (2006). Induction of pluripotent stem cells from mouse embryonic and adult fibroblast cultures by defined factors. *Cell*, *126*(4), 663–676.

Takizawa, H., Schanz, U., et al. (2011). Ex vivo expansion of hematopoietic stem cells: Mission accomplished? *Swiss Medical Weekly*, *141*, w13316.

Tang, Y., Harrington, A., et al. (2010). The contribution of the Tie2+ lineage to primitive and definitive hematopoietic cells. *Genesis*, *48*(9), 563–567.

Taoudi, S., Bee, T., et al. (2011). ERG dependence distinguishes developmental control of hematopoietic stem cell maintenance from hematopoietic specification. *Genes & Development*, *25*(3), 251–262.

Taoudi, S., Gonneau, C., et al. (2008). Extensive hematopoietic stem cell generation in the AGM region via maturation of VE-cadherin+CD45+ pre-definitive HSCs. *Cell Stem Cell*, *3*(1), 99–108.

Taoudi, S., & Medvinsky, A. (2007). Functional identification of the hematopoietic stem cell niche in the ventral domain of the embryonic dorsal aorta. *Proceedings of the National Academy of Sciences of the United States of America*, *104*(22), 9399–9403.

Tavian, M., Robin, C., et al. (2001). The human embryo, but not its yolk sac, generates lympho-myeloid stem cells: Mapping multipotent hematopoietic cell fate in intraembryonic mesoderm. *Immunity*, *15*(3), 487–495.

Taylor, E., Taoudi, S., et al. (2010). Hematopoietic stem cell activity in the aorta-gonad-mesonephros region enhances after mid-day 11 of mouse development. *The International Journal of Developmental Biology*, *54*(6-7), 1055–1060.

Tober, J., Yzaguirre, A. D., et al. (2013). Distinct temporal requirements for Runx1 in hematopoietic progenitors and stem cells. *Development, 140*(18), 3765–3776.

Tsai, F. Y., Keller, G., et al. (1994). An early haematopoietic defect in mice lacking the transcription factor GATA-2. *Nature, 371*(6494), 221–226.

Visvader, J. E., Fujiwara, Y., et al. (1998). Unsuspected role for the T-cell leukemia protein SCL/tal-1 in vascular development. *Genes & Development, 12*(4), 473–479.

Wilkinson, R. N., Pouget, C., et al. (2009). Hedgehog and Bmp polarize hematopoietic stem cell emergence in the zebrafish dorsal aorta. *Developmental Cell, 16*(6), 909–916.

Wilson, N. K., Foster, S. D., et al. (2010). Combinatorial transcriptional control in blood stem/progenitor cells: Genome-wide analysis of ten major transcriptional regulators. *Cell Stem Cell,* 7(4), 532–544.

Wilson, N. K., Timms, R. T., et al. (2010). Gfi1 expression is controlled by five distinct regulatory regions spread over 100 kilobases, with Scl/Tal1, Gata2, PU.1, Erg, Meis1, and Runx1 acting as upstream regulators in early hematopoietic cells. *Molecular and Cellular Biology, 30*(15), 3853–3863.

Wood, H. B., May, G., et al. (1997). CD34 expression patterns during early mouse development are related to modes of blood vessel formation and reveal additional sites of hematopoiesis. *Blood, 90*(6), 2300–2311.

Yamamoto, R., Morita, Y., et al. (2013). Clonal analysis unveils self-renewing lineage-restricted progenitors generated directly from hematopoietic stem cells. *Cell, 154*(5), 1112–1126.

Yokomizo, T., & Dzierzak, E. (2010). Three-dimensional cartography of hematopoietic clusters in the vasculature of whole mouse embryos. *Development, 137*(21), 3651–3661.

Zhang, P., He, Q., et al. (2015). G protein-coupled receptor 183 facilitates endothelial-to-hematopoietic transition via Notch1 inhibition. *Cell Research, 25*(10), 1093–1107.

Zhu, H., Traver, D., et al. (2005). Regulation of the lmo2 promoter during hematopoietic and vascular development in zebrafish. *Developmental Biology, 281*(2), 256–269.

Zovein, A. C., Hofmann, J. J., et al. (2008). Fate tracing reveals the endothelial origin of hematopoietic stem cells. *Cell Stem Cell, 3*(6), 625–636.

CHAPTER TWO

Hematopoietic Stem Cell and Its Bone Marrow Niche

V.W.C. Yu*,†,‡, D.T. Scadden*,†,‡,1

*Massachusetts General Hospital, Boston, MA, United States
†Harvard Stem Cell Institute, Cambridge, MA, United States
‡Harvard University, Cambridge, MA, United States
[1]Corresponding author: e-mail address: david_scadden@harvard.edu

Contents

Abstract

Stem cells do not thrive without their niche. The bone marrow microenvironment is where hematopoietic stem cells maintain their cell state while receiving physiological input to modify their activity in response to changing physiological demands. The complexity of the bone marrow microenvironment is being unraveled and indicates that multiple different cell types contribute to the regulation of stem and progenitor cells. Further, it is becoming evident that the bone marrow represents a composite of niches with different components and different functional roles in hematopoiesis. It is now evident that alterations in specific stromal cells that comprise the bone marrow microenvironment can contribute to hematologic pathology. In this chapter, we will review the history of the niche concept, evolving information about its components and how niche dysfunction may contribute to disease.

Current Topics in Developmental Biology, Volume 118
ISSN 0070-2153
http://dx.doi.org/10.1016/bs.ctdb.2016.01.009

ABBREVIATIONS

AML acute myelogenous leukemia
BM bone marrow
BMPRIA bone morphogenic protein receptor type IA
CAR CXCL12-abundant reticular
CFU-S colony-forming unit-spleen cell
CLP common lymphoid progenitor
CML chronic myelogenous leukemia
Col(I)a2.3 2.3 kb promoter of collagen Iα
CXCL12 CXC chemokine ligand (CXCL) 12
CXCR4 receptor for CXC chemokine ligand (CXCL) 12
FAP fibroblast activation protein
FDA US Food and Drug Administration
G-CSF granulocyte colony-stimulating factor
HSC hematopoietic stem cell
HSCT hematopoietic stem cell transplantation
HSPC hematopoietic stem and progenitor cell
HSPG heparan sulfate proteoglycan
LEPR leptin receptor
LKS $Lineage^{Lo}Sca^{+}c\text{-}Kit^{+}$
MMP metalloproteinase
NE norepinephrine
Ocn osteocalcin
OSM oncostatin M
Osx osterix
PPARγ peroxisome proliferator-activated receptor gamma
PTH parathyroid hormone
RARγ retinoic acid receptor gamma
RB retinoblastoma protein
SCF stem cell factor
SDF-1 stromal cell-derived factor 1
SLAM markers Lineage, c-Kit, Sca, CD150, CD48, CD41
SNS sympathetic nervous system
TGF-β1 transforming growth factor beta 1
VCAM-1 vascular cell adhesion molecule 1
VLA-4 very late antigen 4

1. INTRODUCTION

In the human body, the softest organ of all, the blood, is ironically encapsulated by the hardest—the skeleton. The bone marrow (BM) microenvironment with its specialized anatomy and interconnected vasculature provides a sanctuary where hematopoietic stem cells (HSCs) reside, are maintained, and differentiate into multiple blood lineages. The bone marrow niche is a critical microenvironment that regulates many stem cell activities including self-renewal, mobilization, engraftment, and lineage differentiation. The importance of the hematopoietic niche is highlighted by evidence, showing that mutations of the nonhematopoietic cells of the marrow microenvironment are sufficient to cause hematopoietic neoplasia. This chapter will provide an overview of the niche concept, the functional and anatomical relationships of cells within the bone marrow, and summarize the recent literature of the hematopoietic niche in blood diseases.

2. EVOLUTION OF THE STEM CELL NICHE CONCEPT

Formulation of the niche hypothesis indirectly relates to the first experimental demonstration of tissue stem cells. Exploring how ionizing radiation affects mammalian cells, Till and McCulloch laid out the first experimental demonstration of the "self-renewing unit" in the hematopoietic system (Till & McCulloch, 1961). They accomplished this by irradiating mice with a dose that would kill the animals within 30 days if the mice did not receive a transplant of fresh cells. Transplantation of donor bone marrow cells conferred radioprotection. Not only did the donor cells reconstitute the bone marrow of the recipients, but they also gave rise to nodules in the spleen. Using chromosome breaks as durable genetic markers of individual transplanted cells, Till, McCulloch, and colleagues elegantly demonstrated that these spleen nodules were myeloid, erythroid, and lymphoid cell containing colonies derived from single BM donor cells. They proposed that these "self-renewing units" must be the primitive cell source that were capable of giving rise to multiple lineages and regenerated the whole hematopoietic system, and therefore they hypothesized these cells to be stem cells (Worton, McCulloch, & Till, 1969). Although it was not known until much later that these cells were not stem cells but progenitors, this breakthrough experiment laid the experimental groundwork that led to the discovery of the long-term repopulating HSC many years later.

Following the study of Till and McCulloch, Schofield was puzzled by the fact that transplantation of bone marrow cells derived from either young or old wild-type mice into W/Wv mice (having a c-kit mutation) was able to reconstitute hematopoiesis indefinitely. However, cells that formed colonies in the spleen upon transplantation, the ones that Till and McCulloch defined as stem cells and named colony-forming units-spleen cells (CFU-Ss), could not reconstitute W/Wv mice and had a limited serial passage capacity. Schofield hypothesized that CFU-Ss were not stem cells but reflected a different cell state due to the spleen in which they resided. Around that time, Dexter and colleagues published a landmark paper describing the requirement of a bone marrow stromal feeder layer in order to sustain primitive hematopoietic cells in ex vivo cultures (Dexter, Allen, & Lajtha, 1977). Driven by his own experimental observation and the findings of his laboratory neighbor and colleague, Dexter, Schofield articulated the stem cell niche concept in 1978. He concluded that stem cells needed to reside in the bone marrow to retain their stemness. Once they left the niche, they could become CFU-Ss, but at the expense of their immortality (Fig. 1). He proposed that when these cells reoccupied the niche, they could regain their stemness (Schofield, 1978). Schofield's proposal presented the basic concepts of a stem cell niche: (1) a defined anatomical site, (2) a location where stem cells could be maintained and reproduce, (3) a place where stem cell differentiation was inhibited, and (4) a defined space that limited the number of stem cells. He had no experimental evidence to prove these new concepts and was challenged by McCulloch and others, but he was correct.

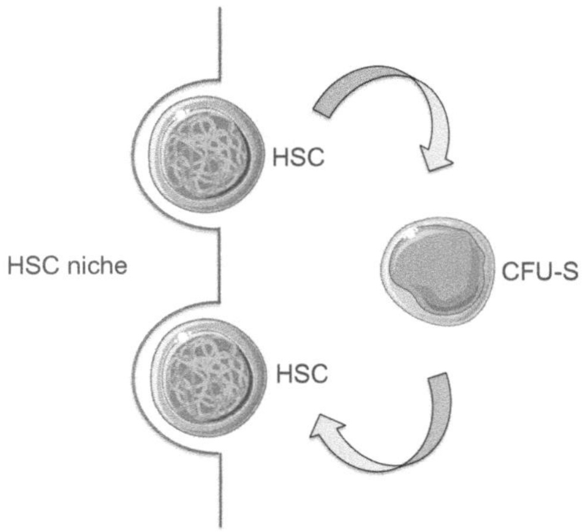

Fig. 1 The hypothetical view of the stem cell niche from R. Schofield. The stem cell can become the daughter cell, CFU-S, once the stem cell leaves its niche. But if the stem cell finds and reoccupies the niche, it will itself return to its stem cell state. *Adopted from Schofield, R. (1978). The relationship between the spleen colony-forming cell and the haemopoietic stem cell.* Blood Cells, 4, *7–25.* (See the color plate.)

3. LOCATION MATTERS

Schofield proposed that the bone marrow microenvironment was special, providing an HSC niche. His concept was well supported by comparative biology where it is clear that land animals with bones have hematopoiesis in those bones. Therefore, the quest for defining HSC niches reasonably began with assessment of how bone cells influenced blood cell production. This was first shown by two studies. One where osteoblastic cells were constitutively expressing an activated parathyroid hormone receptor in the mouse. The animals had an increase in hematopoietic stem and progenitor cell (HSPC) (Calvi et al., 2003). The other where the bone morphogenic protein receptor type IA (BMPRIA) was conditionally depleted in hematopoietic cells using the Mx1-Cre promotor (Zhang et al., 2003). In both studies, they observed increased number of osteolineage cells, correlated with increased number of HSPCs (Lineage^{Lo}Sca^{+}c-Kit^{+} cells). When interpreting these data, one must be mindful that at the time when these studies were conducted, Lineage^{Lo}Sca^{+}c-Kit^{+} (LKS) cells were used to immunophenotypically define HSCs. We now know that within the LKS population, progenitor cells represent the major population, while HSCs only constitute a minority. Second, in these studies, hematopoietic reconstitution of transplanted hematopoietic cells was tracked up to 12 weeks. It is now known that progenitors contribute to the hematopoietic system up to 12 weeks, and therefore to assess the function of long-term HSCs, a longer period of reconstitution is needed. Nonetheless, these studies were the first in vivo demonstration that a specific microenvironmental cell could modulate HSPC, thereby validated Schofield's niche concept.

Pursuit of more specific definition of niche components remains a main driving force in the field. In addition to selected genetic changes in cell populations to assess the role of specific cells and specific genes in the HSPC niche, imaging experiments provided important information. They have defined certain principles. Under homeostatic conditions, HSCs reside in close proximity to blood vessels. Following radiation conditioning, bone marrow microvessels are disrupted and transplanted HSCs engraft in periendosteal sites.

Identification of transplanted cell localization was conducted by a number of laboratories. In vivo monitoring of HSPC after transplant over time has been particularly useful for fully defining that infused cells arrive at particular CXCL12 abundant positions in the microvasculature (Sipkins et al., 2005). The cells then either remain there or migrate to periendosteal position with more immature HSPCs in more proximate relationship to osteoblastic cells and more actively cycling cells at a greater distance (Lo Celso et al., 2009). These events occur within minutes to hours of HSPC infusion in the mouse conditioned with lethal irradiation.

Under homeostatic conditions, similar live animal in vivo imaging has not been possible due to lack of HSPC-specific fluorescent tags. Immunohistochemistry has revealed, however, that $CD150^+$, $CD48^-$, and $CD41^-$ (SLAM markers) HSCs localize adjacent to microvascular sinusoids and are scattered diffusely throughout the marrow (Kiel, Yilmaz, Iwashita, Terhorst, & Morrison, 2005). Further, other studies reported that HSCs were close to cells expressing high amounts of CXCL12, also known as the CXCL12-abundant reticular (CAR) cells. These cells surround the sinusoidal endothelial cells located near the endosteum (Sugiyama, Kohara, Noda, & Nagasawa, 2006).

Laser scanning cytometry then enabled imaging of HSCs throughout the longitude of the long bones in mice. Early application of this technology revealed that HSPCs preferentially localize in endosteal zones (Nombela-Arrieta et al., 2013), where the majority of cells closely interact with sinusoidal and nonsinusoidal BM microvessels. Combined whole-mount confocal immunofluorescence imaging and computational modeling have been used to demonstrate a three-dimensional association in the mouse BM among vascular structures, stromal cells, and HSCs. Quiescent HSCs were found to associate with small arterioles preferentially found in the periendosteal bone marrow (Kunisaki et al., 2013). These arterioles are ensheathed by $NG2^+$ pericytes, distinct from the sinusoid-associated leptin receptor ($LEPR^+$) cells. Interestingly, when HSCs enter into cell cycle, they were reported to redistribute from $NG2^+$ periarteriolar niches to $LEPR^+$ perisinusoidal sites. Depletion of $NG2^+$ cells caused HSC cycling and reduced functional long-term repopulating HSCs in the BM, implicating that the arteriolar $NG2^+$ cells control HSC quiescence. However, the debate continues as where precisely HSCs reside and whether there are specific locations that provide quiescence signals.

Other issues are more clear based on recent studies using probes detecting elemental oxygen. The marrow space is hypoxic with progressively decreasing O_2 levels from endosteum to central sinus (Spencer et al., 2014). Further, time-lapse and knockout mouse models have revealed that HSCs are highly dynamic and transit between different niches upon stimulation. They are mobile, regularly entering and exiting the circulation (Wright, Wagers, Gulati, Johnson, & Weissman, 2001). While much prior work has provided single snapshots of HSC within the BM microenvironment, it is likely that we will soon be able to acquire a dynamic view of HSC interactions with regulatory partners in the BM under homeostatic and, ultimately, disease conditions.

However, one should be mindful that most in vivo data are obtained through mouse models. There are major differences that exist between rodents and humans with regard to location of hematopoiesis. For instance, in mice, all bones support hematopoiesis, and the long bones are the sites in which the BM hematopoiesis is studied. In adult humans, the axial skeleton (the cranium, sternum, ribs, vertebra, and ilium) is the major site of blood cell production, and the red marrow in the long bones is replaced by hematopoietically inactive yellow marrow between 5 and 7 years of age, with the exception of the proximal regions of the long bones (Kricun, 1985). Second, the distinction between bone marrow under homeostasis vs postconditioning must be emphasized. Total body irradiation or myeloablative cytotoxic reagents are necessary in a transplant setting to permit HSC engraftment. However, these interventions are known to markedly disrupt the marrow environment. In particular, they destroy sinusoidal vessels and lead to marked hemorrhage within the marrow space. Therefore, the architecture and cellular composition of the marrow microenvironment are very different postconditioning compared with homeostasis. Third, the niche is a dynamic entity. Many of the experimental approaches focused on a static moment of cell–cell interaction may not be reflective of changes that occur over time or under different physiological contexts. This is more problematic if molecular modification of cells is driven by constitutively active Cre recombinases, as these models almost certainly associate with compensatory changes and changes in additional cell types than those putatively expressing the Cre at the point of study.

4. CELLULAR PARTICIPANTS IN THE BONE MARROW MICROENVIRONMENT

A number of cell populations have been examined for their roles in the HSPC bone marrow niche.

4.1 Osteolineage Cells

Early HSCs and bone marrow stromal cell coculture experiments suggested that osteoblasts provide support for primitive hematopoietic cells. The first in vivo demonstration of osteolineage cell requirement for hematopoiesis came from two groups simultaneously using different mouse models. One group conditionally deleted the BMPRIA in hematopoietic cells using the Mx1-Cre promotor (Zhang et al., 2003), while another group used a 2.3 kb promoter of collagen Iα (Col(I)a2.3) to target osteoblast-specific expression of the parathyroid hormone protein (PTH) or PTH-related protein receptor (Calvi et al., 2003). While these studies illustrate genetic modification of osteolineage cells alters HSPC number and downstream hematopoiesis, it was recently shown that certain osteolineage subtypes (including the Osx^+ and Ocn^+ cells) are partially dispensable (50–70% cell ablation) in adult BM without compromising LT-HSC maintenance under homeostatsis (Yu et al., 2015). Nevertheless, different osteolineage cells play distinct roles in several important hematopoietic processes.

First, mature osteolineage cells appear to be important in granulocyte colony-stimulating factor (G-CSF)-mediated HSPC mobilization. In studies where osteocalcin (Ocn)-expressing cells were depleted, G-CSF induced mobilization was markedly compromised (Ferraro et al., 2011). Further, different osteolineage subtypes create distinct lymphopoietic niches. Specifically, it was described that the deletion of early osteolineage cells (including the osterix-expressing (Osx^+) and Col(I)a2.3-expressing osteolineage cell populations) or of CXCL12 expression from these cells impacted B progenitor cell maturation (Greenbaum et al., 2013; Visnjic et al., 2004; Zhu et al., 2007) and reduced mature B and T cell numbers (Ding & Morrison, 2013). Targeted deletion of mature osteocalcin-expressing (Ocn^+) osteolineage cells resulted in a loss of T lineage cells. These animals have decreased T competent common lymphoid progenitors (CLPs) with a minimal effect on B cell biased CLP ($Ly6D^+$) (Yu et al., 2015) due to a defective generation of C–C chemokine receptor type 7 dependent thymic-seeding progenitor cells. The T cell lymphopenic effect seen in the Ocn^+ cell-depleted animals could be recapitulated by selective deletion of the Notch ligand, delta-like protein 4 expression in Ocn^+ cells, or of its receptor and downstream signaling molecules in primitive hematopoietic cells. Another study using genetic mouse model to achieve endogenous depletion of osteocytes demonstrated systemic disruption of metabolism and similar loss of T cells (Sato et al., 2013). Altogether, these studies show that early- and late-stage osteolineage cells play very different roles in supporting B and T lymphopoiesis. While Osx^+ osteoprogenitors have a

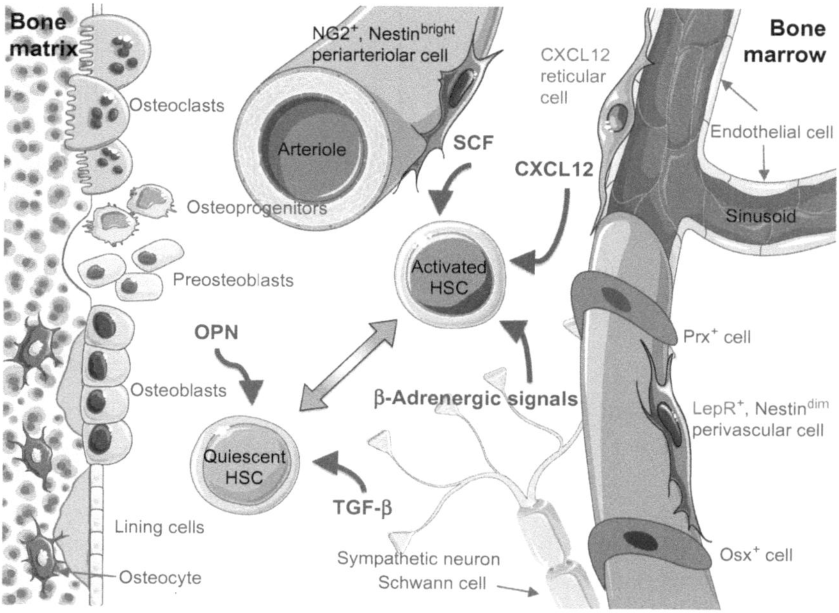

Fig. 2 The hematopoietic stem cell niche. Extrinsic signals including growth factors, cytokines, morphogens, extracellular matrix proteins, and adhesion molecules regulate the behavior of hematopoietic stem cells (HSCs) and protect them from exhaustion, although few have been pinpointed to derive from a specific cell source. Recent advances in flow cytometry, transgenic mouse models, and intravital microscopy have revolutionized our understanding of the BM niche and how it controls various stem cell behaviors including self-renewal, proliferation, differentiation, lineage commitment, and mobilization. Enlisted in the diagram are cellular and molecular participants that have been experimentally shown to impact different HSC activities through promoter-specific Cre deletion or overexpression. But the diagram is far from complete and represents an active area of ongoing investigation. (See the color plate.)

more pronounced role in the maturation of B cell progenitors, Ocn^+ cells modestly alter B cell production while distinctly affecting T cell specification. These data redefine bone cells as important immune participants that regulate specific production of cells of the adaptive immune system. The emerging model is one of a highly interrelated system with "intermediate" populations of both skeleton and blood having very specific interactions (Fig. 2). Whether these interactions are perturbed is of particular relevance in settings where specific subsets of cells are deficient as in particular blood disorders, or most problematically in the lack of T cell generation postallogeneic bone marrow transplantation. Future directions will involve exploring these heterologous cell interactions in malignant processes such as lymphoma, leukemia, and bone metastatic processes.

4.2 Perivascular Cells

The frequent localization of HSCs near blood vessels led to the hypothesis that perivascular regions may promote HSC maintenance. Mesenchymal stromal cells (MSCs) are cells that localize adjacent to blood vessels in bone marrow and are thought to be primitive cells capable of differentiation into osteolineage cells, chondrocytes, and adipocytes. Different markers have been used to define MSCs, including CD146 (Sacchetti et al., 2007) and CD271 (Matsuoka et al., 2015) in humans, CXCL12-GFP (Sugiyama et al., 2006), Nestin-GFP (Mendez-Ferrer et al., 2010), LEPR (Ding, Saunders, Enikolopov, & Morrison, 2012), Prx-1-Cre (Greenbaum et al., 2013), and Mx-1-Cre (Park et al., 2012). These cells are capable of differentiation into osteolineage cells in mice and express factors that promote HSC maintenance.

CAR cells adjacent to sinusoids were first shown to colocalize with HSCs by immunohistochemistry on bone marrow sections (Sugiyama et al., 2006). Ablation of these CAR cells resulted in severe impairment of adipogenic and osteogenic capacity of bone marrow cells, leading to reduced production of cytokines SCF (stem cell factor) and CXCL12 in the bone marrow, and eventually decreased cycling lymphoid and erythroid progenitors and HSCs (Omatsu et al., 2010). Using a different marker to identify MSCs, Frenette and colleagues found that cells that express a Nestin-GFP transgene similarly localize around blood vessels throughout the bone marrow, express high levels of SCF and CXCL12, and are indispensible for HSC maintenance (Mendez-Ferrer et al., 2010). Fibroblast activation protein (FAP) is expressed by stromal cells with many MSC characteristics, including expression of SCF, CXCL12, PDGFRα, Nestin, Sca-1, and CD51 (Morikawa et al., 2009; Pinho et al., 2013). Genetic depletion of FAP$^+$ cells led to reduction of osteolineage cells, impaired B lymphopoiesis and erythropoiesis, and eventually cachexia and anemia. Despite our incomplete characterization of the heterogeneity of the different mesenchymal stem cell subtypes, their relationships, and the respective markers to identify these cells, these studies provide strong evidence that mesenchymal cells are critical components that maintain HSPCs in the BM niche.

4.3 Endothelial Cells

Endothelial cells are known to secrete specific paracrine growth factors, cytokines (Kobayashi et al., 2010), and adhesion molecules such as

E-selectin, P-selectin, vascular cell adhesion molecule 1 (VCAM-1), and intercellular adhesion molecule 1 (Mazo et al., 1998; Rafii, Mohle, Shapiro, Frey, & Moore, 1997) to regulate the homeostasis of HSPCs. For instance, bone marrow sinusoidal endothelial cells had been shown to express Notch ligands to promote HSPC expansion in culture (Butler et al., 2010). E-selectin has been suggested to be exclusively expressed by endothelial cells in the bone marrow and E-selectin blockade in mice improved HSC survival upon chemotherapeutic agents or irradiation treatment (Winkler et al., 2012). One pathway by which endothelial cells balance the rate of proliferation vs differentiation of HSPCs is possibly through recruiting mTOR and activating AKT signaling. Selective activation of AKT in the endothelial cells of adult mice increased the number of colony-forming units in the spleen and $CD34^{-}Flt3^{-}$ LKS HSPCs with LT-HSC activity in the bone marrow and accelerated hematopoietic recovery (Kobayashi et al., 2010).

Another approach to study the niche cells was to examine which cell populations were the key sources of factors that promote HSC maintenance in vivo. Conditional deletion of the SCF from mesenchymal stem cells (LEPR-expressing) or endothelial cells (Tie2-expressing) depleted endogenous HSCs (Ding et al., 2012). Likewise, conditional deletion of CXCL12 from $Tie2^{+}$ endothelial cells reduced HSC number without inducing stem cell mobilization (Greenbaum et al., 2013).

Therefore, endothelial cell is another regulatory component of the HSC niche and appear to have pleotropic effect on HSC properties, although these studies did not address whether the effect was direct or indirect in vivo.

4.4 Adipocytes

In mouse and human, the percentage of adipocytic cells within the bone marrow, the so-called "fatty marrow," gradually increases as the organism ages, but the function of HSC decreases inversely. This observation led scientists to question whether BM adipocytes influence hematopoiesis. In a mouse model of lipoatrophy, which was genetically incapable of forming adipocytes, and in another model that inhibits adipogenesis by treating mice with the peroxisome proliferator-activated receptor gamma (PPARγ) inhibitor bisphenol-A-diglycidyl-ether, one group demonstrated that bone marrow engraftment by HSCs was accelerated in the fatless mice or in the PPARγ inhibitor-treated mice compared to wild type or untreated controls (Naveiras et al., 2009). These data implicate adipocytes as negative regulator of hematopoiesis, although the resultant alteration in marrow space or the number of other stromal cells in those mouse models could have indirectly contributed to this result.

4.5 Macrophages

Macrophage was recently found to be another player in the niche mediating HSC mobilization through signaling the MSCs. In studying the mobilization of HSCs into the bloodstream in response to G-CSF, scientists found that there was concomitant loss of macrophages and reduced bone formation (Winkler et al., 2010). In vivo depletion of macrophages, in either macrophage Fas-induced apoptosis (Mafia) transgenic mice or by administration of clodronate-loaded liposomes into wild-type mice, led to marked reduction of endosteal osteoblasts and HSC mobilization into blood, mimicking the phenotype observed during G-CSF administration. Specifically, it is the CD169$^+$ macrophage but not the Gr-1^{hi} monocyte that promotes the retention of HSCs in the endosteal niche. CD169$^+$ macrophage secretes oncostatin M (OSM), which in turn induces Nestin$^+$ cells and possibly other MSCs to express CXCL12 via a mitogen-activated protein kinase kinase-p38-signal transducer and activator of a transcription 3-dependent pathway (Albiero et al., 2015; Chow et al., 2011). CXCL12 engagement with CXCR4 receptor on HSC cell surface is crucial for HSC retention in the BM niche. In mice and patients with diabetes, there were increased CD169$^+$ macrophages and elevated OSM in the BM, and reduced HSC mobilization. OSM neutralization (Albiero et al., 2015) and CXCR4 antagonist (Chow et al., 2011) in diabetic mice have been shown to improve G-CSF-induced HSC mobilization, suggesting that targeting CD169$^+$ macrophages or antagonizing OSM may be a strategy to restore niche function, particularly in diabetic settings.

Interestingly, CD169$^+$ macrophage also seems to be a participant that supports late erythroid lineage development. Specific depletion of CD169$^+$ macrophages in mice reduced the number of erythroblasts in the BM but did not lead to overt anemia under homeostatic conditions. However, in the case of hemolytic anemia, CD169$^+$ macrophage depletion led to myeloablation, impaired erythropoietic recovery, and eventually acute blood loss. In contrast, in a JAK2(V617F)-driven mouse model of polycythemia vera, macrophage deletion normalized the erythroid compartment, suggesting that erythropoiesis in polycythemia vera is mediated through macrophages in the BM microenvironment.

Altogether, these studies show that the macrophage is an indispensible cellular participant needed for HSC retention in the BM, through signaling other bone marrow stromal cells, and it also serves as a niche cell that supports erythropoiesis through a yet unknown pathway. The macrophage studies are excellent examples illustrating the concept of one cell type playing multiple roles in supporting different hematopoietic cell processes in the BM, through very different mechanisms.

5. THE SYMPATHETIC NERVOUS SYSTEM

Sensory and autonomic innervation of the bone marrow not only regulates physiological homeostatic trafficking of hematopoietic cells but also enables them to respond to acute stress signals. HSCs and progenitors circulate in the bloodstream under homeostatic conditions, exhibiting robust circadian fluctuations in antiphase with the expression of the chemokine CXCL12 by stromal cells in the BM niche. Sympathetic tone delivered by nerves in the bone marrow is transmitted to stromal cells through the beta(3)-adrenergic receptor, leading to reduced Sp1 transcription factor and the rapid downregulation of CXCL12 expressed by these stromal cells (Mendez-Ferrer, Lucas, Battista, & Frenette, 2008). As such, the cyclical circadian release of HSCs into the bloodstream and the concomitant reduction of CXCL12 in the BM microenvironment are maintained by core genes of the molecular clock through noradrenaline secretion from the sympathetic nervous system (SNS).

The concept that signals from the SNS are critical for HSPC egress from the bone marrow was further illustrated by several studies using complementary mouse models targeting this pathway. In one study, nonmyelinating Schwann cells were proven to be responsible for maintaining HSC hibernation through inducing TGF-β production from stromal cells. Mice with autonomic nerve denervation, reduced number of TGF-β-producing cells, or TGF-β type II receptor-deficient HSCs all exhibited reduced level of Smad activation in HSCs, loss of HSCs from BM, and impaired long-term repopulation activity (Yamazaki et al., 2011). In a different study, UDP-galactose ceramide:galactosyltransferase-deficient ($Cgt^{-/-}$) mice exhibited aberrant nerve conduction and displayed no HSPC egress from the BM following G-CSF or fucoidan administration (Katayama et al., 2006). Pharmacological or genetic ablation of adrenergic neurotransmission in mice indicates that norepinephrine (NE) signaling controls G-CSF-induced osteolineage cell suppression, CXCL12 downregulation from stromal cells, and HSPC mobilization. In contrast, administration of a β(2) adrenergic agonist enhances mobilization in both NE-deficient and control mice, suggesting that the SNS is responsible for the attraction of stem cells to their niche. It was found in later studies that the cells that relay the adrenergic signals from the SNS to HSPCs appear to be nestin$^+$ MSCs. First, nestin$^+$ cells display a number of genes that regulate HSPC mobilization, including CXCL12, Kitl, Angpt1, IL7, and Vcam-1, and the expression of these genes decreases upon G-CSF stimulation or β(3) adrenoreceptor activation. Second, purified HSCs were found to home near nestin$^+$ MSCs in the bone marrow of lethally irradiated mice. Third, administration of parathyroid hormone, an agent that favors HSPC engraftment in the BM (Adams et al., 2007), doubles the number of HSC-associated nestin$^+$ MSCs and their osteoblastic differentiation. Most importantly, in vivo nestin$^+$ cell depletion significantly reduces BM homing of HSPCs (Mendez-Ferrer et al., 2010).

It is known that some diabetic patients fail to achieve sufficient CD34$^+$ HSPCs upon G-CSF stimulation. Using mouse models of type 1 (streptozotocin-induced) and type 2 (db/db) diabetes, scientists found that HSPCs that were unable to egress upon G-CSF treatment and localized aberrantly in the BM niche with altered pattern of catecholaminergic nerve termini (Ferraro et al., 2011). This altered sympathetic innervation was associated with decreased sensitization of beta-adrenergic receptor in nestin-GFP$^+$ cells, which was in turn unable to downmodulate CXCL12 in response to G-CSF treatment. The HPSC mobilization defect in diabetic mouse models can be rescued by pharmacological inhibition of the interaction of CXCL12 with its receptor CXCR4 using the receptor antagonist AMD3100.

In summary, these studies described an unprecedented pathway of mesenchymal cell control of stem cell trafficking—through the SNS. Sympathetic nerve fiber was an indispensible component in relaying signals from the niche cells to the HSCs and regulates stem cell trafficking during hibernation, circadian oscillation, and G-CSF-induced mobilization. These data offer new intervention opportunities to overcome poor HPSC mobilization in clinical and disease settings including diabetes.

6. HSC MOBILIZATION FROM THE BM NICHE

HSCs for HSC transplantation (HSCT) can be derived from BM, peripheral blood, or umbilical cord blood. Harvesting adult HSCs from peripheral blood is relatively easy and much less painful than from the

BM, but the frequency of HSCs in the peripheral blood is low. Mobilization of HSCs from the BM niche has thus been explored as a mechanism to enhance HSC frequency in the donor's peripheral blood for the purpose of HSCT.

G-CSF was first described as an agent that causes HSC mobilization in 1988 (Socinski et al., 1988) and has since been the mainstay in the clinic for HSC mobilization, yet its mechanisms of action are still not fully understood today. Given that HSCs themselves do not express the G-CSF receptor but rather cells of the myelomonocytic series, including macrophages and osteomacs, express the receptor, it was hypothesized that mobilization through G-CSF is indirect and several mechanisms have been proposed. Increasing evidence suggests that one of the pathways involves activation of osteomacs and macrophages, which in turn regulates neighboring stromal cells to decrease SDF-1 production (Albiero et al., 2015; Chow et al., 2011). Reduced stromal SDF-1 leads to attenuated anchoring of HSCs to the BM stroma through the CXCR4 receptor expressed on HSCs. Additional pathways involve degradation of VCAM-1, fibronectin, and OPN, leading to reduced cellular adhesion of HSC to stromal cells through very late antigen 4 (VLA-4) adhesion receptor (Bonig & Papayannopoulou, 2012; Levesque, Takamatsu, Nilsson, Haylock, & Simmons, 2001). Recently, two independent studies show that G-CSF-induced mobilization signal is relayed through the SNS, as pharmacological intervention or sympathectomy with β-blockers results in impaired HSC mobilization in mice (Ferraro et al., 2011; Katayama et al., 2006). Therefore, the mechanism of G-CSF-induced HSC mobilization is indirect, multifactorial, and likely involves multiple stromal components.

SDF-1 (CXCL12) is a CXC chemokine, secreted by various BM stromal cells, including CAR, nestin$^+$ cells, osteolineage cells, and endothelial cells. The interaction of SDF-1 with its receptor CXCR4 on HSC plays a pivotal role in HSC retention to the niche, as treatment of human cells with a CXCR4 antibody prevented their engraftment in a xenotransplantation model (Peled et al., 1999), and treatment of mice with neutralizing anti-SDF-1 or anti-CXCR4 antibodies inhibited HSC mobilization (Petit et al., 2002). These discoveries led to the development of plerixafor (AMD3100), a US Food and Drug Administration (FDA) approved CXCR4 mobilizing agent, that disrupts the SDF-1/CXCR4 axis in a synergistic way to G-CSF.

The β1 integrin, VLA-4, is expressed by HSCs and facilitates their adhesion on BM stroma through interaction with vascular cell adhesion molecule

1 (VCAM-1), fibronectin, and OPN. Studies in mice, primates, and humans have shown that administration of the anti-VLA-4 monoclonal antibody (natalizumab) or blockade of its ligands leads to a potent mobilization effect (Bonig & Papayannopoulou, 2012; Bonig & Papayannopoulou, 2013; Zohren et al., 2008). The disruption of the VLA-4/VCAM-1 axis, however, is dependent upon induction by G-CSF and CXCR4 inhibitors.

The interaction between CD44 on HSC and hyaluronic acid, a component of the extracellular matrix protein, is essential for the retention of HSC in the BM. The most hyaluronic acid-rich regions of the bone marrow are the endosteum and the sinusoidal endothelium. CD44 cleavage on HSC membrane is associated with mobilization and is dependent on the abundance of its proteolytic enzyme, metalloproteinase (MMP). G-CSF administration has been documented to increase MMP on $CD34^+$ cells, resulting in cleavage of CD44 on HSC cell surface and HSC egress from the BM (Avigdor et al., 2004; Vagima et al., 2009). In addition, homing of HSC is impaired upon administration of anti-CD44 antibodies in mice, highlighting the importance of CD44 in HSC retention at the niche.

Heparan sulfate proteoglycans (HSPGs) are thought to serve as extracellular binding partners for secreted signaling molecules. In hematopoiesis, HSPG has been theorized a potential role in bone marrow compartmentalization, by forming matrices that retain the right cytokines and/or morphogens to the vicinity of HSPCs. It has been hypothesized that gradients of cytokine and morphogen maintained by interaction with locally secreted matrix proteins are essential in sustaining the HSPC niche. Indeed, conditional deletion of the *Ext1* gene, a glycosyltransferase essential for the synthesis of heparin sulfate, in $Mx1^+$ BM stromal cells affected HSPC localization and retention in the BM, in part by modulating VCAM-1 (Saez et al., 2014). This mechanistic pathway was further confirmed by the data that showed pharmacologic inhibition of endogenous heparin sulfate enhanced the mobilization efficacy of G-CSF, including in the setting of mobilization resistance in a murine diabetes model. HSPCs mobilized by heparin sulfate inhibition were shown to have improved reconstitution ability in primary and secondary transplanted mice compared to G-CSF-mobilized HSPCs. Finally, HSPCs engrafted efficiently in the BM of *Ext1*-deficient mice without cytotoxic conditioning. These findings suggest that targeting heparin sulfate or the enzyme, *Ext1*, may provide novel means to mobilize HSCs of improved quality or to achieve noncytotoxic conditioning, both are of critical value for clinical HSC transplantation.

7. NICHE OF HEMATOPOIETIC MALIGNANCIES

In analogy to normal HSCs, cancer-initiating cells also require a proper niche to thrive and expand. Recent discoveries of the niche's critical involvement in the induction and development of hematologic neoplasia further stressed the importance of defining and characterizing the stromal subpopulations that support these pathologic processes.

It is well documented that in diseased and aged conditions, not only the hematopoietic cells exhibit genetic alterations, but massive reorganization of the BM niche also takes place. These alterations within the BM microenvironment include osteoporosis in aged individuals and increased vascularization in hematopoietic neoplasms and leukemias. In the case of multiple myeloma, it is well established that myeloma patients experience acute bone lesions. There was activation of osteoclastogenesis and suppression of osteoblastic activity due to an unbalanced RANK/OPG ratio, and dysregulation of a variety of inflammatory cytokines such as MIP-1α, MIP-1β, IL-3, IL-6, IL-1β, TNF-α, HGF, VEGF, OPN, and SDF-1α in the myeloma niche (Reagan, Liaw, Rosen, & Ghobrial, 2015). While these were thought to be evidence of disease induced changes in the microenvironment, it was not known until the first two reports that showed genetic mutation in the BM niche itself can induce irreversible intrinsic changes in the hematopoietic cells. In these studies, mice deficient for the retinoblastoma protein (RB) (Walkley, Shea, Sims, Purton, & Orkin, 2007a) or the retinoic acid receptor gamma (RARγ) (Walkley et al., 2007b) developed myelodysplasia. Unexpectedly, transplant studies revealed that the disease was not intrinsic to the hematopoietic cells themselves because BM from wild-type mice transplanted into an $RB^{-/-}$ or $RARg^{-/-}$ recipient mice developed the myeloproliferative disease, suggesting for the first time that mutation in the BM microenvironment can confer neoplastic development of hematopoietic cells. Following these studies, our laboratory discovered that deletion of the RNase III endonuclease, Dicer1, in BM osteoprogenitor cells but not in mature osteoblasts resulted in myelodysplasia and predisposition to acute myelogenous leukemia (AML) (Raaijmakers et al., 2010). Interestingly, mice with osteoprogenitor-specific Dicer1 deletion had reduced expression of Sbds, the gene mutated in Shwachman–Bodian–Diamond syndrome, corroborating with a human disease condition characterized by bone marrow failure and leukemia predisposition. Deletion of Sbds in mouse osteoprogenitors further confirmed the development of myelodysplasia, proving the perturbation of specific subsets of mesenchymal cells can disorient the differentiation, proliferation, and apoptosis of parenchymal hematopoietic cells. Confirming the concept of niche-induced leukemia, another group found that activating mutation of β-catenin in osteoblasts led to the development of AML with cell autonomous disease progression (Kode et al., 2014), likely through elevated Notch signaling in hematopoietic progenitor cells. Collectively, these in vivo mouse studies indicate that mutation in the niche can be the sole cause of hematopoietic disorders, and that primary stromal cell dysfunction can result in secondary hematologic pathologies.

It appears that the BM niche not only capable of inducing leukemia, but during disease state, it remodels itself to further support the growth of specific leukemias. One of the factors secreted by osteoblasts that differentially regulate the development of chronic myelogenous leukemia (CML) and AML is transforming growth factor beta 1 (TGF-β1). Osteoblast-specific activation of the PTH receptor attenuated BCR-ABL1-induced CML-like myeloproliferative neoplasia while enhanced MLL-AF9-induced AML in mouse models, possibly through opposing effects of increased TGF-β1 on the respective leukemia initiating cells (Krause et al., 2013). This study illustrates the distinct niche requirement for CML and AML and suggests that niche modulation may have very specific outcome depending on the type of the disease.

Aberrant or malignant hematopoietic cells can also confer signals to the stromal cells to initiate niche remodeling that reinforces leukemic growth. In the context of myeloproliferative neoplasia, leukemic blasts make direct contact with MSCs and produce TPO and CCL3 to stimulate MSCs to overproduce functionally altered osteolineage cells (Schepers et al., 2013). These myeloproliferative neoplasia-expanded osteolineage cells have altered gene expression profile and accumulate in the BM cavity as myelofibrotic cells. They exhibit impaired ability to support normal HSCs, yet produce inflammatory signals that effectively support leukemic development. In a BCR/ABL-induced model of CML, leukemic cells produce proinflammatory cytokines to alter the microenvironment such that it favors disease development (Welner et al., 2015). Interestingly, the normal bystander cells acquired gene expression profiles resembling their malignant counterparts. IL-6 is a strong candidate responsible for most of these changes. IL-6 produced by CML cells induces normal hematopoietic progenitors to suppress lymphoid lineage differentiation but expand myeloid lineage cells (Reynaud et al., 2011; Welner et al., 2015), a phenomenon highly reinforcing disease progression. Mice deficient of IL-6 receptor on hematopoietic cells or treated with anti-IL-6 neutralizing antibody rescued the skewed lymphoid and myeloid cell ratio and attenuated the CML (Welner et al., 2015), suggesting that the differentiation bias caused by IL-6 is reversible and blockade of IL-6 may be a potential mechanism to treat CML.

As mentioned in the earlier section, SNS regulates normal HSC retention in the niche by delivering sympathetic tone to stromal cells in the bone marrow through the β(3)-adrenergic receptor, causing a downregulation of CXCL12 expressed by stromal cells and attenuated engagement of CXCR4 receptor on HSCs. The end result is reduced lodgment of HSCs in the BM microenvironment. SNS also participates in AML development, although through a different mechanism. In a mouse model of SNS neuropathy, BM infiltration with MLL-AF9 AML cells was significantly enhanced. Development of AML was found to disrupt SNS nerves, decrease the number of HSC-maintaining $NG2^+$ periarteriolar niche cells, reduce production of cytokines that support normal HSC maintenance, and prime $nestin^+$ niche cells for osteoblastic differentiation. But most importantly, transplantation of MLL-AF9 leukemic cells into mice treated with adrenergic receptor β2 but not β3 antagonist, or mice deficient of adrenergic receptor β2, significantly augmented leukemic cell engraftment and number in the bone marrow in comparison to control mice. These studies indicate that malignant cells are capable of reorganizing its surrounding niche architecture and supportive cellular neighbors to transform the normal HSC niche into a cancer-supporting niche. The current developing concept is that there is a coevolution that exists between cancer cells and the niche, and that there are dynamic interactions between the two. This implies that the BM environment adapts according to the different stages of the disease. These data stress the importance of considering the corresponding changes that occur in the niche while targeting cancer. Cutting off environmental support may improve the efficacy to eradicate cancer cells.

8. PERSPECTIVES

Decades of experimentation have validated the importance of the niche in stem cell regulation and revealed some fundamental insights about the cellular and molecular participants involved. Studies of mutated BM niches have revealed unprecedented information on niche-induced and niche-maintained hematologic neoplasia and leukemias. Yet the complete list of the stromal participants in the normal and leukemic hematopoietic niches remains incomplete and many outstanding questions regarding the biological nature of HSC under homeostasis and in aberrant conditions are unanswered. Here, we will provide our perspectives on some future directions.

Dissecting the complexity of the niche components within the BM remains a high order priority. To date, our knowledge of the regulatory cell types that exist in the bone marrow "stroma" is limited to a few that can be targeted using existing cell markers, and the list of niche participants is far from complete. Single-cell sequencing and single-cell proteomic technologies will enable the study of cellular heterogeneity in the stroma independent of any existing knowledge about these cells.

HSCs in the fetal liver expand on a daily basis but maintained at a steady level under homeostasis in the bone marrow. Differences in the cellular components of these two sites may provide insights to achieve HSC expansion for transplantation therapies. Much can be learned from parallel stem cell systems in different tissues or even species. Comparison of stromal components among different organs from the adult may inform us about shared characteristics or functions among these niches. As aforementioned, the physiological site, the cellular, and the molecular components of the BM niche in mouse and human do not directly translate. Comparing homologous niches between human and rodent may answer why certain leukemias develop in one species but not the other, and also aid our effort in translating therapeutic molecular targets from mice to human.

New technologies will enable us to pinpoint stem cell engagement with particular cellular components in the niche with increased precision and study their dynamic interactions over time. For instance, we can now achieve a highly precise, short-range cell perturbation using laser capture (unpublished). Gene expression can be manipulated in single cells in vivo using nanowave or fluorescent protein conversion laser technologies. Multifluorescent transgenic mouse models will allow simultaneous labeling and tracing of multiple stromal subtypes in an animal. Conventional gene knockout studies using homologous recombination enable the knockout of one gene at a time. However, if the gene is involved in a compensatory network, this strategy may not create a visible phenotype. The emerging CRISPR/Cas9 technology can disrupt multiple genes simultaneously in a cell or in a mouse, allowing the study of multifactorial pathways that will only manifest a phenotype when multiple members of the gene network are perturbed.

Recent literature reporting niche involvement in myelodysplasia and leukemias only revealed the tip of the iceberg. How the niche contributes to hematologic malignancies or bone marrow diseases remains largely unexplored. Is there competition of anatomical space or niche resources between normal and malignant cells? Is there a hierarchy of components in the niche that confers cancer cell survival? Can specific intervention of these niche interactions improve therapeutic outcome?

The promises of unraveling niche contribution to normal and disease physiology are warranted with new tools at hand. It is anticipated that with better understanding of what specific subsets of hematopoietic cells are governing by what stromal elements, the ability to engineer a particular hematopoietic outcome will become feasible. Further, gaining a "systems" level understanding of the hematopoietic bone marrow may provide useful paradigms for studying other tissue niches with the long-term goal of ultimately manipulating better regenerative processes in contexts of disease.

REFERENCES

Adams, G. B., et al. (2007). Therapeutic targeting of a stem cell niche. *Nature Biotechnology*, *25*, 238–243.

Albiero, M., et al. (2015). Bone marrow macrophages contribute to diabetic stem cell mobilopathy by producing Oncostatin M. *Diabetes*, *64*, 2957–2968.

Avigdor, A., et al. (2004). CD44 and hyaluronic acid cooperate with SDF-1 in the trafficking of human CD34+ stem/progenitor cells to bone marrow. *Blood*, *103*, 2981–2989.

Bonig, H., & Papayannopoulou, T. (2012). Mobilization of hematopoietic stem/progenitor cells: General principles and molecular mechanisms. *Methods in Molecular Biology*, *904*, 1–14.

Bonig, H., & Papayannopoulou, T. (2013). Hematopoietic stem cell mobilization: Updated conceptual renditions. *Leukemia*, *27*, 24–31.

Butler, J. M., et al. (2010). Endothelial cells are essential for the self-renewal and repopulation of Notch-dependent hematopoietic stem cells. *Cell Stem Cell*, *6*, 251–264.

Calvi, L. M., et al. (2003). Osteoblastic cells regulate the haematopoietic stem cell niche. *Nature*, *425*, 841–846.

Chow, A., et al. (2011). Bone marrow CD169+ macrophages promote the retention of hematopoietic stem and progenitor cells in the mesenchymal stem cell niche. *The Journal of Experimental Medicine*, *208*, 261–271.

Dexter, T. M., Allen, T. D., & Lajtha, L. G. (1977). Conditions controlling the proliferation of haemopoietic stem cells in vitro. *Journal of Cellular Physiology*, *91*, 335–344.

Ding, L., & Morrison, S. J. (2013). Haematopoietic stem cells and early lymphoid progenitors occupy distinct bone marrow niches. *Nature*, *495*, 231–235.

Ding, L., Saunders, T. L., Enikolopov, G., & Morrison, S. J. (2012). Endothelial and perivascular cells maintain haematopoietic stem cells. *Nature*, *481*, 457–462.

Ferraro, F., et al. (2011). Diabetes impairs hematopoietic stem cell mobilization by altering niche function. *Science Translational Medicine*, *3*, 104ra101.

Greenbaum, A., et al. (2013). CXCL12 in early mesenchymal progenitors is required for haematopoietic stem-cell maintenance. *Nature*, *495*, 227–230.
Katayama, Y., et al. (2006). Signals from the sympathetic nervous system regulate hematopoietic stem cell egress from bone marrow. *Cell*, *124*, 407–421.
Kiel, M. J., Yilmaz, O. H., Iwashita, T., Terhorst, C., & Morrison, S. J. (2005). SLAM family receptors distinguish hematopoietic stem and progenitor cells and reveal endothelial niches for stem cells. *Cell*, *121*, 1109–1121.
Kobayashi, H., et al. (2010). Angiocrine factors from Akt-activated endothelial cells balance self-renewal and differentiation of haematopoietic stem cells. *Nature Cell Biology*, *12*, 1046–1056.
Kode, A., et al. (2014). Leukaemogenesis induced by an activating beta-catenin mutation in osteoblasts. *Nature*, *506*, 240–244.
Krause, D. S., et al. (2013). Differential regulation of myeloid leukemias by the bone marrow microenvironment. *Nature Medicine*, *19*, 1513–1517.
Kricun, M. E. (1985). Red-yellow marrow conversion: Its effect on the location of some solitary bone lesions. *Skeletal Radiology*, *14*, 10–19.
Kunisaki, Y., et al. (2013). Arteriolar niches maintain haematopoietic stem cell quiescence. *Nature*, *502*, 637–643.
Levesque, J. P., Takamatsu, Y., Nilsson, S. K., Haylock, D. N., & Simmons, P. J. (2001). Vascular cell adhesion molecule-1 (CD106) is cleaved by neutrophil proteases in the bone marrow following hematopoietic progenitor cell mobilization by granulocyte colony-stimulating factor. *Blood*, *98*, 1289–1297.
Lo Celso, C., et al. (2009). Live-animal tracking of individual haematopoietic stem/progenitor cells in their niche. *Nature*, *457*, 92–96.
Matsuoka, Y., et al. (2015). Prospectively isolated human bone marrow cell-derived MSCs support primitive human CD34-negative hematopoietic stem cells. *Stem Cells*, *33*, 1554–1565.
Mazo, I. B., et al. (1998). Hematopoietic progenitor cell rolling in bone marrow microvessels: Parallel contributions by endothelial selectins and vascular cell adhesion molecule 1. *The Journal of Experimental Medicine*, *188*, 465–474.
Mendez-Ferrer, S., Lucas, D., Battista, M., & Frenette, P. S. (2008). Haematopoietic stem cell release is regulated by circadian oscillations. *Nature*, *452*, 442–447.
Mendez-Ferrer, S., et al. (2010). Mesenchymal and haematopoietic stem cells form a unique bone marrow niche. *Nature*, *466*, 829–834.
Morikawa, S., et al. (2009). Prospective identification, isolation, and systemic transplantation of multipotent mesenchymal stem cells in murine bone marrow. *The Journal of Experimental Medicine*, *206*, 2483–2496.
Naveiras, O., et al. (2009). Bone-marrow adipocytes as negative regulators of the haematopoietic microenvironment. *Nature*, *460*, 259–263.
Nombela-Arrieta, C., et al. (2013). Quantitative imaging of haematopoietic stem and progenitor cell localization and hypoxic status in the bone marrow microenvironment. *Nature Cell Biology*, *15*, 533–543.
Omatsu, Y., et al. (2010). The essential functions of adipo-osteogenic progenitors as the hematopoietic stem and progenitor cell niche. *Immunity*, *33*, 387–399.
Park, D., et al. (2012). Endogenous bone marrow MSCs are dynamic, fate-restricted participants in bone maintenance and regeneration. *Cell Stem Cell*, *10*, 259–272.
Peled, A., et al. (1999). Dependence of human stem cell engraftment and repopulation of NOD/SCID mice on CXCR4. *Science*, *283*, 845–848.
Petit, I., et al. (2002). G-CSF induces stem cell mobilization by decreasing bone marrow SDF-1 and up-regulating CXCR4. *Nature Immunology*, *3*, 687–694.
Pinho, S., et al. (2013). PDGFRalpha and CD51 mark human nestin+ sphere-forming mesenchymal stem cells capable of hematopoietic progenitor cell expansion. *The Journal of Experimental Medicine*, *210*, 1351–1367.

Raaijmakers, M. H., et al. (2010). Bone progenitor dysfunction induces myelodysplasia and secondary leukaemia. *Nature, 464*, 852–857.
Rafii, S., Mohle, R., Shapiro, F., Frey, B. M., & Moore, M. A. (1997). Regulation of hematopoiesis by microvascular endothelium. *Leukemia & Lymphoma, 27*, 375–386.
Reagan, M. R., Liaw, L., Rosen, C. J., & Ghobrial, I. M. (2015). Dynamic interplay between bone and multiple myeloma: Emerging roles of the osteoblast. *Bone, 75*, 161–169.
Reynaud, D., et al. (2011). IL-6 controls leukemic multipotent progenitor cell fate and contributes to chronic myelogenous leukemia development. *Cancer Cell, 20*, 661–673.
Sacchetti, B., et al. (2007). Self-renewing osteoprogenitors in bone marrow sinusoids can organize a hematopoietic microenvironment. *Cell, 131*, 324–336.
Saez, B., et al. (2014). Inhibiting stromal cell heparan sulfate synthesis improves stem cell mobilization and enables engraftment without cytotoxic conditioning. *Blood, 124*, 2937–2947.
Sato, M., et al. (2013). Osteocytes regulate primary lymphoid organs and fat metabolism. *Cell Metabolism, 18*, 749–758.
Schepers, K., et al. (2013). Myeloproliferative neoplasia remodels the endosteal bone marrow niche into a self-reinforcing leukemic niche. *Cell Stem Cell, 13*, 285–299.
Schofield, R. (1978). The relationship between the spleen colony-forming cell and the haemopoietic stem cell. *Blood Cells, 4*, 7–25.
Sipkins, D. A., et al. (2005). In vivo imaging of specialized bone marrow endothelial microdomains for tumour engraftment. *Nature, 435*, 969–973.
Socinski, M. A., et al. (1988). Granulocyte-macrophage colony stimulating factor expands the circulating haemopoietic progenitor cell compartment in man. *Lancet, 1*, 1194–1198.
Spencer, J. A., et al. (2014). Direct measurement of local oxygen concentration in the bone marrow of live animals. *Nature, 508*, 269–273.
Sugiyama, T., Kohara, H., Noda, M., & Nagasawa, T. (2006). Maintenance of the hematopoietic stem cell pool by CXCL12-CXCR4 chemokine signaling in bone marrow stromal cell niches. *Immunity, 25*, 977–988.
Till, J. E., & McCulloch, C. E. (1961). A direct measurement of the radiation sensitivity of normal mouse bone marrow cells. *Radiation Research, 14*, 213–222.
Vagima, Y., et al. (2009). MT1-MMP and RECK are involved in human CD34+ progenitor cell retention, egress, and mobilization. *The Journal of Clinical Investigation, 119*, 492–503.
Visnjic, D., et al. (2004). Hematopoiesis is severely altered in mice with an induced osteoblast deficiency. *Blood, 103*, 3258–3264.
Walkley, C. R., Shea, J. M., Sims, N. A., Purton, L. E., & Orkin, S. H. (2007a). Rb regulates interactions between hematopoietic stem cells and their bone marrow microenvironment. *Cell, 129*, 1081–1095.
Walkley, C. R., et al. (2007b). A microenvironment-induced myeloproliferative syndrome caused by retinoic acid receptor gamma deficiency. *Cell, 129*, 1097–1110.
Welner, R. S., et al. (2015). Treatment of chronic myelogenous leukemia by blocking cytokine alterations found in normal stem and progenitor cells. *Cancer Cell, 27*, 671–681.
Winkler, I. G., et al. (2010). Bone marrow macrophages maintain hematopoietic stem cell (HSC) niches and their depletion mobilizes HSCs. *Blood, 116*, 4815–4828.
Winkler, I. G., et al. (2012). Vascular niche E-selectin regulates hematopoietic stem cell dormancy, self renewal and chemoresistance. *Nature Medicine, 18*, 1651–1657.
Worton, R. G., McCulloch, E. A., & Till, J. E. (1969). Physical separation of hemopoietic stem cells differing in their capacity for self-renewal. *The Journal of Experimental Medicine, 130*, 91–103.
Wright, D. E., Wagers, A. J., Gulati, A. P., Johnson, F. L., & Weissman, I. L. (2001). Physiological migration of hematopoietic stem and progenitor cells. *Science, 294*, 1933–1936.
Yamazaki, S., et al. (2011). Nonmyelinating Schwann cells maintain hematopoietic stem cell hibernation in the bone marrow niche. *Cell, 147*, 1146–1158.

Yu, V. W., et al. (2015). Specific bone cells produce DLL4 to generate thymus-seeding progenitors from bone marrow. *The Journal of Experimental Medicine, 212*, 759–774.

Zhang, J., et al. (2003). Identification of the haematopoietic stem cell niche and control of the niche size. *Nature, 425*, 836–841.

Zhu, J., et al. (2007). Osteoblasts support B-lymphocyte commitment and differentiation from hematopoietic stem cells. *Blood, 109*, 3706–3712.

Zohren, F., et al. (2008). The monoclonal anti-VLA-4 antibody natalizumab mobilizes CD34+ hematopoietic progenitor cells in humans. *Blood, 111*, 3893–3895.

CHAPTER THREE

The Hematopoietic Stem and Progenitor Cell Cistrome: GATA Factor-Dependent *cis*-Regulatory Mechanisms

K.J. Hewitt*,†, K.D. Johnson*,†, X. Gao*,†, S. Keles‡, E.H. Bresnick*,†,1
*UW-Carbone Cancer Center, University of Wisconsin School of Medicine and Public Health, Madison, WI, United States
†UW-Madison Blood Research Program, Madison, WI, United States
‡University of Wisconsin School of Medicine and Public Health, Madison, WI, United States
[1]Corresponding author: e-mail address: ehbresni@wisc.edu

Contents

Abstract

Transcriptional regulators mediate the genesis and function of the hematopoietic system by binding complex ensembles of *cis*-regulatory elements to establish genetic networks. While thousands to millions of any given *cis*-element resides in a genome, how transcriptional regulators select these sites and how site attributes dictate functional output is not well understood. An instructive system to address this problem involves the GATA family of transcription factors that control vital developmental and physiological processes and are linked to multiple human pathologies. Although GATA factors bind DNA motifs harboring the sequence GATA, only a very small subset

Current Topics in Developmental Biology, Volume 118
ISSN 0070-2153
http://dx.doi.org/10.1016/bs.ctdb.2016.01.002

of these abundant motifs are occupied in genomes. Mechanistic studies revealed a unique configuration of a GATA factor-regulated *cis*-element consisting of an E-box and a downstream GATA motif separated by a short DNA spacer. GATA-1- or GATA-2-containing multiprotein complexes at these composite elements control transcription of genes critical for hematopoietic stem cell emergence in the mammalian embryo, hematopoietic progenitor cell regulation, and erythroid cell maturation. Other constituents of the complex include the basic helix–loop–loop transcription factor Scl/TAL1, its heterodimeric partner E2A, and the Lim domain proteins LMO2 and LDB1. This chapter reviews the structure/function of E-box-GATA composite *cis*-elements, which collectively constitute an important sector of the hematopoietic stem and progenitor cell cistrome.

1. *CIS*-REGULATORY MECHANISMS

Many proteins and protein networks are implicated in controlling the generation and function of hematopoietic stem and progenitor cells (HSPCs) and their hematopoietic cell progeny (Rossi et al., 2012). However, only recently have DNA regulatory motifs (*cis*-elements) in the genome been described that nonredundantly control cell fate decisions during hematopoiesis (Gao et al., 2013; Johnson et al., 2012, 2015). As a single *cis*-element integrates activities of a limited ensemble of regulatory proteins and signals, it provides a unique path to discover important pathways and regulatory networks. While *cis*-elements reside at distinct gene locations, including promoters, enhancers, and insulators, and there can be both similarities and differences in their functions at these locations, the focus of this review is on enhancer mechanisms.

Enhancers are short (~50–500 bp) DNA sequences residing distal to genes or in introns that function in *cis* to increase transcription (Plank & Dean, 2014). Typical enhancers contain clustered *cis*-elements, each binding one or more transcription factors (TF). Mechanistic steps in enhancer function include TF recruitment of coregulators that modify or remodel chromatin, thereby regulating chromatin template access to additional regulators and RNA Polymerase II (Pol II). Whereas Pol II recruitment to promoters is commonly stimulated by enhancers, enhancers also recruit Pol II directly to the enhancer (Johnson, Christensen, Zhao, & Bresnick, 2001; Johnson et al., 2003; Szutorisz, Dillon, & Tora, 2005). Consequences of Pol II recruitment to an enhancer include generation of enhancer-derived RNA transcripts (Lai & Shiekhattar, 2014), the exact function of which remains controversial, and Pol II transfer to the promoter, a poorly understood reaction (Johnson et al., 2001; Vieira et al., 2004). In principle, transfer may involve processive Pol II tracking along the template (Ling, Ainol, et al., 2004) or DNA/chromatin looping in which the enhancer complex physically associates with promoter components (Cullen, Kladde, & Seyfred, 1993; Dekker, Rippe, Dekker, & Kleckner, 2002; Schleif, 1992; Willis & Seyfred, 1996). Finally, enhancers control locus subnuclear localization, which can expel a locus from a nuclear microenvironment unfavorable for high-level transcription, while positioning it in a microenvironment permissive for high-level transcription (Fraser & Bickmore, 2007; Jackson & Cook, 1995; Kosak et al., 2002; Lee, Johnson, Boyer, & Bresnick, 2011; Lee et al., 2009; Rajapakse & Groudine, 2011; Schubeler et al., 2000; Zullo et al., 2012). Enhancer mechanisms are vital determinants of cell phenotype in a wide swath of biological and pathological contexts.

2. GATA FACTOR INTERACTIONS WITH DNA AND CHROMATIN

Enhancers involved in the development and function of the hematopoietic system are often activated by members of the dual zinc finger GATA TF family (GATA-1 to GATA-6) (Bresnick, Katsumura, Lee, Johnson, & Perkins, 2012; Charron & Nemer, 1999; Molkentin, 2000), which recognize naked DNA containing the consensus (A/T)GATA (A/G) (Evans, Reitman, & Felsenfeld, 1988; Ko & Engel, 1993; Martin & Orkin, 1989; Merika & Orkin, 1993). While certain TFs commonly localize to promoter complexes, GATA factors have a considerably higher propensity to occupy enhancers (Cheng et al., 2009; Fujiwara et al., 2009; Kang et al., 2012; Linnemann, O'Geen, Keles, Farnham, & Bresnick, 2011; Yu et al., 2009). GATA-1, -2, and -3 are related to GATA factors that control stem and progenitor cell transitions that give rise to the hematopoietic system (Bresnick et al., 2012). These proteins also control important functions of differentiated hematopoietic cells.

This review will focus on GATA-1 (Evans & Felsenfeld, 1989; Tsai et al., 1989) and GATA-2 (Dorfman, Wilson, Bruns, & Orkin, 1992; Tsai et al., 1994), key regulators of HSPCs and myeloerythroid lineage cells in which considerable mechanistic/biological insights and principles have accrued. We will not discuss GATA-3, an important regulator of lymphopoiesis (Ho et al., 1991; Ko et al., 1991; Scripture-Adams et al., 2014; Ting, Olson, Barton, & Leiden, 1996). GATA-2 is expressed in HSPCs and promotes their generation and function (de Pater et al., 2013;

Gao et al., 2013; Johnson et al., 2012; Ling, Ottersbach, et al., 2004; Rodrigues et al., 2005; Tsai et al., 1994). The targeted deletion of *Gata2* in the mouse is embryonic lethal at E10.5 due to lack of hematopoietic development (Tsai et al., 1994). GATA-2 is also expressed in early erythroid precursors, megakaryocytes, and mast cells (Huang et al., 2009; Mouthon et al., 1993; Tsai & Orkin, 1997; Weiss, Keller, & Orkin, 1994; Zon et al., 1991). GATA-2 expression in HSPCs requires an intronic enhancer (+9.5) containing an E-box–GATA composite element (Gao et al., 2013; Grass et al., 2006; Johnson et al., 2012; Sanalkumar et al., 2014). The +9.5 site resides 9.5 kb downstream of the *Gata2* transcription start site, and its deletion abrogates hematopoietic stem cell (HSC) generation in the aorta-gonad-mesonephros during embryogenesis (Gao et al., 2013). The GATA-2 concentration must be maintained within a critical physiological window, as ectopically elevated GATA-2 blocks erythroid cell differentiation (Briegel et al., 1993; Persons et al., 1999), and HSC expansion and function are defective in $Gata2^{+/-}$ mice (Ling, Ottersbach, et al., 2004; Rodrigues et al., 2005). Consistent with the tight control of *Gata2* transcription, GATA-2 protein has a short half-life of <1 h, which provides opportunities to rapidly change its cellular concentration (Lurie, Boyer, Grass, & Bresnick, 2008). GATA-2 deficiency induces a wide spectrum of human disorders (Dickinson et al., 2014; Spinner et al., 2013). Familial and sporadic heterozygous mutations in *GATA2* cause a complex immunodeficiency syndrome (MonoMAC) that often progresses to myelodysplastic syndrome and acute myeloid leukemia (AML; Dickinson et al., 2011; Hahn et al., 2011; Hsu et al., 2011). Mutation of the *GATA2* +9.5 *cis*-element also yields MonoMAC syndrome (Hsu et al., 2013; Johnson et al., 2012). Furthermore, high GATA-2 correlates with poor prognosis of pediatric and adult AML (Luesink et al., 2012; Vicente et al., 2012).

Unlike GATA-2, GATA-1 is not required for HSPC genesis and function. GATA-1 is expressed predominantly in megakaryocyte and erythroid lineage cells and regulates red blood cell, platelet, mast cell, and eosinophil differentiation (Fujiwara, Browne, Cunniff, Goff, & Orkin, 1996; Pevny et al., 1995, 1991; Shivdasani, Fujiwara, McDevitt, & Orkin, 1997; Simon et al., 1992; Weiss et al., 1994). Mutational disruption of *GATA1* yields familial dyserythropoietic anemia and thrombocytopenia (Nichols et al., 2000) and megakaryoblastic leukemia (Wechsler et al., 2002). In many contexts, GATA-1 requires a multi-zinc finger coregulator Friend of GATA-1 (FOG-1) (Tsang, Fujiwara, Hom, & Orkin, 1998; Tsang et al., 1997) for both activation and repression of target gene transcription

(Crispino, Lodish, MacKay, & Orkin, 1999). Though FOG-1 lacks DNA-binding activity, it promotes GATA-1 chromatin occupancy (Letting, Chen, Rakowski, Reedy, & Blobel, 2004; Pal et al., 2004) and recruits the chromatin-remodeling NuRD complex to GATA-1-bound sites (Gao et al., 2010; Hong et al., 2005; Miccio et al., 2010; Ross, Mavoungou, Bresnick, & Milot, 2012). During erythropoiesis, GATA-1 replaces GATA-2 at many of its chromatin sites (Bresnick, Lee, Fujiwara, Johnson, & Keles, 2010; Grass et al., 2003). This "GATA-switch" mechanism has been proposed to control the transition of early erythroid precursors to terminally differentiating erythroid cells (Bresnick et al., 2012, 2010; Bresnick, Martowicz, Pal, & Johnson, 2005; DeVilbiss, Sanalkumar, Johnson, Keles, & Bresnick, 2014; Grass et al., 2003). However, GATA switches do not occur at all GATA-2-occupied loci (May et al., 2013). The often different GATA factor expression patterns in various cell types reflect their qualitatively distinct biological roles, and GATA switching illustrates one example by which distinct GATA factors function differentially through a common chromatin site. In addition, in certain contexts, GATA factors appear to function redundantly. Yolk sac-derived primitive erythroblasts express both GATA-1 and GATA-2. In these cells, targeted deletion of both factors is required to ablate a survival mechanism (Fujiwara, Chang, Williams, & Orkin, 2004).

GATA factors colocalize on chromatin with the E-box-binding basic helix–loop–helix (bHLH) protein Scl/TAL1 (Aplan, Nakahara, Orkin, & Kirsch, 1992; Begley et al., 1989; Kallianpur, Jordan, & Brandt, 1994). Scl/TAL1 heterodimerizes with a class A bHLH E-box protein, typically E12 or E47, and the heterodimer binds with high affinity to the E-box motif (Hsu, Cheng, Chen, & Baer, 1991; Hsu, Wadman, & Baer, 1994). The Scl/TAL1–GATA complex occupies chromosomal sites containing a GATA motif, with or without a neighboring E-box (Cheng et al., 2009; Kassouf et al., 2010; Tripic et al., 2008; Wozniak et al., 2008; Yu et al., 2009). In ChIP-seq studies, approximately 20% of Scl/TAL1 occupancy does not coincide with an E-box motif (Kassouf et al., 2010). At an E-box–GATA element, CRISPR/Cas9-mediated deletion of only the E-box motif abrogates GATA-2 occupancy and reduces transcription of the endogenous loci (Hewitt et al., 2015). Thus, the E-box is required for complex formation at certain E-box–GATA elements. ChIP studies indicate that Scl/TAL1 has a higher propensity for occupying E-box–GATA elements in a transcriptionally active state in comparison with a repressed state (Tripic et al., 2008; Yu et al., 2009).

Resembling GATA-2, Scl/TAL1 is crucial for definitive hematopoiesis, since targeted deletion of *Scl/Tal1* in mice severely impairs hematopoiesis (Porcher et al., 1996; Shivdasani, Mayer, & Orkin, 1995). The conditional ablation of *Scl/Tal1* in mice provided evidence for an Scl/TAL1 function to generate HSCs, but not for their function (Mikkola et al., 2003). As expected for such a critical hematopoietic regulator, dysregulated Scl/TAL1 expression is linked to leukemogenesis, specifically T-cell acute lymphocytic leukemia (Aplan, Lombardi, et al., 1992; Aplan, Raimondi, & Kirsch, 1992; Bash et al., 1993, 1995; Ferrando et al., 2004; Gerby et al., 2014; Sanda et al., 2012; Tremblay et al., 2010). A non-DNA-binding Scl/TAL1 mutant delays embryonic lethality in a *Scl/Tal1*$^{-/-}$ context, suggesting that intrinsic DNA-binding activity is not essential for hematopoietic development (Kassouf, Chagraoui, Vyas, & Porcher, 2008). However, since these partially rescued mice have severe survival defects and are grossly anemic, the DNA-binding function likely contributes to Scl/TAL1 activity. Since Scl/TAL1 occupies certain chromatin sites lacking an E-box, the partial rescue may also reflect activity via tethering to a GATA factor at these target sites.

2.1 Sequence Requirements

Analysis of GATA-2 chromatin occupancy in erythroid cells reiterated the importance of the naked DNA consensus (A/T)GATA(A/G) and revealed a more complex consensus (C/G)(A/T)GATAA(G/A/C)(G/A/C) enriched at chromatin occupancy sites (Fujiwara et al., 2009). While sequences conforming to this extended consensus (297,124 copies in the human genome) mediate GATA factor chromatin occupancy, these sequences are insufficient for chromatin occupancy and enhancer activity. Based on chromatin immunoprecipitation (ChIP) analysis, the vast majority of the sequences (>99%) are not GATA factor-occupied in cells (Fujiwara et al., 2009; Kang et al., 2012; Linnemann et al., 2011), presumably due to their occlusion by inaccessible chromatin and/or the absence of neighboring *cis*-elements that confer stable complex formation on chromatin. GATA factor occupancy is not restricted to enhancers, however, and occupancy at certain promoters correlates with transcriptional regulation (Johnson et al., 2007).

A pioneering study discovered an E-box (CANNTG) followed by a GATA motif, termed an E-box–GATA composite element, that mediates multimeric protein complex assembly in GATA-1-expressing erythroid cells (Wadman et al., 1997). The precise orientation (CANNTG upstream of GATA) and spacing (8–10 bp) support maximal protein complex assembly on naked DNA templates and enhancer activity in transfection assays (Wadman et al., 1997; Wozniak, Boyer, Grass, Lee, & Bresnick, 2007). Either E-box or GATA motif mutations prevent multimeric protein complex formation in vitro (Wadman et al., 1997). Subsequent work demonstrated the importance of E-box–GATA elements as GATA-2-regulated elements in GATA-2-expressing HSPCs and erythroid precursor cells. GATA-1 and GATA-2 occupy several hundred E-box–GATA elements (Fujiwara et al., 2009). More than 300 E-box–GATA elements are GATA-2-occupied in human K562 erythroleukemia cells. Using a more restricted sequence, CATCTG-8 bp-AGATAA (corresponding to the *Gata2* +9.5 site), 62 of 797 sites are GATA-2-occupied (Hewitt et al., 2015). Many of these sites are GATA-1- or GATA-2-occupied in a context-dependent manner; multimeric complexes are GATA-2-occupied in HSPCs and GATA-1-occupied in erythroid cells (Fujiwara et al., 2009; Li et al., 2011; Wilson et al., 2010).

How widespread are E-box–GATA motifs in the human genome? Over 1×10^5 human motifs contain some permutation of the CANNTG-(N8-N14)-AGATAA sequence. Using the E-box motif specific to the *Gata2* +9.5 site, CATCTG-(N8-N14)-AGATAA, nearly 9000 E-box–GATA motifs exist, and 25% are evolutionarily conserved in mouse by location analysis. Further narrowing the spacer sequence constraints yields nearly 800 mouse sequences that conform to CATCTG-8 bp-AGATAA, and ~5% of these sequences are GATA-2- and Scl/TAL1-occupied, based on ChIP-seq analysis (Hewitt et al., 2015).

It is instructive to consider how GATA factors function through E-box–GATA elements, in comparison with GATA motifs lacking E-boxes. Whereas this relationship is not understood, E-box–GATA composite elements have been functionally analyzed at several hematopoietic loci. At *Epb4.2,* encoding an erythroid cytoskeletal component, tandem E-box–GATA motifs are each required for *Epb4.2* transcriptional activation in a transient transfection assay (Xu, Huang, Chang, Agulnick, & Brandt, 2003). At the *Klf1* locus, an E-box–GATA motif is required for transcriptional activation in a transient transfection assay (Anderson, Crable, & Lingrel, 1998) and with a reporter-containing transgene (Anderson, Crable, & Lingrel, 2000). At the *Gata2* +9.5 E-box–GATA composite element, the 7–9 bp distance between the E-box and GATA motifs is an important determinant of enhancer activity in a transient transfection assay

(Wozniak et al., 2007). The specific spacer sequence can be altered without functional consequences, at least in this transfection context (Wozniak et al., 2007). The spacer length of E-box–GATA elements varies at different loci. E-box–GATA elements consisting of longer spacer sequences at the *Kit* promoter (Lecuyer et al., 2002), *Runx1* intron (Bee et al., 2009), *Smad1* intron, and *Smad5* promoter (Marks-Bluth et al., 2015) exhibit E-box- and GATA motif-dependent enhancer activity in transient transfection and transgenic mouse assays.

The principle of a restricted spacer length segregating the E-box and GATA motif has been reinforced by chromatin occupancy studies. GATA-1 and GATA-2 occupy E-box–GATA elements with a spacer length of 9 bp (Fujiwara et al., 2009; Hewitt et al., 2015; Li et al., 2013). Naked DNA-binding studies indicate that Scl/TAL1, a GATA-2 target gene and essential determinant of hematopoiesis, prefers E-box sequences CAGGTG (Wadman et al., 1997) or CAGATG (Hsu, Huang, et al., 1994) at composite elements. In erythroid cells, however, Scl/TAL1 occupies E-boxes with a CATCTG consensus sequence at composite elements (Hewitt et al., 2015). GATA-1 and GATA-2 occupy E-box–GATA elements in K562 erythroleukemia cells with a more degenerate E-box sequence (CANNTG) (Fujiwara et al., 2009). Scl/TAL1 chromatin occupancy has been detected at an E-box–GATA motif with a 13-bp spacer, suggesting potential flexibility in the spacer length in certain contexts (Cohen-Kaminsky et al., 1998). In some cases, a degenerate half E-box motif (CTG) can supplant the requisite E-box motif at hematopoietic enhancers with a spacing upstream of the GATA motif identical to that of the full E-box–GATA element (Bauer et al., 2013; Kassouf et al., 2010; Tallack et al., 2010). E-twenty six (ETS) motifs (TTCC), mediating ETS TF DNA binding, are often associated with GATA-factor-occupied E-box–GATA motifs (Dore, Chlon, Brown, White, & Crispino, 2012; Linnemann et al., 2011), revealing another layer of sequence complexity at composite enhancers. Furthermore, mutations of an ETS motif immediately downstream of the *Gata2* +9.5 E-box–GATA element are associated with MonoMAC syndrome, and mutation of the ETS motif reduces E-box–GATA element enhancer activity in a transient transfection assay (Hsu et al., 2013).

Despite high-affinity GATA factor binding to naked DNA, most E-box–GATA elements are not GATA factor-occupied in mouse and human hematopoietic cell genomes (Fujiwara et al., 2009; Hewitt et al., 2015; Linnemann et al., 2011; Wozniak et al., 2007). Although multifactorial mechanisms governing *cis*-element occupancy in chromatin are complex, a rudimentary principle is that contextual or "epigenetic" parameters irrespective of DNA sequence impact DNA structure in chromatin and *cis*-element accessibility. These parameters include histone modifications (DeVilbiss, Boyer, & Bresnick, 2013), DNA methylation (Jeong & Goodell, 2014; Jones, 2012), and DNA hydroxymethylation (Koh et al., 2011; Madzo et al., 2014; Tahiliani et al., 2009). In addition, mechanisms that expel or attract loci to nuclear microenvironments favor or preclude regulatory modes (Kosak et al., 2002; Lee et al., 2011, 2009; Ragoczy, Bender, Telling, Byron, & Groudine, 2006; Ragoczy, Telling, Sawado, Groudine, & Kosak, 2003; Rajapakse & Groudine, 2011; Schubeler et al., 2000; Zullo et al., 2012). Further complicating the matter, while GATA factor occupancy is required for transcriptional regulation of endogenous loci, occupancy does not suffice to generate a measurable functional output. Many occupancy sites are not physically linked to GATA factor-regulated genes. In principle, occupied *cis*-elements might control neighboring genes in three-dimensional space or generate regulatory RNAs that act in *trans* to control distinct loci. Deciphering parameters dictating TF binding at endogenous loci and endowing noncoding sequences with *cis*-element activity remains challenging, and there is no universally accepted methodology to achieve this goal.

Is evolutionary conservation the ultimate parameter for predicting the functional importance of GATA factor-regulated enhancers? The current evidence suggests that while conservation has utility to winnow and sift through large ensembles of GATA-bound sequences, as a sole parameter, it lacks predictive power. Though essential *cis*-elements are likely to be retained through speciation events (Pennacchio et al., 2006), human to mouse conservation of E-box–GATA sequences is below 1% (Hewitt et al., 2015), far lower than the ~85% conservation of protein-coding sequences. Furthermore, there is little to no evidence that positional conservation of enhancers universally equates to functional importance. Orthologous alignments of candidate enhancers reveal that most GATA-occupied sites are poorly conserved, and only a small fraction of conserved elements are GATA factor-occupied in mouse and human (Cheng et al., 2014; Hewitt et al., 2015). Thus, mouse and human genomes might be regulated by distinct enhancer mechanisms, or similar enhancer mechanisms in these species might exhibit considerable fluidity in genomic positioning of the *cis*-elements. Of course, these two possibilities are not mutually exclusive. By using less-stringent location analysis to identify similar sequence motifs at orthologous gene loci, more than 25% of human E-box–GATA sequences are detected in mouse (Hewitt et al., 2015). While it is not known whether all of these elements are functionally conserved, there is greater conservation of E-box–GATA composite elements at loci than defined by strict positional alignments. A noteworthy example of enhancer conservation at distinct locations involves *Gata1*, where an upstream GATA-occupied mouse enhancer corresponds to an intronic enhancer in human (Valverde-Garduno et al., 2004). Enhancer diversity and locus organization are features that almost certainly contribute to mammalian phenotypic diversity and interspecies trait variability.

2.2 Genomic and Bioinformatics Analysis

ChIP coupled with next-generation sequencing (ChIP-seq) permits detailed mapping of TF interactions and chromatin attributes genome-wide. Enormous efforts have established TF and histone occupancy profiles genome-wide (Giresi, Kim, McDaniell, Iyer, & Lieb, 2007; Hannah, Joshi, Wilson, Kinston, & Gottgens, 2011; Heintzman et al., 2009; Li et al., 2013; May et al., 2013; Shen et al., 2012). These studies suggest the following parameters to predict enhancers: chromatin occupancy, chromatin accessibility, histone modifications (which can regulate accessibility and occupancy), and DNA sequence attributes (including disease mutations and polymorphisms). Statistical tools to integrate multiple ChIP-seq datasets in parallel have revealed combinatorial patterns of protein–DNA interactions genome-wide in diverse cell types (Zeng et al., 2013). These efforts have led to the annotation of thousands of putative enhancers and have devised innovative and powerful methods when combined with direct functional analysis.

While chromatin occupancy and accessibility have merit for localizing enhancers in complex genomes (Kellis et al., 2014; Yue et al., 2014), these parameters are insufficient to predict enhancers at endogenous loci. Prospective enhancers with promising attributes can be deleted from the genome with little to no impact on target gene expression (Sanalkumar et al., 2014; Snow et al., 2010, 2011). By contrast, enhancers with essential non-redundant functions have been identified (Johnson et al., 2012, 2015). Recently, "super-enhancers" have been described as a subclass of large, clustered enhancers with particularly high ChIP-seq signals that can span several thousand base pairs in length (Hnisz et al., 2013). The attributes of these complex sequences recapitulate those of conventional enhancers, albeit super-enhancers are considerably longer and therefore contain many more prospective *cis*-elements. The mechanistic relationship of super-enhancers to locus control regions (LCRs) (Forrester, Takegawa, Papayannopoulou, Stamatoyannopoulos, & Groudine, 1987; Grosveld, van Assendelft, Greaves, & Kollias, 1987) and conventional enhancers is unclear. The β-globin LCR confers high-level transcription to all of the β-like globin genes at different stages of development (Bender, Bulger, Close, & Groudine, 2000; Bender et al., 2001; Fiering et al., 1995; Forrester et al., 1987; Grosveld et al., 1987), and this activity involves mechanistic steps shared with enhancers, namely chromatin looping, coregulator recruitment, Pol II recruitment to the enhancer and promoter, and control of target loci subnuclear localization (Bresnick, Johnson, Kim, & Im, 2006; Katsumura, DeVilbiss, Pope, Johnson, & Bresnick, 2013).

Despite limitations to deriving definitive functional insights, chromatin attribute datasets permit comparisons of multiple parameters at cohorts of predicted enhancers. Enhancer sequences can be stratified genome-wide based on their contextual parameters at endogenous loci (Hannah et al., 2011). Enhancer activity in transient transfection assays with reporter genes in plasmids and/or in transgenic mouse harboring reporter genes integrated at ectopic chromosomal loci can be utilized as training sets for algorithms that may improve upon existing approaches to predict enhancers. Comparison of the attributes (eg, chromatin occupancy, accessibility, and function) to sequence motifs, such as E-box–GATA elements, genome-wide allows one to classify these elements based on shared parameters, which may infer functional similarity (Hewitt et al., 2015). Methodological development continues to improve with the goal to predict the full complement of genomic enhancers with activity at endogenous loci. As functional validation at endogenous loci has lagged behind the development of enhancer predictive utilities, the picture remains murky regarding the extent to which the thousands of predicted enhancers increase promoter activity at their endogenous loci. The advent of facile gene-editing technologies has revolutionized validation strategies, which can now rigorously test function at endogenous loci in cultured cells and in mice.

2.3 Discovery and Validation of Enhancers Controlling HSPCs

Historically, enhancers were defined based on activity to increase reporter gene activity in a position- and orientation-independent manner in transiently transfected plasmids (Khoury & Gruss, 1983; Wasylyk, Wasylyk, & Chambon, 1984). Electrophoretic mobility shift assays (Garner & Revzin, 1981) were used to discover proteins that bind with sequence specificity to the functionally important sequences. This conceptually straightforward approach was extended to transgenic mice in which one or more copies of reporter genes are integrated at ectopic chromosomal sites (Pennacchio et al., 2006). While these approaches can identify enhancers, enhancer activity at endogenous loci almost invariably differs from that at ectopic chromosomal sites. In part, this reflects the differential chromatin environments at endogenous loci that create unique mechanistic requirements for conferring activation or repression. DNA sequences may elicit promiscuous enhancer activity at ectopic chromosomal sites due to the absence of endogenous locus sequences that occlude the respective *cis*-elements. Distinct ensembles of neighboring *cis*-elements may be permissive, stimulatory, or antagonistic for enhancer function.

Technologies to discover and analyze enhancers fall into two groups. Correlative methods infer function from DNA–protein interaction and chromatin attribute patterns. Causative methods involve disrupting or deleting the endogenous sequence, which can provide a definitive assessment of functional importance, at least in the specific context analyzed. However, as enhancers often have cell type- and developmental-specific activities, deletion analysis in any single cell would not be expected to identify all enhancers.

Multiple technologies have emerged to establish the chromatin accessibility state and TF composition of enhancers. DNase hypersensitivity, formaldehyde-assisted isolation of regulatory elements (FAIRE) (Giresi et al., 2007), or ATAC-seq (Buenrostro, Wu, Chang, & Greenleaf, 2015) is used to identify accessible or "open" chromatin regions that often reflect sites lacking nucleosomes or containing altered nucleosomes. Such sites frequently harbor clusters of *cis*-regulatory elements bound by TFs and can demarcate enhancers. ChIP relies on formaldehyde cross-linking of proteins to DNA on chromosomes to measure TF, coregulator, and modified histone occupancy of specific chromatin sites (DeVilbiss et al., 2014). A variation of ChIP, chromatin conformation capture (3C) (Dekker et al., 2002), is used to detect higher-order chromatin structure, specifically long-range interactions between chromosomal segments, and powerful high-throughput permutations of 3C have been developed (Dekker, Marti-Renom, & Mirny, 2013). This is important, as enhancer localization near or in a gene does not definitively establish that the enhancer is committed to this particular gene. The enhancer could regulate a gene residing in close proximity in three-dimensional space, which might be distant along the DNA template. Alternatively, the target gene might reside on another chromosome. Proteins bound at distal sites that interact with promoters via chromatin looping might cross-link to the promoter, even though the factors are bound directly to distal DNA sequences. Whereas the aggregate data have painted detailed pictures of chromatin landscapes and pinpointed many prospective enhancers, these assays need to be used in concert with direct functional analyses to identify enhancers.

Enhancer deletions at endogenous loci represent the acid test of functional significance, and this approach can yield results contrasting with those inferred from correlations. DNase hypersensitivity sites 2 (HS2) and 3 (HS3) upstream of *Hbb* locus have all the hallmarks of a classic enhancer: DNase hypersensitivity, reporter activity, and multi-TF occupancy including GATA-1. However, the targeted deletion of either HS2 or HS3 individually in mice only slightly decreases globin gene expression (Bender et al., 2001; Fiering et al., 1995; Hug et al., 1996). The targeted deletion of a major DNase hypersensitivity site at *Gata1* does not influence GATA-1 expression (Guyot et al., 2006). Knockout of a candidate enhancer in the Scl/TAL1 locus, exhibiting DNase hypersensitivity, activity in a reporter gene assay, and TF occupancy, does not impact Scl/TAL1 expression (Gottgens et al., 2004). As enhancers can be cell type- and developmental-stage specific, minimal or no activity might reflect an inappropriate context for the analysis. Alternatively, certain enhancers may modulate promoter activity, without inducing quantitatively large changes. In principle, incremental regulation may have critical physiological consequences. Finally, certain enhancers may not be particularly crucial in homeostasis, but might function in contexts of stress or other pathological insults. It can therefore be difficult to comprehend negative or incremental results obtained from endogenous locus gene editing.

The issues discussed earlier are well illustrated by consideration of enhancers controlling *Gata2* expression. Five GATA-1-/GATA-2-occupied sites at *Gata2* (named for their distance in kilobases from the *Gata2* transcription start site) are referred to as GATA-switch sites due to the replacement of GATA-2 by GATA-1 at these sites during erythroid differentiation (Bresnick et al., 2010). Mutant mouse strains lacking individual sites from the endogenous *Gata2* locus have been generated. As the targeted deletion of *Gata2* yields a severe block in definitive hematopoiesis and embryonic lethality, one expectation was that deletion of enhancers essential for conferring *Gata2* expression in HSPCs would also block hematopoiesis and be embryonic lethal. However, −1.8, −2.8, and −3.9 sites were not individually required for hematopoietic development, despite GATA factor occupancy and enhancer activity in a transient transfection assay (Grass et al., 2006; Martowicz, Grass, Boyer, Guend, & Bresnick, 2005). The −1.8 site has a unique activity to maintain *Gata2* repression in late-stage erythroblasts, but its deletion has little to no influence on hematopoiesis (Snow et al., 2010). The −2.8 site confers maximal *Gata2* expression, but is not required for *Gata2* expression or repression (Snow et al., 2011). Moreover, the −3.9 does not influence *Gata2* expression or repression (Sanalkumar et al., 2014). By contrast, the intronic +9.5 enhancer, containing an E-box–GATA element, is required for HSC genesis (Johnson et al., 2012), establishment of HSPCs in the fetal liver, and embryonic development. A distal −77 *cis*-element (Grass et al., 2006) is also required for embryonic development, but operates distinctly from the +9.5 site (Johnson et al., 2015). Whereas HSCs are generated in $-77^{-/-}$ mouse embryos, myeloid progenitors accumulate and are defective in their activity to generate diverse myeloerythroid progeny. Consistent with this impairment of myeloid progenitor cell function, a sector of the myeloid progenitor transcriptome is altered. Thus, multiple *Gata2* sites bearing enhancer attributes exhibit remarkably distinct functions at the endogenous locus in vivo. Elucidating mechanisms that endow these distinct functions may permit improved predictions of enhancer function.

Simple gene-editing technologies, such as TALENs and CRISPR/Cas9, permit enhancer deletions in cultured cells (Cho, Kim, Kim, & Kim, 2013; Kim et al., 2013). Thus, it has become relatively straightforward to validate enhancer function at endogenous loci. Moreover, both TALENs and CRISPR/Cas9 have greatly reduced the time required to make mutant mouse strains (Kim & Kim, 2014; Yang et al., 2013). Genome-wide screens using CRISPR libraries can be used to elucidate details of complex genetic networks (Zhou et al., 2014), and this technology can be adapted to interrogate candidate enhancers. Despite lacking enhancer activity in a transient transfection assay in a biologically relevant cell type, CRISPR/Cas9-mediated deletion of an E-box–GATA element at the *Bcl2l1* locus indicated its importance at the endogenous locus. This result further illustrates a lack of concordance between correlative and causative data (Hewitt et al., 2015). Predicted disease-associated loci can be deleted in cell lines to elucidate their molecular mechanisms and to inform disease mechanisms (Bauer et al., 2013). These gene-editing technologies have transformed efforts to discover functionally critical *cis*-elements, to dissect their sequence requirements, and to establish their roles in cell function and phenotype.

3. COMPONENTS OF SCL/TAL1 AND GATA FACTOR COMPLEXES AT E-BOX–GATA ELEMENTS

Scores of TFs that function directly or indirectly through enhancers are implicated in development and function of the hematopoietic system. The scope of this discussion is restricted to TFs that function through E-box–GATA elements. Whereas it has been known for more than a decade that multimeric protein complexes assemble at these elements (Wadman et al., 1997), many questions remain unanswered regarding the exact contribution of the constituents involved and the underlying mechanisms. Given the critical functions of the established components, eg, GATA-1, GATA-2, Scl/TAL1, LDB1, and LMO2, the future development of work on this problem promises to yield vital biological, mechanistic, and pathological insights.

3.1 LMO2

The Scl/TAL1–GATA complex contains the non-DNA-binding components LIM-only (LMO) proteins 1 and 2 (Love, Warzecha, & Li, 2014; Osada, Grutz, Axelson, Forster, & Rabbitts, 1995). The LIM domain is a cysteine-rich zinc finger-like motif that acts as a docking site for proteins and mediates multiprotein complex assembly. As with GATA-1, GATA-2, and Scl/TAL1, LMO2 promotes erythropoiesis (Warren et al., 1994). Though LMO2 has little to no transcriptional activation activity with reporter genes in transient transfection assays, LMO2 bridges the GATA factor and Scl/TAL1 at E-box–GATA elements and is believed to function as a scaffold for additional components of the complex (Lecuyer et al., 2007; Wadman et al., 1997). Disruption of *LMO2*, eg, due to retroviral insertion in a gene therapy context, can cause human myeloid leukemia (Hacein-Bey-Abina et al., 2003). Conditional overexpression of LMO2 in T-cells induces T-cell acute lymphoblastic lymphoma (T-ALL) (Fisch et al., 1992). The LMO2-containing DNA-binding complex in this mouse model of T-ALL forms at a bipartite E-box–E-box sequence, rather than an E-box–GATA sequence (Grutz, Forster, & Rabbitts, 1998), which highlights a potential new configuration of Scl/TAL1-responsive *cis*-elements lacking a GATA motif.

3.2 LDB1

Another well-studied component of the Scl/TAL1–GATA complex at E-box–GATA elements is the non-DNA-binding protein LIM-domain-binding protein 1 (LDB1). LDB1 binds the LMO2 LIM domain and contributes to the assembly of the Scl/TAL1–GATA complex. LDB1 occupies chromatin at many erythroid-specific genes (Li et al., 2013). *Ldb1* knockout embryos are severely anemic and die at ~E9.5 (Mukhopadhyay et al., 2003). LDB1 promotes formation of higher-order chromatin loops, eg, at *Hbb* (Song, Hou, & Dean, 2007), and these long-range interactions are believed to promote transcriptional activation. LDB1 is also required for subnuclear localization of *Hbb* to sites of active transcription (Song et al., 2010), which may or may not reflect its capacity to induce chromatin looping; the relationship between chromatin looping and alteration of subnuclear localization is elusive. The mechanism for the LDB1-induced long-range looping involves LDB1 interactions at promoters and distal enhancers and its capacity to homodimerize, thus tethering together LBD1-bound sites (Soler et al., 2010). While LMO2 provides a scaffold connecting Scl/TAL1 to LDB1, LDB1 self-association appears to underlie promoter–enhancer interactions in certain contexts. This self-association capacity has been exploited to engineer fusion proteins that dock at specific chromosomal sites and have the capacity to interact. These proteins represent innovative tools to study long-range chromatin interactions (Deng et al., 2012).

3.3 KLF1

Erythroid Kruppel-like factor 1/Kruppel-like factor 1 (EKLF or KLF1) (Miller & Bieker, 1993) promotes erythropoiesis (Perkins, Sharpe, & Orkin, 1995) by regulating a large cohort of erythroid genes (Tallack et al., 2012). KLF1$^{-/-}$ mice die of anemia in late-embryonic development, primarily due to defective terminal erythroid maturation (Nuez, Michalovich, Bygrave, Ploemacher, & Grosveld, 1995; Perkins et al., 1995). Nearly 50% of KLF1-occupied sites have neighboring GATA-1-occupancy in erythroid cells (Tallack et al., 2010). A half E-box–GATA motif, with identical spacing to the canonical E-box–GATA element, is frequently associated with KLF1-occupied chromatin sites (Tallack et al., 2010). At LDB1-occupied chromatin, the KLF1-binding motif is also enriched (Li et al., 2013). While this proximity suggests that KLF1 may interact or functionally interface with Scl/TAL1–GATA-1 complexes to coordinate gene expression, deciphering the functional relationship between Scl/TAL1–GATA-1 complexes and KLF1 requires additional investigation. Whereas KLF1 is undoubtedly crucial for establishment/maintenance of the erythroid cell transcriptome (Nilson, Sabatino, Bodine, & Gallagher, 2006; Pilon et al., 2008;

Tallack et al., 2012, 2010), many questions remain regarding how it functions with GATA-1 and Scl/TAL1, whether it functions via qualitatively distinct mechanisms, or whether it predominantly amplifies GATA-1- and Scl/TAL1-dependent transcriptional outputs.

3.4 SWI/SNF Chromatin-Remodeling Complex

Brahma-related gene 1 (BRG1) is a catalytic subunit of the SWI/SNF complex, which has a broad function in diverse cell types to remodel nucleosome structure (Bevilacqua, Willis, & Bultman, 2014; Hargreaves & Crabtree, 2011). By altering nucleosome structure and/or positioning, which may expose or occlude *cis*-elements and/or alter higher-order chromatin transitions, the SWI/SNF complex can promote or inhibit transcriptional activation. In hematopoietic cells, BRG1 confers chromatin accessibility at enhancers (Hu et al., 2011; Kim, Bresnick, & Bultman, 2009; Kim, Bultman, Kiefer, Dean, & Bresnick, 2009). Studies using *Brg1*$^{-/-}$ cells demonstrated that BRG1 regulates chromatin looping at the α- and β-globin loci (Kim, Bresnick, et al., 2009; Kim, Bultman, et al., 2009). While BRG1 lacks intrinsic sequence specificity for chromatin site occupancy, it is recruited to chromatin via protein–protein interactions with DNA-bound factors. For example, it is recruited by the Scl/TAL1–GATA-1 complex at an E-box–GATA composite element in the *Epb4.2* promoter (Xu, Meng, Cai, Koury, & Brandt, 2006). In erythroid cells, BRG1 also occupies the E-box–GATA element at *Gata2* +9.5 (Sanalkumar et al., 2014). In leukemic cells, BRG1 colocalizes with LMO2 at an enhancer regulating *Myc* expression (Shi et al., 2013). Given the broadly important BRG1 functions, BRG1-dependent mechanisms at E-box–GATA elements are likely to be crucial in both physiological and pathological contexts. However, elucidating the full complement of coregulators that trigger or suppress E-box–GATA element activity requires major additional work.

3.5 Single-Stranded DNA-Binding Proteins

Single-stranded DNA-binding proteins (SSBPs) bind LDB1 and regulate its function (Chen et al., 2002). This family of proteins exerts important roles in embryonic developmental patterning, and hypomorphic alleles of SSBPs have developmental phenotypes resembling that of LDB1 and LMO2 mutants (van Meyel, Thomas, & Agulnick, 2003). SSBP proteins Ssbp2 and Ssbp3 occupy chromatin at the Epb4.2 E-box–GATA site and modestly increase its transcriptional activity in a transfection assay (Xu et al., 2007). Their association with LDB1 and LMO2 is essential for maintaining the abundance of both proteins in erythroid precursor cells by inhibiting ubiquitin-mediated protein degradation (Xu et al., 2007). This mechanism sustains LDB1 levels and promotes multimeric complex assembly at E-box–GATA sites. The human *SSBP2* locus is disrupted in a subset of therapy-induced AML patients, and *Ssbp2*$^{-/-}$ mice are susceptible to lymphoma and other carcinomas, supporting its role as a tumor suppressor (Wang et al., 2010). Ssbp2 also plays a critical role in the maintenance and stress response of HSPC (Li et al., 2014).

4. HSPC CISTROME

A common pattern emerges when examining each factor of the multimeric complex at E-box–GATA elements. Each TF is an obligate regulator of hematopoiesis during embryonic development, and all are implicated in hematopoietic pathologies. Presumably, these factors synergize through E-box–GATA elements to control hematopoietic development and homeostasis. It is reasonable to assume that alterations in any of the mechanistic steps involving these factors will dysregulate E-box–GATA element function, thereby impairing regulation of the important cohort of genes harboring this element.

E-box–GATA elements regulate gene expression at hematopoietic loci involved in critical cellular processes (Fig. 1). An E-box–GATA element controls transcription of *Gata1* (Vyas et al., 1999). As described earlier, the *Gata2* +9.5 E-box–GATA element controls *Gata2* transcription in HSPCs and is essential for HSC genesis and function (Gao et al., 2013; Johnson et al., 2012; Sanalkumar et al., 2014). At *Epb4.2*, an E-box–GATA element controls expression of a key membrane protein in erythrocytes critical for maintaining red blood cell size and cation transport (Peters et al., 1999; Xu et al., 2003). A *Klf1* E-box–GATA element is implicated in *Klf1* expression (Anderson et al., 2000). E-box–GATA elements have been functionally validated using gene-editing techniques at the *Bcl2l1*, *Dapp1*, and *Samd14* loci (Hewitt et al., 2015). *Bcl2l1* +12.2 controls expression of Bcl-xL, an essential constituent of apoptotic pathways and survival determinant in HSPCs and other hematopoietic cells (Chao & Korsmeyer, 1998; Gregory et al., 1999). *Dapp1* +23.5 controls expression of Bam32, which regulates germinal center progression, mast cell

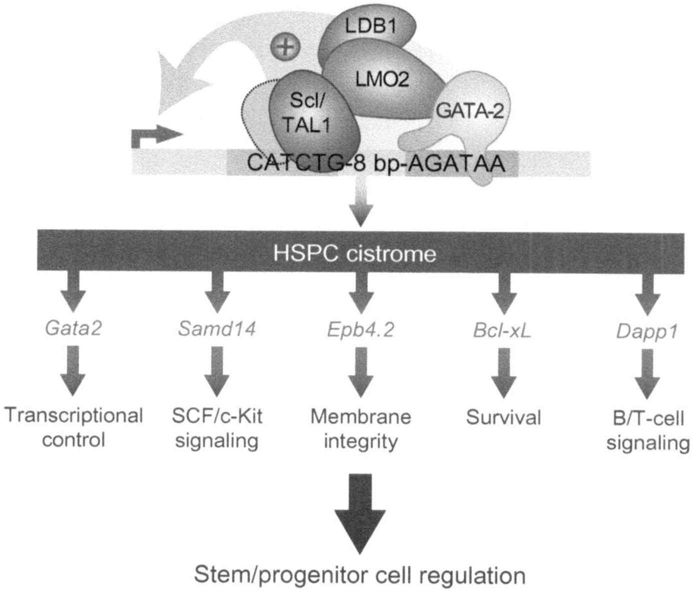

Fig. 1 Composite E-box–GATA elements regulate diverse aspects of hematopoietic stem/progenitor cell development and function. A multimeric complex, consisting of GATA-2, Scl/TAL-1, E2A, LMO2, and LDB1, and almost certainly additional components, forms at a highly restricted subset of E-box–GATA elements in the genome. The GATA-2-activated E-box–GATA elements are important constituents of the HSPC cistrome, which orchestrates vital steps in the stem and progenitor cell transitions that give rise to the full complement of blood cells. Examples of genes controlled by GATA-2 through these elements include *Gata2*, *Samd14*, *Epb4.2*, *Bcl2l1*, and *Dapp1*. (See the color plate.)

activation, and B- and T-cell receptor signaling (Han, Saijo, Mecklenbrauker, Tarakhovsky, & Nussenzweig, 2003).

Samd14 +2.5 contains an E-box–GATA element conserved in mouse and humans. GATA-2 occupies the +2.5 site and regulates *Samd14* transcription in HSPCs (Hewitt et al., 2015). While E-box sequences are not believed to be essential for multimeric complex assembly in all contexts, deletion of E-box sequences from the *Samd14* E-box–GATA composite element severely impairs *Samd14* +2.5 function. +2.5 site function requires LDB1, since LDB1 knockdown in *Samd14* $+2.5^{+/-}$cells only affects the wild-type *Samd14* allele. *Samd14* promotes c-Kit signaling in fetal liver HSPCs, and shRNA-mediated downregulation of *Samd14* reduces fetal liver hematopoietic progenitor levels/activity. As GATA-2 also activates *Kit* expression (Jing et al., 2008; Munugalavadla et al.,

2005), *Gata2*, *Samd14*, and *Kit* constitute a type I coherent feed-forward loop that controls the function of hematopoietic progenitors (Hewitt et al., 2015). One would assume that this axis would also regulate HSCs, but this has not yet been described.

5. PERSPECTIVE

Dissecting mammalian genome function is complicated by the extraordinarily diverse and large ensemble of *cis*-elements, a subset of which comprise enhancers. E-box–GATA elements nucleate chromatin assembly of a multiprotein complex that regulates expression of key hematopoietic genes. Given the thousands of E-box–GATA elements in the genome, identifying those with critical functions at endogenous loci could be considered to be formidable. However, recent efforts utilizing sequence and chromatin attributes of essential enhancer elements have identified essential components of the HSPC cistrome—importantly, components with a quantitatively large contribution to endogenous locus transcription (Hewitt et al., 2015). By linking these enhancers to the cognate genes, this approach has led to the discovery of unknown regulators of HSPCs, including a facilitator of SCF-c-Kit signaling (Hewitt et al., 2015). One can assemble a systems-wide network of functionally critical enhancers and their respective genes, rather than factor-occupied sites that might or might not reflect functionally critical sites. This global network derived from the HSPC cistrome will almost certainly inform fundamental processes including, but not limited to, stem cell genesis, self-renewal, and differentiation.

There are many uncertainties involved with using DNA and chromatin attributes to assign functional importance of *cis*-elements, without the requisite direct functional analysis at endogenous loci. We have not attempted to review the numerous studies that have rigorously mapped chromatin parameters genome-wide in HSPCs. It will be extremely interesting to establish the full complement of bona fide enhancers resembling the *Gata2* +9.5 and −-77 sites in controlling hematopoiesis nonredundantly. Do these enhancers parse into subgroups based on their functions to control distinct sectors of the hematopoietic cellular hierarchy? Do they parse into subgroups based on distinct molecular mechanisms? Clearly, major mechanistic and biological questions remain unanswered, but given existing systems and tools, it has become tractable to sift through genomes with considerably greater ease than ever before possible. Similar approaches can be deployed to analyze enhancers at loci encoding master regulators that control distinct biological processes.

REFERENCES

Anderson, K. P., Crable, S. C., & Lingrel, J. B. (1998). Multiple proteins binding to a GATA-E box-GATA motif regulate the erythroid Kruppel-like factor (EKLF) gene. *The Journal of Biological Chemistry*, *273*, 14347–14354.

Anderson, K. P., Crable, S. C., & Lingrel, J. B. (2000). The GATA-E-box motif in the EKLF promoter is required for in vivo expression. *Blood*, *95*, 1652–1655.

Aplan, P. D., Lombardi, D. P., Reaman, G. H., Sather, H. N., Hammond, G. D., & Kirsch, I. R. (1992). Involvement of the putative hematopoietic transcription factor SCL in T-cell acute lymphoblastic leukemia. *Blood*, *79*, 1327–1333.

Aplan, P. D., Nakahara, K., Orkin, S. H., & Kirsch, I. R. (1992). The SCL gene product: A positive regulator of erythroid differentiation. *The EMBO Journal*, *11*, 4073–4081.

Aplan, P. D., Raimondi, S. C., & Kirsch, I. R. (1992). Disruption of the SCL gene by a t(1;3) translocation in a patient with T cell acute lymphoblastic leukemia. *The Journal of Experimental Medicine*, *176*, 1303–1310.

Bash, R. O., Crist, W. M., Shuster, J. J., Link, M. P., Amylon, M., Pullen, J., et al. (1993). Clinical features and outcome of T-cell acute lymphoblastic leukemia in childhood with respect to alterations at the TAL1 locus: A Pediatric Oncology Group study. *Blood*, *81*, 2110–2117.

Bash, R. O., Hall, S., Timmons, C. F., Crist, W. M., Amylon, M., Smith, R. G., et al. (1995). Does activation of the TAL1 gene occur in a majority of patients with T-cell acute lymphoblastic leukemia? A pediatric oncology group study. *Blood*, *86*, 666–676.

Bauer, D. E., Kamran, S. C., Lessard, S., Xu, J., Fujiwara, Y., Lin, C., et al. (2013). An erythroid enhancer of BCL11A subject to genetic variation determines fetal hemoglobin level. *Science*, *342*, 253–257.

Bee, T., Liddiard, K., Swiers, G., Bickley, S. R., Vink, C. S., Jarratt, A., et al. (2009). Alternative Runx1 promoter usage in mouse developmental hematopoiesis. *Blood Cells, Molecules & Diseases*, *43*, 35–42.

Begley, C. G., Aplan, P. D., Denning, S. M., Haynes, B. F., Waldmann, T. A., & Kirsch, I. R. (1989). The gene SCL is expressed during early hematopoiesis and encodes a differentiation-related DNA-binding motif. *Proceedings of the National Academy of Sciences of the United States of America*, *86*, 10128–10132.

Bender, M. A., Bulger, M., Close, J., & Groudine, M. (2000). Beta-globin gene switching and DNaseI sensitivity of the endogenous beta-globin locus in mice do not require the locus control region. *Molecular Cell*, *5*, 387–393.

Bender, M. A., Roach, J. N., Halow, J., Close, J., Alami, R., Bouhassira, E. E., et al. (2001). Targeted deletion of 5'HS1 and 5'HS4 of the beta-globin locus control region reveals additive activity of the DNaseI hypersensitive sites. *Blood*, *98*, 2022–2027.

Bevilacqua, A., Willis, M. S., & Bultman, S. J. (2014). SWI/SNF chromatin-remodeling complexes in cardiovascular development and disease. *Cardiovascular Pathology: The Official Journal of the Society for Cardiovascular Pathology*, *23*, 85–91.

Bresnick, E. H., Johnson, K. D., Kim, S.-I., & Im, H. (2006). Establishment and regulation of chromatin domains: Mechanistic insights from studies of hemoglobin synthesis. *Progress in Nucleic Acid Research and Molecular Biology*, *81*, 435–471.

Bresnick, E. H., Katsumura, K. R., Lee, H. Y., Johnson, K. D., & Perkins, A. S. (2012). Master regulatory GATA transcription factors: Mechanistic principles and emerging links to hematologic malignancies. *Nucleic Acids Research*, *40*, 5819–5831.

Bresnick, E. H., Lee, H. Y., Fujiwara, T., Johnson, K. D., & Keles, S. (2010). GATA switches as developmental drivers. *The Journal of Biological Chemistry*, *285*, 31087–31093.

Bresnick, E. H., Martowicz, M. L., Pal, S., & Johnson, K. D. (2005). Developmental control via GATA factor interplay at chromatin domains. *Journal of Cellular Physiology*, *205*, 1–9.

Briegel, K., Lim, K. C., Plank, C., Beug, H., Engel, J. D., & Zenke, M. (1993). Ectopic expression of a conditional GATA-2/estrogen receptor chimera arrests erythroid differentiation in a hormone-dependent manner. *Genes & Development*, *7*, 1097–1109.

Buenrostro, J. D., Wu, B., Chang, H. Y., & Greenleaf, W. J. (2015). ATAC-seq: A method for assaying chromatin accessibility genome-wide. *Current Protocols in Molecular Biology*, *109*, 21.29.1–21.29.9.

Chao, D. T., & Korsmeyer, S. J. (1998). BCL-2 family: Regulators of cell death. *Annual Review of Immunology*, *16*, 395–419.

Charron, F., & Nemer, M. (1999). GATA transcription factors and cardiac development. *Seminars in Cell & Developmental Biology*, *10*, 85–91.

Chen, L., Segal, D., Hukriede, N. A., Podtelejnikov, A. V., Bayarsaihan, D., Kennison, J. A., et al. (2002). Ssdp proteins interact with the LIM-domain-binding protein Ldb1 to regulate development. *Proceedings of the National Academy of Sciences of the United States of America*, *99*, 14320–14325.

Cheng, Y., Ma, Z., Kim, B. H., Wu, W., Cayting, P., Boyle, A. P., et al. (2014). Principles of regulatory information conservation between mouse and human. *Nature*, *515*, 371–375.

Cheng, Y., Wu, W., Kumar, S. A., Yu, D., Deng, W., Tripic, T., et al. (2009). Erythroid GATA1 function revealed by genome-wide analysis of transcription factor occupancy, histone modifications, and mRNA expression. *Genome Research*, *19*, 2172–2184.

Cho, S. W., Kim, S., Kim, J. M., & Kim, J. S. (2013). Targeted genome engineering in human cells with the Cas9 RNA-guided endonuclease. *Nature Biotechnology*, *31*, 230–232.

Cohen-Kaminsky, S., Maouche-Chretien, L., Vitelli, L., Vinit, M. A., Blanchard, I., Yamamoto, M., et al. (1998). Chromatin immunoselection defines a TAL-1 target gene. *The EMBO Journal*, *17*, 5151–5160.

Crispino, J. D., Lodish, M. B., MacKay, J. P., & Orkin, S. H. (1999). Use of altered specificity mutants to probe a specific protein-protein interaction in differentiation: The GATA-1: FOG complex. *Molecular Cell*, *3*, 219–228.

Cullen, K. E., Kladde, M. P., & Seyfred, M. A. (1993). Interaction between transcriptional regulatory elements of prolactin chromatin. *Science*, *261*, 203–206.

Dekker, J., Marti-Renom, M. A., & Mirny, L. A. (2013). Exploring the three-dimensional organization of genomes: Interpreting chromatin interaction data. *Nature Reviews. Genetics*, *14*, 390–403.

Dekker, J., Rippe, K., Dekker, M., & Kleckner, N. (2002). Capturing chromosome conformation. *Science*, *295*, 1306–1311.

Deng, W., Lee, J., Wang, H., Miller, J., Reik, A., Gregory, P. D., et al. (2012). Controlling long-range genomic interactions at a native locus by targeted tethering of a looping factor. *Cell*, *149*, 1233–1244.

de Pater, E., Kaimakis, P., Vink, C. S., Yokomizo, T., Yamada-Inagawa, T., van der Linden, R., et al. (2013). Gata2 is required for HSC generation and survival. *The Journal of Experimental Medicine*, *210*, 2843–2850.

DeVilbiss, A. W., Boyer, M. E., & Bresnick, E. H. (2013). Establishing a hematopoietic genetic network through locus-specific integration of chromatin regulators. *Proceedings of the National Academy of Sciences of the United States of America*, *110*, E3398–E3407.

DeVilbiss, A. W., Sanalkumar, R., Johnson, K. D., Keles, S., & Bresnick, E. H. (2014). Hematopoietic transcriptional mechanisms: From locus-specific to genome-wide vantage points. *Experimental Hematology*, *42*, 618–629.

Dickinson, R. E., Griffin, H., Bigley, V., Reynard, L. N., Hussain, R., Haniffa, M., et al. (2011). Exome sequencing identifies GATA-2 mutation as the cause of dendritic cell, monocyte, B and NK lymphoid deficiency. *Blood*, *118*, 2656–2658.

Dickinson, R. E., Milne, P., Jardine, L., Zandi, S., Swierczek, S. I., McGovern, N., et al. (2014). The evolution of cellular deficiency in GATA2 mutation. *Blood*, *123*, 863–874.
Dore, L. C., Chlon, T. M., Brown, C. D., White, K. P., & Crispino, J. D. (2012). Chromatin occupancy analysis reveals genome-wide GATA factor switching during hematopoiesis. *Blood*, *119*, 3724–3733.
Dorfman, D. M., Wilson, D. B., Bruns, G. A., & Orkin, S. H. (1992). Human transcription factor GATA-2. Evidence for regulation of preproendothelin-1 gene expression in endothelial cells. *The Journal of Biological Chemistry*, *267*, 1279–1285.
Evans, T., & Felsenfeld, G. (1989). The erythroid-specific transcription factor Eryf1: A new finger protein. *Cell*, *58*, 877–885.
Evans, T., Reitman, M., & Felsenfeld, G. (1988). An erythrocyte-specific DNA-binding factor recognizes a regulatory sequence common to all chicken globin genes. *Proceedings of the National Academy of Sciences of the United States of America*, *85*, 5976–5980.
Ferrando, A. A., Herblot, S., Palomero, T., Hansen, M., Hoang, T., Fox, E. A., et al. (2004). Biallelic transcriptional activation of oncogenic transcription factors in T-cell acute lymphoblastic leukemia. *Blood*, *103*, 1909–1911.
Fiering, S., Epner, E., Robinson, K., Zhuang, Y., Telling, A., Hu, M., et al. (1995). Targeted deletion of 5'HS2 of the murine beta-globin LCR reveals that it is not essential for proper regulation of the beta-globin locus. *Genes & Development*, *9*, 2203–2213.
Fisch, P., Boehm, T., Lavenir, I., Larson, T., Arno, J., Forster, A., et al. (1992). T-cell acute lymphoblastic lymphoma induced in transgenic mice by the RBTN1 and RBTN2 LIM-domain genes. *Oncogene*, *7*, 2389–2397.
Forrester, W. C., Takegawa, S., Papayannopoulou, T., Stamatoyannopoulos, G., & Groudine, M. (1987). Evidence for a locus activation region: The formation of developmentally stable hypersensitive sites in globin-expressing hybrids. *Nucleic Acids Research*, *15*, 10159–10177.
Fraser, P., & Bickmore, W. A. (2007). Nuclear organization of the genome and the potential for gene regulation. *Nature*, *447*, 413–417.
Fujiwara, Y., Browne, C. P., Cunniff, K., Goff, S. C., & Orkin, S. H. (1996). Arrested development of embryonic red cell precursors in mouse embryos lacking transcription factor GATA-1. *Proceedings of the National Academy of Sciences of the United States of America*, *93*, 12355–12358.
Fujiwara, Y., Chang, A. N., Williams, A. M., & Orkin, S. H. (2004). Functional overlap of GATA-1 and GATA-2 in primitive hematopoietic development. *Blood*, *103*, 583–585.
Fujiwara, T., O'Geen, H., Keles, S., Blahnik, K., Linnemann, A. K., Kang, Y. A., et al. (2009). Discovering hematopoietic mechanisms through genome-wide analysis of GATA factor chromatin occupancy. *Molecular Cell*, *36*, 667–681.
Gao, Z., Huang, Z., Olivey, H. E., Gurbuxani, S., Crispino, J. D., & Svensson, E. C. (2010). FOG-1-mediated recruitment of NuRD is required for cell lineage re-enforcement during haematopoiesis. *The EMBO Journal*, *29*, 457–468.
Gao, X., Johnson, K. D., Chang, Y. I., Boyer, M. E., Dewey, C. N., Zhang, J., et al. (2013). Gata2 cis-element is required for hematopoietic stem cell generation in the mammalian embryo. *The Journal of Experimental Medicine*, *210*, 2833–2842.
Garner, M. M., & Revzin, A. (1981). A gel electrophoresis method for quantifying the binding of proteins to specific DNA regions: Application to components of the *Escherichia coli* lactose operon regulatory system. *Nucleic Acids Research*, *9*, 3047–3060.
Gerby, B., Tremblay, C. S., Tremblay, M., Rojas-Sutterlin, S., Herblot, S., Hebert, J., et al. (2014). SCL, LMO1 and Notch1 reprogram thymocytes into self-renewing cells. *PLoS Genetics*, *10*, e1004768.
Giresi, P. G., Kim, J., McDaniell, R. M., Iyer, V. R., & Lieb, J. D. (2007). FAIRE (Formaldehyde-Assisted Isolation of Regulatory Elements) isolates active regulatory elements from human chromatin. *Genome Research*, *17*, 877–885.

Gottgens, B., Broccardo, C., Sanchez, M. J., Deveaux, S., Murphy, G., Gothert, J. R., et al. (2004). The scl +18/19 stem cell enhancer is not required for hematopoiesis: Identification of a 5' bifunctional hematopoietic-endothelial enhancer bound by Fli-1 and Elf-1. *Molecular and Cellular Biology, 24*, 1870–1883.

Grass, J. A., Boyer, M. E., Pal, S., Wu, J., Weiss, M. J., & Bresnick, E. H. (2003). GATA-1-dependent transcriptional repression of GATA-2 via disruption of positive autoregulation and domain-wide chromatin remodeling. *Proceedings of the National Academy of Sciences of the United States of America, 100*, 8811–8816.

Grass, J. A., Jing, H., Kim, S.-I., Martowicz, M. L., Pal, S., Blobel, G. A., et al. (2006). Distinct functions of dispersed GATA factor complexes at an endogenous gene locus. *Molecular and Cellular Biology, 26*, 7056–7067.

Gregory, T., Yu, C., Ma, A., Orkin, S. H., Blobel, G. A., & Weiss, M. J. (1999). GATA-1 and erythropoietin cooperate to promoter erythroid cell survival by regulating bcl-xl expression. *Blood, 94*, 87–96.

Grosveld, F., van Assendelft, G. B., Greaves, D. R., & Kollias, G. (1987). Position-independent high-level expression of the human b-globin gene in transgenic mice. *Cell, 51*, 975–985.

Grutz, G., Forster, A., & Rabbitts, T. H. (1998). Identification of the LMO4 gene encoding an interaction partner of the LIM-binding protein LDB1/NLI1: A candidate for displacement by LMO proteins in T cell acute leukaemia. *Oncogene, 17*, 2799–2803.

Guyot, B., Murai, K., Fujiwara, Y., Valverde-Garduno, V., Hammett, M., Wells, S., et al. (2006). Characterization of a megakaryocyte-specific enhancer of the key hemopoietic transcription factor GATA1. *The Journal of Biological Chemistry, 281*, 13733–13742.

Hacein-Bey-Abina, S., Von Kalle, C., Schmidt, M., McCormack, M. P., Wulffraat, N., Leboulch, P., et al. (2003). LMO2-associated clonal T cell proliferation in two patients after gene therapy for SCID-X1. *Science, 302*, 415–419.

Hahn, C. N., Chong, C. E., Carmichael, C. L., Wilkins, E. J., Brautigan, P. J., Li, X. C., et al. (2011). Heritable GATA2 mutations associated with familial myelodysplastic syndrome and acute myeloid leukemia. *Nature Genetics, 43*, 1012–1017.

Han, A., Saijo, K., Mecklenbrauker, I., Tarakhovsky, A., & Nussenzweig, M. C. (2003). Bam32 links the B cell receptor to ERK and JNK and mediates B cell proliferation but not survival. *Immunity, 19*, 621–632.

Hannah, R., Joshi, A., Wilson, N. K., Kinston, S., & Gottgens, B. (2011). A compendium of genome-wide hematopoietic transcription factor maps supports the identification of gene regulatory control mechanisms. *Experimental Hematology, 39*, 531–541.

Hargreaves, D. C., & Crabtree, G. R. (2011). ATP-dependent chromatin remodeling: Genetics, genomics and mechanisms. *Cell Research, 21*, 396–420.

Heintzman, N. D., Hon, G. C., Hawkins, R. D., Kheradpour, P., Stark, A., Harp, L. F., et al. (2009). Histone modifications at human enhancers reflect global cell-type-specific gene expression. *Nature, 459*, 108–112.

Hewitt, K. J., Kim, D. H., Devadas, P., Prathibha, R., Zuo, C., Sanalkumar, R., et al. (2015). Hematopoietic signaling mechanism revealed from a stem/progenitor cell cistrome. *Molecular Cell, 59*, 62–74.

Hnisz, D., Abraham, B. J., Lee, T. I., Lau, A., Saint-Andre, V., Sigova, A. A., et al. (2013). Super-enhancers in the control of cell identity and disease. *Cell, 155*, 934–947.

Ho, I. C., Vorhees, P., Marin, N., Oakley, B. K., Tsai, S. F., Orkin, S. H., et al. (1991). Human GATA-3: A lineage-restricted transcription factor that regulates the expression of the T cell receptor alpha gene. *The EMBO Journal, 10*, 1187–1192.

Hong, W., Nakazawa, M., Chen, Y. Y., Kori, R., Vakoc, C. R., Rakowski, C., et al. (2005). FOG-1 recruits the NuRD repressor complex to mediate transcriptional repression by GATA-1. *The EMBO Journal, 24*, 67–78.

Hsu, H. L., Cheng, J. T., Chen, Q., & Baer, R. (1991). Enhancer-binding activity of the tal-1 oncoprotein in association with the E47/E12 helix-loop-helix proteins. *Molecular and Cellular Biology*, *11*, 3037–3042.

Hsu, H. L., Huang, L., Tsan, J. T., Funk, W., Wright, W. E., Hu, J. S., et al. (1994). Preferred sequences for DNA recognition by the TAL1 helix-loop-helix proteins. *Molecular and Cellular Biology*, *14*, 1256–1265.

Hsu, A. P., Johnson, K. D., Falcone, E. L., Sanalkumar, R., Sanchez, L., Hickstein, D. D., et al. (2013). GATA2 haploinsufficiency caused by mutations in a conserved intronic element leads to MonoMAC syndrome. *Blood*, *121*, 3830–3837. S3831–S3837.

Hsu, A. P., Sampaio, E. P., Khan, J., Calvo, K. R., Lemieux, J. E., Patel, S. Y., et al. (2011). Mutations in GATA2 are associated with the autosomal dominant and sporadic monocytopenia and mycobacterial infection (MonoMAC) syndrome. *Blood*, *118*, 2653–2655.

Hsu, H. L., Wadman, I., & Baer, R. (1994). Formation of in vivo complexes between the TAL1 and E2A polypeptides of leukemic T-cells. *Proceedings of the National academy of Sciences of the United States of America*, *91*, 3181–3185.

Hu, G., Schones, D. E., Cui, K., Ybarra, R., Northrup, D., Tang, Q., et al. (2011). Regulation of nucleosome landscape and transcription factor targeting at tissue-specific enhancers by BRG1. *Genome Research*, *21*, 1650–1658.

Huang, Z., Dore, L. C., Li, Z., Orkin, S. H., Feng, G., Lin, S., et al. (2009). GATA-2 reinforces megakaryocyte development in the absence of GATA-1. *Molecular and Cellular Biology*, *29*, 5168–5180.

Hug, B. A., Wesselschmidt, R. L., Fiering, S., Bender, M. A., Epner, E., Groudine, M., et al. (1996). Analysis of mice containing a targeted deletion of beta-globin locus control region 5′ hypersensitive site 3. *Molecular and Cellular Biology*, *16*, 2906–2912.

Jackson, D. A., & Cook, P. R. (1995). The structural basis of nuclear function. *International Review of Cytology*, *162A*, 125–149.

Jeong, M., & Goodell, M. A. (2014). New answers to old questions from genome-wide maps of DNA methylation in hematopoietic cells. *Experimental Hematology*, *42*, 609–617.

Jing, H., Vakoc, C. R., Ying, L., Mandat, S., Wang, H., Zheng, X., et al. (2008). Exchange of GATA factors mediates transitions in looped chromatin organization at a developmentally regulated gene locus. *Molecular Cell*, *29*, 232–242.

Johnson, K. D., Boyer, M. E., Kang, J. A., Wickrema, A., Cantor, A. B., & Bresnick, E. H. (2007). Friend of GATA-1-independent transcriptional repression: A novel mode of GATA-1 function. *Blood*, *109*, 5230–5233.

Johnson, K. D., Christensen, H. M., Zhao, B., & Bresnick, E. H. (2001). Distinct mechanisms control RNA polymerase II recruitment to a tissue-specific locus control region and a downstream promoter. *Molecular Cell*, *8*, 465–471.

Johnson, K. D., Grass, J. A., Im, H., Park, C., Choi, K., & Bresnick, E. H. (2003). Highly restricted localization of RNA polymerase II to the hypersensitive site cores of a tissue-specific locus control region. *Molecular and Cellular Biology*, *23*, 6468–6493.

Johnson, K. D., Hsu, A. P., Ryu, M. J., Wang, J., Gao, X., Boyer, M. E., et al. (2012). Cis-element mutated in GATA2-dependent immunodeficiency governs hematopoiesis and vascular integrity. *The Journal of Clinical Investigation*, *122*, 3692–3704.

Johnson, K. D., Kong, G., Gao, X., Chang, Y.-I., Hewitt, K. J., Sanalkumar, R., et al. (2015). Cis-regulatory mechanisms governing stem and progenitor cell transitions. *Science Advances*, *1*, e1500503.

Jones, P. A. (2012). Functions of DNA methylation: Islands, start sites, gene bodies and beyond. *Nature Reviews. Genetics*, *13*, 484–492.

Kallianpur, A. R., Jordan, J. E., & Brandt, S. J. (1994). The SCL/TAL-1 gene is expressed in progenitors of both the hematopoietic and vascular systems during embryogenesis. *Blood*, *83*, 1200–1208.

Kang, Y. A., Sanalkumar, R., O'Geen, H., Linnemann, A. K., Chang, C. J., Bouhassira, E. E., et al. (2012). Autophagy driven by a master regulator of hematopoiesis. *Molecular and Cellular Biology, 32*, 226–239.

Kassouf, M. T., Chagraoui, H., Vyas, P., & Porcher, C. (2008). Differential use of SCL/TAL-1 DNA-binding domain in developmental hematopoiesis. *Blood, 112*, 1056–1067.

Kassouf, M. T., Hughes, J. R., Taylor, S., McGowan, S. J., Soneji, S., Green, A. L., et al. (2010). Genome-wide identification of TAL1's functional targets: Insights into its mechanisms of action in primary erythroid cells. *Genome Research, 20*, 1064–1083.

Katsumura, K. R., DeVilbiss, A. W., Pope, N. J., Johnson, K. D., & Bresnick, E. H. (2013). Transcriptional mechanisms underlying hemoglobin synthesis. *Cold Spring Harbor Perspectives in Medicine, 3*, a015412.

Kellis, M., Wold, B., Snyder, M. P., Bernstein, B. E., Kundaje, A., Marinov, G. K., et al. (2014). Defining functional DNA elements in the human genome. *Proceedings of the National Academy of Sciences of the United States of America, 111*, 6131–6138.

Khoury, G., & Gruss, P. (1983). Enhancer elements. *Cell, 33*, 313–314.

Kim, S. I., Bresnick, E. H., & Bultman, S. J. (2009). BRG1 directly regulates nucleosome structure and chromatin looping of the alpha globin locus to activate transcription. *Nucleic Acids Research, 37*, 6019–6027.

Kim, S.-I., Bultman, S. J., Kiefer, C. M., Dean, A., & Bresnick, E. H. (2009). BRG1 requirement for long-range interaction of a locus control region with a downstream promoter. *Proceedings of the National Academy of Sciences of the United States of America, 106*, 2259–2264.

Kim, H., & Kim, J. S. (2014). A guide to genome engineering with programmable nucleases. *Nature Reviews. Genetics, 15*, 321–334.

Kim, Y., Kweon, J., Kim, A., Chon, J. K., Yoo, J. Y., Kim, H. J., et al. (2013). A library of TAL effector nucleases spanning the human genome. *Nature Biotechnology, 31*, 251–258.

Ko, L. J., & Engel, J. D. (1993). DNA-binding specificities of the GATA transcription factor family. *Molecular and Cellular Biology, 13*, 4011–4022.

Ko, L. J., Yamamoto, M., Leonard, M. W., George, K. M., Ting, P., & Engel, J. D. (1991). Murine and human T-lymphocyte GATA-3 factors mediate transcription through a cis-regulatory element within the human T-cell receptor delta gene enhancer. *Molecular and Cellular Biology, 11*, 2778–2784.

Koh, K. P., Yabuuchi, A., Rao, S., Huang, Y., Cunniff, K., Nardone, J., et al. (2011). Tet1 and Tet2 regulate 5-hydroxymethylcytosine production and cell lineage specification in mouse embryonic stem cells. *Cell Stem Cell, 8*, 200–213.

Kosak, S. T., Skok, J. A., Medina, K. L., Riblet, R., Le Beau, M. M., Fisher, A. G., et al. (2002). Subnuclear compartmentalization of immunoglobulin loci during lymphocyte development. *Science, 296*, 158–162.

Lai, F., & Shiekhattar, R. (2014). Enhancer RNAs: The new molecules of transcription. *Current Opinion in Genetics & Development, 25*, 38–42.

Lecuyer, E., Herblot, S., Saint-Denis, M., Martin, R., Begley, C. G., Porcher, C., et al. (2002). The SCL complex regulates c-kit expression in hematopoietic cells through functional interaction with Sp1. *Blood, 100*, 2430–2440.

Lecuyer, E., Lariviere, S., Sincennes, M. C., Haman, A., Lahlil, R., Todorova, M., et al. (2007). Protein stability and transcription factor complex assembly determined by the SCL-LMO2 interaction. *The Journal of Biological Chemistry, 282*, 33649–33658.

Lee, H. Y., Johnson, K. D., Boyer, M. E., & Bresnick, E. H. (2011). Relocalizing genetic loci into specific subnuclear neighborhoods. *The Journal of Biological Chemistry, 286*, 18834–18844.

Lee, H.-Y., Johnson, K. D., Fujiwara, T., Boyer, M. E., Kim, S.-I., & Bresnick, E. H. (2009). Controlling hematopoiesis through sumoylation-dependent regulation of a GATA factor. *Molecular Cell, 36*, 984–995.

Letting, D. L., Chen, Y. Y., Rakowski, C., Reedy, S., & Blobel, G. A. (2004). Context-dependent regulation of GATA-1 by friend of GATA-1. *Proceedings of the National Academy of Sciences of the United States of America, 101*, 476–481.

Li, L., Freudenberg, J., Cui, K., Dale, R., Song, S. H., Dean, A., et al. (2013). Ldb1-nucleated transcription complexes function as primary mediators of global erythroid gene activation. *Blood, 121*, 4575–4585.

Li, L., Jothi, R., Cui, K., Lee, J. Y., Cohen, T., Gorivodsky, M., et al. (2011). Nuclear adaptor Ldb1 regulates a transcriptional program essential for the maintenance of hematopoietic stem cells. *Nature Immunology, 12*, 129–136.

Li, J., Kurasawa, Y., Wang, Y., Clise-Dwyer, K., Klumpp, S. A., Liang, H., et al. (2014). Requirement for ssbp2 in hematopoietic stem cell maintenance and stress response. *Journal of Immunology, 193*, 4654–4662.

Ling, J., Ainol, L., Zhang, L., Yu, X., Pi, W., & Tuan, D. (2004). HS2 enhancer function is blocked by a transcriptional terminator inserted between the enhancer and the promoter. *The Journal of Biological Chemistry, 279*, 51704–51713.

Ling, K. W., Ottersbach, K., van Hamburg, J. P., Oziemlak, A., Tsai, F. Y., Orkin, S. H., et al. (2004). GATA-2 plays two functionally distinct roles during the ontogeny of hematopoietic stem cells. *The Journal of Experimental Medicine, 200*, 871–882.

Linnemann, A. K., O'Geen, H., Keles, S., Farnham, P. J., & Bresnick, E. H. (2011). Genetic framework for GATA factor function in vascular biology. *Proceedings of the National Academy of Sciences of the United States of America, 108*, 13641–13646.

Love, P. E., Warzecha, C., & Li, L. (2014). Ldb1 complexes: The new master regulators of erythroid gene transcription. *Trends in Genetics, 30*, 1–9.

Luesink, M., Hollink, I. H., van der Velden, V. H., Knops, R. H., Boezeman, J. B., de Haas, V., et al. (2012). High GATA2 expression is a poor prognostic marker in pediatric acute myeloid leukemia. *Blood, 120*, 2064–2075.

Lurie, L. J., Boyer, M. E., Grass, J. A., & Bresnick, E. H. (2008). Differential GATA factor stabilities: Implications for chromatin occupancy by structurally similar transcription factors. *Biochemistry, 47*, 859–869.

Madzo, J., Liu, H., Rodriguez, A., Vasanthakumar, A., Sundaravel, S., Caces, D. B., et al. (2014). Hydroxymethylation at gene regulatory regions directs stem/early progenitor cell commitment during erythropoiesis. *Cell Reports, 6*, 231–244.

Marks-Bluth, J., Khanna, A., Chandrakanthan, V., Thoms, J., Bee, T., Eich, C., et al. (2015). SMAD1 and SMAD5 expression is coordinately regulated by FLI1 and GATA2 during endothelial development. *Molecular and Cellular Biology, 35*, 2165–2172.

Martin, D. I., & Orkin, S. H. (1989). An erythroid-specific DNA binding factor mediates increased gamma-globin expression in hereditary persistence of fetal hemoglobin (HPFH). *Progress in Clinical and Biological Research, 316A*, 217–228.

Martowicz, M. L., Grass, J. A., Boyer, M. E., Guend, H., & Bresnick, E. H. (2005). Dynamic GATA factor interplay at a multicomponent regulatory region of the GATA-2 locus. *The Journal of Biological Chemistry, 280*, 1724–1732.

May, G., Soneji, S., Tipping, A. J., Teles, J., McGowan, S. J., Wu, M., et al. (2013). Dynamic analysis of gene expression and genome-wide transcription factor binding during lineage specification of multipotent progenitors. *Cell Stem Cell, 13*, 754–768.

Merika, M., & Orkin, S. H. (1993). DNA-binding specificity of GATA family transcription factors. *Molecular and Cellular Biology, 13*, 3999–4010.

Miccio, A., Wang, Y., Hong, W., Gregory, G. D., Wang, H., Yu, X., et al. (2010). NuRD mediates activating and repressive functions of GATA-1 and FOG-1 during blood development. *The EMBO Journal, 29*, 442–456.

Mikkola, H. K., Klintman, J., Yang, H., Hock, H., Schlaeger, T. M., Fujiwara, Y., et al. (2003). Hematopoietic stem cells retain long-term repopulating activity and multipotency in the absence of stem-cell leukemia SCL/tal-1 gene. *Nature, 421*, 547–551.

Miller, I. J., & Bieker, J. J. (1993). A novel, erythroid cell-specific murine transcription factor that binds to the CACCC element and is related to the Kruppel family of nuclear proteins. *Molecular and Cellular Biology*, *13*, 2776–2786.

Molkentin, J. D. (2000). The zinc finger-containing transcription factors GATA-4, -5, and -6. Ubiquitously expressed regulators of tissue-specific gene expression. *The Journal of Biological Chemistry*, *275*, 38949–38952.

Mouthon, M. A., Bernard, O., Mitjavila, M. T., Romeo, P. H., Vainchenker, W., & Mathieu-Mahul, D. (1993). Expression of Tal-1 and GATA-binding proteins during human hematopoiesis. *Blood*, *81*, 647–655.

Mukhopadhyay, M., Teufel, A., Yamashita, T., Agulnick, A. D., Chen, L., Downs, K. M., et al. (2003). Functional ablation of the mouse Ldb1 gene results in severe patterning defects during gastrulation. *Development*, *130*, 495–505.

Munugalavadla, V., Dore, L. C., Tan, B. L., Hong, L., Vishnu, M., Weiss, M. J., et al. (2005). Repression of c-kit and its downstream substrates by GATA-1 inhibits cell proliferation during erythroid maturation. *Molecular and Cellular Biology*, *25*, 6747–6759.

Nichols, K. E., Crispino, J. D., Poncz, M., White, J. G., Orkin, S. H., Maris, J. M., et al. (2000). Familial dyserythropoietic anaemia and thrombocytopenia due to an inherited mutation in GATA-1. *Nature Genetics*, *24*, 266–270.

Nilson, D. G., Sabatino, D. E., Bodine, D. M., & Gallagher, P. G. (2006). Major erythrocyte membrane protein genes in EKLF-deficient mice. *Experimental Hematology*, *34*, 705–712.

Nuez, B., Michalovich, D., Bygrave, A., Ploemacher, R., & Grosveld, F. (1995). Defective haematopoiesis in fetal liver resulting from inactivation of the EKLF gene. *Nature*, *375*, 316–318.

Osada, H., Grutz, G., Axelson, H., Forster, A., & Rabbitts, T. H. (1995). Association of erythroid transcription factors: Complexes involving the LIM protein RBTN2 and the zinc-finger protein GATA1. *Proceedings of the National Academy of Sciences of the United States of America*, *92*, 9585–9589.

Pal, S., Cantor, A. B., Johnson, K. D., Moran, T., Boyer, M. E., Orkin, S. H., et al. (2004). Coregulator-dependent facilitation of chromatin occupancy by GATA-1. *Proceedings of the National Academy of Sciences of the United States of America*, *101*, 980–985.

Pennacchio, L. A., Ahituv, N., Moses, A. M., Prabhakar, S., Nobrega, M. A., Shoukry, M., et al. (2006). In vivo enhancer analysis of human conserved non-coding sequences. *Nature*, *444*, 499–502.

Perkins, A. C., Sharpe, A. H., & Orkin, S. H. (1995). Lethal beta-thalassaemia in mice lacking the erythroid CACCC-transcription factor EKLF. *Nature*, *375*, 318–322.

Persons, D. A., Allay, J. A., Allay, E. R., Ashmun, R. A., Orlic, D., Jane, S. M., et al. (1999). Enforced expression of the GATA-2 transcription factor blocks normal hematopoiesis. *Blood*, *93*, 488–499.

Peters, L. L., Jindel, H. K., Gwynn, B., Korsgren, C., John, K. M., Lux, S. E., et al. (1999). Mild spherocytosis and altered red cell ion transport in protein 4.2-null mice. *The Journal of Clinical Investigation*, *103*, 1527–1537.

Pevny, L., Lin, C. S., D'Agati, V., Simon, M. C., Orkin, S. H., & Costantini, F. (1995). Development of hematopoietic cells lacking transcription factor GATA-1. *Development*, *121*, 163–172.

Pevny, L., Simon, M. C., Robertson, E., Klein, W. H., Tsai, S. F., D'Agati, V., et al. (1991). Erythroid differentiation in chimaeric mice blocked by a targeted mutation in the gene for transcription factor GATA-1. *Nature*, *349*, 257–260.

Pilon, A. M., Arcasoy, M. O., Dressman, H. K., Vayda, S. E., Maksimova, Y. D., Sangerman, J. I., et al. (2008). Failure of terminal erythroid differentiation in EKLF-deficient mice is associated with cell cycle perturbation and reduced expression of E2F2. *Molecular and Cellular Biology*, *28*, 7394–7401.

Plank, J. L., & Dean, A. (2014). Enhancer function: Mechanistic and genome-wide insights come together. *Molecular Cell*, *55*, 5–14.

Porcher, C., Swat, W., Rockwell, K., Fujiwara, Y., Alt, F. W., & Orkin, S. H. (1996). The T cell leukemia oncoprotein SCL/tal-1 is essential for development of all hematopoietic lineages. *Cell*, *86*, 47–57.

Ragoczy, T., Bender, M. A., Telling, A., Byron, R., & Groudine, M. (2006). The locus control region is required for association of the murine beta-globin locus with engaged transcription factories during erythroid maturation. *Genes & Development*, *20*, 1447–1457.

Ragoczy, T., Telling, A., Sawado, T., Groudine, M., & Kosak, S. T. (2003). A genetic analysis of chromosome territory looping: Diverse roles for distal regulatory elements. *Chromosome Research*, *11*, 513–525.

Rajapakse, I., & Groudine, M. (2011). On emerging nuclear order. *The Journal of Cell Biology*, *192*, 711–721.

Rodrigues, N. P., Janzen, V., Forkert, R., Dombkowski, D. M., Boyd, A. S., Orkin, S. H., et al. (2005). Haploinsufficiency of GATA-2 perturbs adult hematopoietic stem cell homeostasis. *Blood*, *106*, 477–484.

Ross, J., Mavoungou, L., Bresnick, E. H., & Milot, E. (2012). GATA-1 utilizes Ikaros and polycomb repressive complex 2 to suppress Hes1 and to promote erythropoiesis. *Molecular and Cellular Biology*, *32*, 3624–3638.

Rossi, L., Lin, K. K., Boles, N. C., Yang, L., King, K. Y., Jeong, M., et al. (2012). Less is more: Unveiling the functional core of hematopoietic stem cells through knockout mice. *Cell Stem Cell*, *11*, 302–317.

Sanalkumar, R., Johnson, K. D., Gao, X., Boyer, M. E., Chang, Y. I., Hewitt, K. J., et al. (2014). Mechanism governing a stem cell-generating cis-regulatory element. *Proceedings of the National Academy of Sciences of the United States of America*, *111*, E1091–E1100.

Sanda, T., Lawton, L. N., Barrasa, M. I., Fan, Z. P., Kohlhammer, H., Gutierrez, A., et al. (2012). Core transcriptional regulatory circuit controlled by the TAL1 complex in human T cell acute lymphoblastic leukemia. *Cancer Cell*, *22*, 209–221.

Schleif, R. (1992). DNA looping. *Annual Review of Biochemistry*, *61*, 199–223.

Schubeler, D., Francastel, C., Cimbora, D. M., Reik, A., Martin, D. I., & Groudine, M. (2000). Nuclear localization and histone acetylation: A pathway for chromatin opening and transcriptional activation of the human beta-globin locus. *Genes & Development*, *14*, 940–950.

Scripture-Adams, D. D., Damle, S. S., Li, L., Elihu, K. J., Qin, S., Arias, A. M., et al. (2014). GATA-3 dose-dependent checkpoints in early T cell commitment. *Journal of Immunology*, *193*, 3470–3491.

Shen, Y., Yue, F., McCleary, D. F., Ye, Z., Edsall, L., Kuan, S., et al. (2012). A map of the cis-regulatory sequences in the mouse genome. *Nature*, *488*, 116–120.

Shi, J., Whyte, W. A., Zepeda-Mendoza, C. J., Milazzo, J. P., Shen, C., Roe, J. S., et al. (2013). Role of SWI/SNF in acute leukemia maintenance and enhancer-mediated Myc regulation. *Genes & Development*, *27*, 2648–2662.

Shivdasani, R. A., Fujiwara, Y., McDevitt, M. A., & Orkin, S. H. (1997). A lineage-selective knockout establishes the critical role of transcription factor GATA-1 in megakaryocyte growth and platelet development. *The EMBO Journal*, *16*, 3965–3973.

Shivdasani, R. A., Mayer, E. L., & Orkin, S. H. (1995). Absence of blood formation in mice lacking the T-cell leukaemia oncoprotein tal-1/SCL. *Nature*, *373*, 432–434.

Simon, M. C., Pevny, L., Wiles, M. V., Keller, G., Costantini, F., & Orkin, S. H. (1992). Rescue of erythroid development in gene targeted GATA-1-mouse embryonic stem cells. *Nature Genetics*, *1*, 92–98.

Snow, J. W., Trowbridge, J. J., Fujiwara, T., Emambokus, N. E., Grass, J. A., Orkin, S. H., et al. (2010). A single cis element maintains repression of the key developmental regulator Gata2. *PLoS Genetics*, *6*, e1001103.

Snow, J. W., Trowbridge, J. J., Johnson, K. D., Fujiwara, T., Emambokus, N. E., Grass, J. A., et al. (2011). Context-dependent function of "GATA switch" sites in vivo. *Blood*, *117*, 4769–4772.

Soler, E., Andrieu-Soler, C., de Boer, E., Bryne, J. C., Thongjuea, S., Stadhouders, R., et al. (2010). The genome-wide dynamics of the binding of Ldb1 complexes during erythroid differentiation. *Genes & Development*, *24*, 277–289.

Song, S. H., Hou, C., & Dean, A. (2007). A positive role for NL1/Ldb1 in long-range beta-globin locus control region function. *Molecular Cell*, *28*, 810–822.

Song, S. H., Kim, A., Ragoczy, T., Bender, M. A., Groudine, M., & Dean, A. (2010). Multiple functions of Ldb1 required for beta-globin activation during erythroid differentiation. *Blood*, *116*, 2356–2364.

Spinner, M. A., Sanchez, L. A., Hsu, A. P., Shaw, P. A., Zerbe, C. S., Calvo, K. R., et al. (2013). GATA2 deficiency: A protean disorder of hematopoiesis, lymphatics and immunity. *Blood*, *123*, 809–821.

Szutorisz, H., Dillon, N., & Tora, L. (2005). The role of enhancers as centres for general transcription factor recruitment. *Trends in Biochemical Sciences*, *30*, 593–599.

Tahiliani, M., Koh, K. P., Shen, Y., Pastor, W. A., Bandukwala, H., Brudno, Y., et al. (2009). Conversion of 5-methylcytosine to 5-hydroxymethylcytosine in mammalian DNA by MLL partner TET1. *Science*, *324*, 930–935.

Tallack, M. R., Magor, G. W., Dartigues, B., Sun, L., Huang, S., Fittock, J. M., et al. (2012). Novel roles for KLF1 in erythropoiesis revealed by mRNA-seq. *Genome Research*, *22*, 2385–2398.

Tallack, M. R., Whitington, T., Yuen, W. S., Wainwright, E. N., Keys, J. R., Gardiner, B. B., et al. (2010). A global role for KLF1 in erythropoiesis revealed by ChIP-seq in primary erythroid cells. *Genome Research*, *20*, 1052–1063.

Ting, C. N., Olson, M. C., Barton, K. P., & Leiden, J. M. (1996). Transcription factor GATA-3 is required for development of the T-cell lineage. *Nature*, *384*, 474–478.

Tremblay, M., Tremblay, C. S., Herblot, S., Aplan, P. D., Hebert, J., Perreault, C., et al. (2010). Modeling T-cell acute lymphoblastic leukemia induced by the SCL and LMO1 oncogenes. *Genes & Development*, *24*, 1093–1105.

Tripic, T., Deng, W., Cheng, Y., Vakoc, C. R., Gregory, G. D., Hardison, R. C., et al. (2008). SCL and associated protein distinguish active from repressive GATA transcription factor complexes. *Blood*, *113*, 2191–2201.

Tsai, F. Y., Keller, G., Kuo, F. C., Weiss, M., Chen, J., Rosenblatt, M., et al. (1994). An early haematopoietic defect in mice lacking the transcription factor GATA-2. *Nature*, *371*, 221–226.

Tsai, S. F., Martin, D. I., Zon, L. I., D'Andrea, A. D., Wong, G. G., & Orkin, S. H. (1989). Cloning of cDNA for the major DNA-binding protein of the erythroid lineage through expression in mammalian cells. *Nature*, *339*, 446–451.

Tsai, F.-Y., & Orkin, S. H. (1997). Transcription factor GATA-2 is required for proliferation/survival of early hematopoietic cells and mast cell formation, but not for erythroid and myeloid terminal differentiation. *Blood*, *89*, 3636–3643.

Tsang, A. P., Fujiwara, Y., Hom, D. B., & Orkin, S. H. (1998). Failure of megakaryopoiesis and arrested erythropoiesis in mice lacking the GATA-1 transcriptional cofactor FOG. *Genes & Development*, *12*, 1176–1188.

Tsang, A. P., Visvader, J. E., Turner, C. A., Fujiwara, Y., Yu, C., Weiss, M. J., et al. (1997). FOG, a multitype zinc finger protein, acts as a cofactor for transcription factor GATA-1 in erythroid and megakaryocytic differentiation. *Cell*, *90*, 109–119.

Valverde-Garduno, V., Guyot, B., Anguita, E., Hamlett, I., Porcher, C., & Vyas, P. (2004). Differences in the chromatin structure and cis-element organization of the human and mouse GATA1 loci: Implications for cis-element identification. *Blood*, *104*, 3106–3116.

van Meyel, D. J., Thomas, J. B., & Agulnick, A. D. (2003). Ssdp proteins bind to LIM-interacting co-factors and regulate the activity of LIM-homeodomain protein complexes in vivo. *Development*, *130*, 1915–1925.

Vicente, C., Vazquez, I., Conchillo, A., Garcia-Sanchez, M. A., Marcotegui, N., Fuster, O., et al. (2012). Overexpression of GATA2 predicts an adverse prognosis for patients with acute myeloid leukemia and it is associated with distinct molecular abnormalities. *Leukemia*, *26*, 550–554.

Vieira, K. F., Levings, P. P., Hilll, M. A., Cruselle, V. J., Kang, S. H., Engel, J. D., et al. (2004). Recruitment of transcription complexes to the beta-globin gene locus in vivo and in vitro. *The Journal of Biological Chemistry*, *279*, 50350–50357.

Vyas, P., McDevitt, M. A., Cantor, A. B., Katz, S. G., Fujiwara, Y., & Orkin, S. H. (1999). Different sequence requirements for expression in erythroid and megakaryocytic cells within a regulatory element upstream of the GATA-1 gene. *Development*, *126*, 2799–2811.

Wadman, I. A., Osada, H., Grutz, G. G., Agulnick, A. D., Westphal, H., Forster, A., et al. (1997). The LIM-only protein Lmo2 is a bridging molecule assembling an erythroid, DNA-binding complex which includes the TAL1, E47, GATA-1 and Ldb1/NLI proteins. *The EMBO Journal*, *16*, 3145–3157.

Wang, Y., Klumpp, S., Amin, H. M., Liang, H., Li, J., Estrov, Z., et al. (2010). SSBP2 is an in vivo tumor suppressor and regulator of LDB1 stability. *Oncogene*, *29*, 3044–3053.

Warren, A. J., Colledge, W. H., Carlton, M. B., Evans, M. J., Smith, A. J., & Rabbitts, T. H. (1994). The oncogenic cysteine-rich LIM domain protein rbtn2 is essential for erythroid development. *Cell*, *78*, 45–57.

Wasylyk, B., Wasylyk, C., & Chambon, P. (1984). Short and long range activation by the SV40 enhancer. *Nucleic Acids Research*, *12*, 5589–5608.

Wechsler, J., Greene, M., McDevitt, M. A., Anastasi, J., Karp, J. E., LeBeau, M. M., et al. (2002). Acquired mutations in GATA-1 in the megakaryoblastic leukemia of Down syndrome. *Nature Genetics*, *32*, 148–152.

Weiss, M. J., Keller, G., & Orkin, S. H. (1994). Novel insights into erythroid development revealed through in vitro differentiation of GATA-1 embryonic stem cells. *Genes & Development*, *8*, 1184–1197.

Willis, S. D., & Seyfred, M. A. (1996). Pituitary-specific chromatin structure of the rat prolactin distal enhancer element. *Nucleic Acids Research*, *24*, 1065–1072.

Wilson, N. K., Foster, S. D., Wang, X., Knezevic, K., Schutte, J., Kaimakis, P., et al. (2010). Combinatorial transcriptional control in blood stem/progenitor cells: Genome-wide analysis of ten major transcriptional regulators. *Cell Stem Cell*, *7*, 532–544.

Wozniak, R. J., Boyer, M. E., Grass, J. A., Lee, Y., & Bresnick, E. H. (2007). Context-dependent GATA factor function: Combinatorial requirements for transcriptional control in hematopoietic and endothelial cells. *The Journal of Biological Chemistry*, *282*, 14665–14674.

Wozniak, R. J., Keles, S., Lugus, J. J., Young, K., Boyer, M. E., Tran, T. T., et al. (2008). Molecular hallmarks of endogenous chromatin complexes containing master regulators of hematopoiesis. *Molecular and Cellular Biology*, *28*, 6681–6694.

Xu, Z., Huang, S., Chang, L. S., Agulnick, A. D., & Brandt, S. J. (2003). Identification of a TAL1 target gene reveals a positive role for the LIM domain-binding protein Ldb1 in erythroid gene expression and differentiation. *Molecular and Cellular Biology*, *23*, 7585–7599.

Xu, Z., Meng, X., Cai, Y., Koury, M. J., & Brandt, S. J. (2006). Recruitment of the SWI/SNF protein Brg1 by a multiprotein complex effects transcriptional repression in murine erythroid progenitors. *The Biochemical Journal, 399*, 297–304.

Xu, Z., Meng, X., Cai, Y., Liang, H., Nagarajan, L., & Brandt, S. J. (2007). Single-stranded DNA-binding proteins regulate the abundance of LIM domain and LIM domain-binding proteins. *Genes & Development, 21*, 942–955.

Yang, H., Wang, H., Shivalila, C. S., Cheng, A. W., Shi, L., & Jaenisch, R. (2013). One-step generation of mice carrying reporter and conditional alleles by CRISPR/Cas-mediated genome engineering. *Cell, 154*, 1370–1379.

Yu, M., Riva, L., Xie, H., Schindler, Y., Moran, T. B., Cheng, Y., et al. (2009). Insights into GATA-1-mediated gene activation versus repression via genome-wide chromatin occupancy analysis. *Molecular Cell, 36*, 682–695.

Yue, F., Cheng, Y., Breschi, A., Vierstra, J., Wu, W., Ryba, T., et al. (2014). A comparative encyclopedia of DNA elements in the mouse genome. *Nature, 515*, 355–364.

Zeng, X., Sanalkumar, R., Bresnick, E. H., Li, H., Chang, Q., & Keles, S. (2013). jMOSAiCS: Joint analysis of multiple ChIP-seq datasets. *Genome Biology, 14*, R38.

Zhou, Y., Zhu, S., Cai, C., Yuan, P., Li, C., Huang, Y., et al. (2014). High-throughput screening of a CRISPR/Cas9 library for functional genomics in human cells. *Nature, 509*, 487–491.

Zon, L. I., Gurish, M. F., Stevens, R. L., Mather, C., Reynolds, D. S., Austen, K. F., et al. (1991). GATA-binding transcription factors in mast cells regulate the promoter of the mast cell carboxypeptidase A gene. *The Journal of Biological Chemistry, 266*, 22948–22953.

Zullo, J. M., Demarco, I. A., Pique-Regi, R., Gaffney, D. J., Epstein, C. B., Spooner, C. J., et al. (2012). DNA sequence-dependent compartmentalization and silencing of chromatin at the nuclear lamina. *Cell, 149*, 1474–1487.

CHAPTER FOUR

ETS Transcription Factor ETV2/ER71/Etsrp in Hematopoietic and Vascular Development

S. Sumanas*, K. Choi†,[1]
*Cincinnati Children's Hospital Medical Center, Cincinnati, OH, United States
†Washington University, School of Medicine, St. Louis, MO, United States
[1]Corresponding author: e-mail address: kchoi@wustl.edu

Contents

Abstract

Effective establishment of the hematopoietic and vascular systems is prerequisite for successful embryogenesis. The ETS transcription factor *Etv2* has proven to be essential for hematopoietic and vascular development. *Etv2* expression marks the onset of the hematopoietic and vascular development and its deficiency leads to an absolute block in hematopoietic and vascular development. *Etv2* is transiently expressed during development and is mainly expressed in testis in adults. Consistent with its expression pattern, *Etv2* is transiently required for the generation of the optimal levels of the hemangiogenic cell population. Deletion of this gene after the hemangiogenic progenitor formation leads to normal hematopoietic and vascular development. Mechanistically, ETV2 induces the hemangiogenic program by activating blood and endothelial cell lineage specifying genes and enhancing VEGF signaling. Moreover, ETV2 establishes an ETS hierarchy by directly activating other *Ets* genes, which in the face of transient *Etv2* expression, presumably maintain blood and endothelial cell program initiated by ETV2

Current Topics in Developmental Biology, Volume 118
ISSN 0070-2153
http://dx.doi.org/10.1016/bs.ctdb.2016.01.005

through an ETS switching mechanism. Current studies suggest that the hemangiogenic progenitor population is exclusively sensitive to ETV2-dependent FLK1 signaling. Any perturbation in the ETV2, VEGF, and FLK1 balance causing insufficient hemangiogenic progenitor cell generation would lead to defects in hematopoietic and endothelial cell development.

1. INTRODUCTION

Functionally mature blood and endothelial cells are initially detected in the blood islands of the yolk sac during embryogenesis. The close developmental association of hematopoietic and endothelial cells within blood islands has led to the hypothesis that they arise from a common progenitor, termed the hemangioblast (Murray, 1932; Sabin, 1920). As such, yolk sac blood islands are speculated to form from the extraembryonic mesoderm through the intermediary hemangioblast, which differentiate into centrally located embryonic blood cells and endothelial cells that encircle them. It is thus remarkable that many transcription factors and signaling pathways are shared between blood and endothelial cell lineages. Gene-targeting studies have also shown that mutations in any of the shared genes often affect both cell lineages, supporting the notion of the common genetic pathways regulating hematopoietic and endothelial cell lineage development and function. Of the genes that contribute to hematopoietic and endothelial cell development, *Etv2* has emerged as an obligatory factor that functions at the earliest stage in hematopoietic and endothelial cell development.

ETV2 (Ets variant 2), also known as ER71 (ETS-related 71) or Etsrp (Ets-related protein), belongs to the ETS family of transcription factors, which contain a conserved ~85 amino acid domain, termed the ETS domain (Karim et al., 1990). ETS factors recognize the core DNA target sequence GGAA/T and can function as transcriptional activators or repressors (reviewed in Hollenhorst, McIntosh, & Graves, 2011; Sharrocks, 2001). Since the initial identification of the founding member *Ets1*, which was recognized as a viral oncogene in the E26 avian leukemia retrovirus (E26 transformation specific-1) (Leprince et al., 1983), nearly 30 ETS family genes have been identified to date, 28 of which are present in the human genome (reviewed in Adamo & Ladomery, 2015). *Etv2* was initially cloned from the screening of the embryonic (E) day 8.5 mouse embryo cDNA library using degenerate oligonucleotides corresponding to the ETS domain (Brown & McKnight, 1992). Subsequently, microarray analysis of the zebrafish *cloche* mutants, which are defective in blood and endothelial cell formation, identified *etsrp* expression to be greatly reduced in *cloche* mutants (Sumanas, Jorniak, & Lin, 2005). Additionally, microarray analysis of the hemangiogenic progenitor, which generates hematopoietic and endothelial cells, and mouse *Nkx2–5* null hearts identified *Etv2* to be highly expressed in the hemangioblast population compared to its progeny, hematopoietic, and endothelial cells (Lugus et al., 2007), but greatly reduced in *Nkx2–5* null hearts (Ferdous et al., 2009), respectively. These studies suggested that ETV2 might play a role in hematopoietic and cardiovascular cell lineage development. Indeed, recent studies have established the critical and non-redundant function of this factor in hematopoietic and vascular development.

2. Etv2 IS TRANSIENTLY EXPRESSED IN THE HEMANGIOGENIC PROGENITORS

In developing mouse embryo, *Etv2* is expressed in the primitive streak, embryonic mesoderm, amnion, allantois, and the yolk sac blood islands at around E7.5 (Lee et al., 2008). Its expression becomes rapidly downregulated as embryos develop and is mainly detected within major vessels, including the dorsal aorta (DA) and branchial arches, at around E8.25–E9.5 (Lee et al., 2008). *Etv2* expression in adults becomes restricted mainly to testis (Brown & McKnight, 1992). Similarly, *Etv2* expression in differentiating embryonic stem (ES) cells (EBs, embryoid bodies) is initiated at the time of *Brachyury* expression. *Etv2* expression within EB cells is also transient and becomes undetectable after hemangiogenic progenitors are generated (Lee et al., 2008). Consistently, transgenic reporter expression driven by the Etv2 promoter/enhancer region or the endogenous *Etv2* locus can be detected in the extraembryonic mesoderm, lateral mesoderm, and vascular structures in the embryo proper and yolk sac. However, the *Etv2* reporter is no longer expressed in later stages of development (Kataoka et al., 2011; Rasmussen et al., 2011; Wareing et al., 2012).

Similar *Etv2* expression pattern has also been observed in other vertebrates, such as zebrafish and frog *Xenopus laevis*. In both species, *etv2* expression is observed in hematopoietic and vascular endothelial progenitor cells (EPCs) and its expression is downregulated during later stages (Neuhaus, Muller, & Hollemann, 2010; Pham et al., 2007; Salanga, Meadows, Myers, & Krieg, 2010; Sumanas & Lin, 2006). In *Xenopus* embryos, *etv2* expression is observed in the dorsal lateral plate mesoderm (DLP), which gives rise to the DA and hematopoietic stem cells (HSCs), in the ventral blood island, where primitive hematopoietic cells originate, and in the vascular endothelial progenitors that correspond to the vascular plexus and posterior cardinal vein (PCV) (Ciau-Uitz, Pinheiro, Kirmizitas, Zuo, & Patient, 2013; Salanga et al., 2010). In zebrafish, *etv2* expression is observed as early as the 1-somite stage in the anterior lateral plate mesoderm (ALPM) region, which is known to give rise to the cranial vasculature, endocardium, and myeloid progenitors. Slightly later, starting at about 2–3 somite stages, its expression is also initiated in vascular endothelial progenitors of the trunk and tail regions in the posterior lateral plate mesoderm (Sumanas & Lin, 2006). *Etv2* expression in zebrafish precedes that of any previously known hematovascular markers such as *scl* and *fli1*, which makes it the earliest currently known marker of vascular endothelial progenitors. Zebrafish *etv2* displays strong expression in all EPCs throughout the somitogenesis stages. In the trunk and tail regions its expression is initiated first in bilaterally located arterial progenitors, which starting at the 10-somite stage, migrate to the midline and coalesce into the single DA (Kohli, Schumacher, Desai, Rehn, & Sumanas, 2013). Starting at the 10-somite stage, *etv2* expression is also initiated in more laterally located venous progenitors, which migrate to the midline slightly later, starting at about the 15-somite stage, and coalesce into the PCV. *Etv2* expression precedes expression of the zebrafish *flk1* homolog *kdrl* in both arterial and venous progenitors (Kohli et al., 2013). In the ALPM region, intense *etv2* expression is observed in two bilaterally located hematovascular progenitor clusters, rostral, and midbrain organizing centers (ROC and MOC). The ROC gives rise to the most rostral cranial vasculature, while the MOC gives rise to the remainder of the cranial vasculature and myeloid and endocardial progenitors (Proulx, Lu, & Sumanas, 2010). In zebrafish embryos, there is a spatial separation of the primitive hematopoietic regions. ALPM region gives rise to the primitive myeloid progenitors that include macrophages and neutrophilic granulocytes, while PLPM gives rise to the erythroid lineage (Davidson & Zon, 2004; Jin et al., 2012; Warga, Kane, & Ho, 2009). Although zebrafish *etv2* precedes expression of other known hematopoietic markers, it does not colocalize with hematopoietic markers during later stages, and its expression is restricted to the vascular endothelial lineage. Thus, when the earliest myeloid marker *pu.1/spi1* expression is initiated in the ALPM region at the 6-somite stage, *etv2* expression is restricted to vascular endothelial progenitors and does not overlap with *pu.1* expression (Sumanas et al., 2008). However, *etv2:gfp* reporter line labels primitive macrophages and neutrophils (Glenn et al.,

2014; Proulx et al., 2010), which supports the model that myeloid cells originate from *etv2*-expressing cells; because of the long persistence of GFP fluorescence, these cells display GFP fluorescence even after its transcription has been downregulated. Overexpression experiments further demonstrate that *etv2*-expressing cells can give rise to myeloid progenitors, in addition to vascular endothelial cells (Sumanas et al., 2008), further supporting the model that *etv2* is expressed in myeloid progenitors and is downregulated during their differentiation. It is tempting to speculate that *etv2* is expressed in the ALPM region in the hemangioblast, a common precursor of the myeloid and endothelial lineages, and its expression has to be downregulated for the myeloid differentiation to occur. Sustained *etv2* expression in hemangioblast progenitors might lead to vascular endothelial fate commitment. While hemangioblast existence has been previously demonstrated in zebrafish embryos, it is clear that not all cells go through this stage (Vogeli, Jin, Martin, & Stainier, 2006; Warga et al., 2009). Therefore, alternatively, *etv2* may be expressed in distinct pools of myeloid and EPCs and is downregulated only in myeloid cells, while endothelial expression of *etv2* continues throughout vasculogenesis.

In contrast to the ALPM region, *etv2* expression in the zebrafish PLPM is restricted to vascular endothelial cells and does not label erythroid progenitors (Sumanas & Lin, 2006). Similarly, *etv2:gfp* transgene does not label erythroid cells (Proulx et al., 2010), and as discussed further, *etv2* does not appear to be involved in the erythropoiesis in zebrafish. It is currently not clear what accounts for different *etv2* expression in the hematopoietic progenitors within the anterior and posterior LPM.

Based on qPCR analysis, zebrafish *etv2* expression peaks at about 15–20-somite stages, when vasculogenesis actively takes place and is strongly downregulated during later stages (Craig et al., 2015; Moore et al., 2013). Intriguingly, at 20–24 hpf stages, *etv2* shows much stronger expression in the venous vasculature as compared to arterial vessels (Proulx et al., 2010). Functional significance of this pattern is not clear; because venous progenitors originate later than arterial (Kohli et al., 2013), they may retain *etv2* expression until later stages. After 24 hpf, when blood circulation is initiated in zebrafish embryos, *etv2* mRNA displays only weak vascular expression. However, it is still expressed in certain vessels such as PCV or branchial arches, and as discussed further, may have additional functional roles. Based on immunostaining analysis, Etv2 protein is no longer detected after 24 hpf. Micro-RNAs including *let-7* have been implicated in downregulating Etv2 protein expression after 24 hpf (Moore et al., 2013).

Recently, Park et al. (2016) have demonstrated that *Etv2* expression was reactivated in endothelial cells during regenerative angiogenesis and that *Etv2* deficiency in endothelial cells led to impaired vascular regeneration and tissue repair. Upstream signals that activate *Etv2* and its role in pathological angiogenesis such as tumor angiogenesis should be investigated in the future.

3. ETV2 IS SUFFICIENT TO INDUCE HEMATOPOIETIC AND ENDOTHELIAL CELL FORMATION

Developmental stages and cell types can affect the outcome of enforced *Etv2* expression. For example, enforced *Etv2* expression in differentiating mouse ES cells prior to hemangiogenic commitment leads to enhanced hemangiogenic progenitor generation with increased hematopoietic and endothelial cell output (Koyano-Nakagawa et al., 2012; Lee et al., 2008). Similarly, *etv2* mRNA or DNA injection leads to strong expansion of myeloid and vascular endothelial cell lineages in zebrafish embryos (Sumanas et al., 2008; Sumanas & Lin, 2006). All known vascular endothelial markers and many hemangioblast markers, including *scl, fli1, kdrl/flk1, cdh5*, and others are induced by multiple fold in *etv2* RNA or DNA expressing zebrafish embryos. Not only these markers are expressed ectopically, but they are also induced precociously as early as 30% epiboly stage (4.7 hpf), much earlier than their normal expression. This induction is highly specific, and microarray and RNA-seq analysis of *etv2*-overexpressing embryos has been used to discover multiple novel vascular endothelial and blood specific genes (Gomez et al., 2012; Gomez, Veldman, Zhao, Burgess, & Lin, 2009; Wong, Proulx, Rost, & Sumanas, 2009). In addition to vascular endothelial markers, expression of myeloid markers including *pu.1, mpx*, and *lcp1* was also induced in *etv2*-overexpressing embryos (Glenn et al., 2014; Sumanas et al., 2008). However, erythroid markers such as *gata1* were not significantly upregulated upon *etv2* overexpression, consistent with the absence of Etv2 requirement in zebrafish erythropoiesis. Similarly, *etv2* overexpression in *Xenopus* embryos induced ectopic vascular endothelial and myeloid but not erythroid markers (Salanga et al., 2010). By using protein synthesis inhibitor cycloheximide, it was shown that many of these markers, including *flk1, aplnr, runx1*, and *spi1b*, are induced by Etv2 in the absence of protein synthesis, and therefore by direct transcriptional activation. Etv2 was also sufficient to induced hematovascular marker expression in *Xenopus* animal caps (Neuhaus et al., 2010).

Importantly, mouse or human *Etv2* expression in zebrafish can also lead to a robust expansion of hemangioblast and vascular endothelial cell lineages, indicating that ETV2 function in hematopoietic and endothelial cell formation is conserved between mammals and zebrafish (Sumanas et al., 2008). While these studies support the notion that *Etv2* induces hemangiogenic program, enforced *Etv2* expression appears to have a different outcome if targeted at different developmental stages. For instance, transgenic *Tie2-Cre*-mediated constitutive *Etv2* expression results in abnormal yolk sac vascular remodeling and persistent endothelial cell generation with hematopoietic defects in the yolk sac (Hayashi et al., 2012), suggesting an ETV2 dosage dependent effect on endothelial vs hematopoietic differentiation. Moreover, continuous *etv2* expression in zebrafish appears to have an inhibitory effect for vascular and lymphatic development. *Etv2* overexpression under vascular endothelial specific *fli1* promoter resulted in inhibition of intersegmental vessel sprouting in a significant fraction of transgenic zebrafish embryos (Davis, Lubert, Palencia Desai, Winkler, & Sumanas, in review). Furthermore, lymphangiogenesis was inhibited in *fli1:Etv2-mCherry*-overexpressing embryos which also displayed an enlarged PCV. This suggests that *Etv2* expression has to be downregulated for normal angiogenesis to occur and that continued *Etv2* expression would promote vascular development at the expense of lymphatic differentiation.

The outcome of *Etv2* overexpression also depends on the cell type. While *Etv2* overexpression using *Tie2-Cre* leads to endothelialization, somite- or neuronal lineage-specific *Etv2* expression does not lead to any obvious phenotype, suggesting a cell lineage-restricted function of *Etv2* (Hayashi et al., 2012). Consistently, *Etv2* overexpression poorly induces hematopoietic and endothelial cells from undifferentiated ES cells (unpublished data). Intriguingly, however, recent studies have reported that *Etv2*, alone or in combination with other ETS factors, can reprogram somatic cells to functional endothelial cells in vitro (Ginsberg et al., 2012; Han et al., 2014; Morita et al., 2015). Moreover, inducible *etv2* expression using heat-shock promoter in zebrafish embryos was sufficient to convert fast skeletal muscle progenitors into functional vascular endothelium (Veldman et al., 2013). Muscle cells downregulated muscle myosin expression and upregulated expression of vascular endothelial markers. Intriguingly, *etv2* induction at later stages was not sufficient to induce endothelial differentiation in fully differentiated muscle cells. In addition, other differentiated tissues also failed to transdifferentiate into vascular endothelium upon *etv2* overexpression (Veldman et al., 2013). This argues that

Etv2 overexpression alone is not sufficient to override already established epigenetic marks and other factors are necessary for transdifferentiation. Thus, it will be important in the future to further delineate the extent of the ETV2 reprogramming potential. Presumably, there might be a specific chromatin state, transcription factors, regulatory RNA or competency, that needs to be established in order to be susceptible to ETV2 function.

4. NONREDUNDANT AND TRANSIENT ROLE OF ETV2 IN HEMATOPOIETIC AND ENDOTHELIAL CELL FORMATION

In addition to *Etv2*, other ETS transcription factors also contribute to hematopoietic and vascular development (reviewed in Bartel, Higuchi, & Spyropoulos, 2000; Wythe et al., 2013). As such, *Fli1*, *Erg*, *Ets1*, *Ets2*, *Etv6*, and *Elk3* are also expressed in blood and endothelial cells during early stages of development. Moreover, mice deficient in *Fli1*, *Erg*, or *Etv6* have hematopoietic and vascular patterning and remodeling defects (Abedin et al., 2014; Birdsey et al., 2008; Dejana, Taddei, & Randi, 2007; Loughran et al., 2008; McLaughlin et al., 2001; Pham et al., 2007; Spyropoulos et al., 2000). Redundant expression patterns of the *Ets* factors in hematopoietic and endothelial cells suggest that ETS factors might have an overlapping function in hematopoietic and endothelial cell development. Indeed, while single *Ets1* or *Ets2* knockout mice show very little hematopoietic and vascular phenotype, *Ets1* and *Ets2* double knockout mice display severe angiogenesis defects (Wei et al., 2009).

Distinct from these ETS factors, *Etv2*-deficient mice display a complete block in blood and blood vessel formation, indicating that ETV2 performs a nonredundant and indispensable function in hematopoietic and vessel development (Fig. 1; Ferdous et al., 2009; Lee et al., 2008). Similarly, Etv2 functional knockdown in zebrafish using morpholinos, as well as the genetic mutant data, demonstrate critical Etv2 role during embryonic vasculogenesis and hematopoiesis. Vascular endothelial cells fail to differentiate, and expression of all known vascular endothelial markers is lost or downregulated in *etv2* MO knockdown zebrafish embryos (morphants) and *etv2* mutants during early- and mid-somitogenesis stages (Pham et al., 2007; Sumanas & Lin, 2006). This includes a complete or near complete loss of *kdrl/flk1*, *fli1*, *cdh5*, and other marker expression in vascular EPCs (Fig. 2). As observed by *etv2* mRNA expression analysis in *etv2* translation blocking MO-injected embryos, *etv2*-expressing EPCs fail to migrate to the midline,

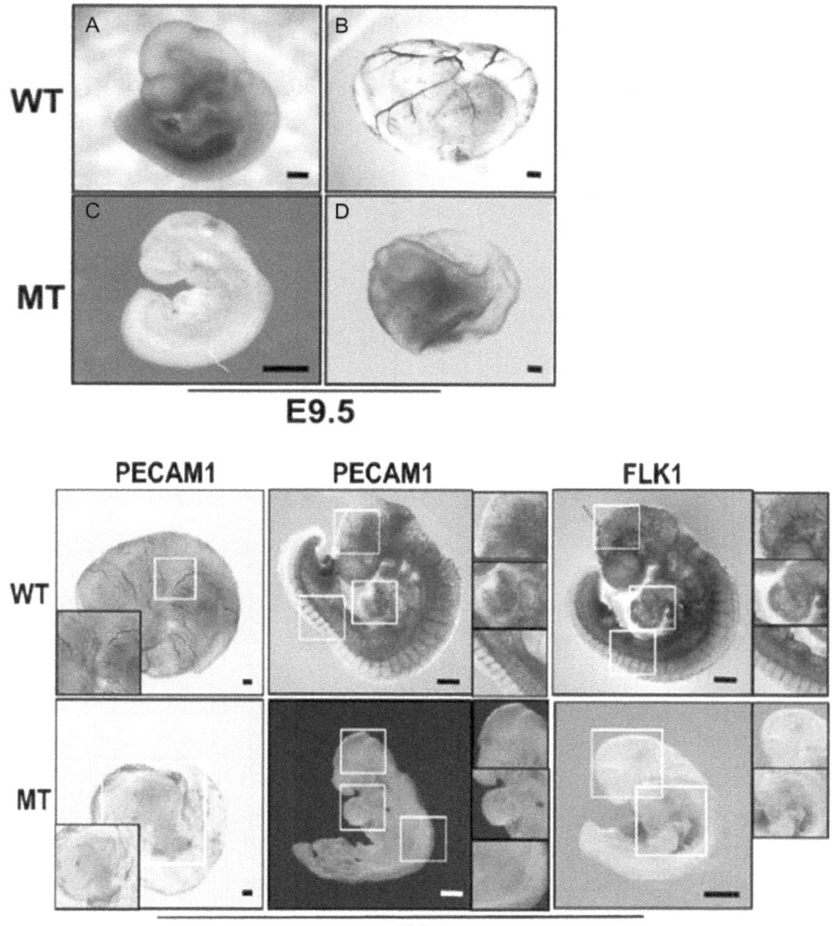

Fig. 1 Etv2 (aka Er71) is required for hematopoietic and endothelial cell development. Gross morphology of *Er71*$^{-/-}$ embryos at E9.5 is shown on the *left* (A and C, embryo proper; B and D, yolk sac), Scale bars, 200 μm. Whole-mount PECAM1 or FLK1 staining of E9.5 embryos is shown on the *right*. The *boxed area* is the vasculature of the yolk sac, brain, heart, and intersomitic regions at higher magnification. Scale bars, 200 μm. *Reprinted from Lee, D. J., Park, C., Lee, H., Lugus, J. J., Kim, S. H., Arentson, E., et al. (2008). ER71 acts downstream of BMP, Notch, and Wnt signaling in blood and vessel progenitor specification.* Cell Stem Cell, 2(5), *497–507, Copyright (2008), with permission from Elsevier.* (See the color plate.)

remain positioned bilaterally, and do not coalesce into blood vessels as they fail to form junctions between adjacent cells (Sumanas & Lin, 2006). Strong upregulation of apoptotic markers in endothelial cells has been observed in *etv2* mutants, arguing that many endothelial progenitors fail to differentiate

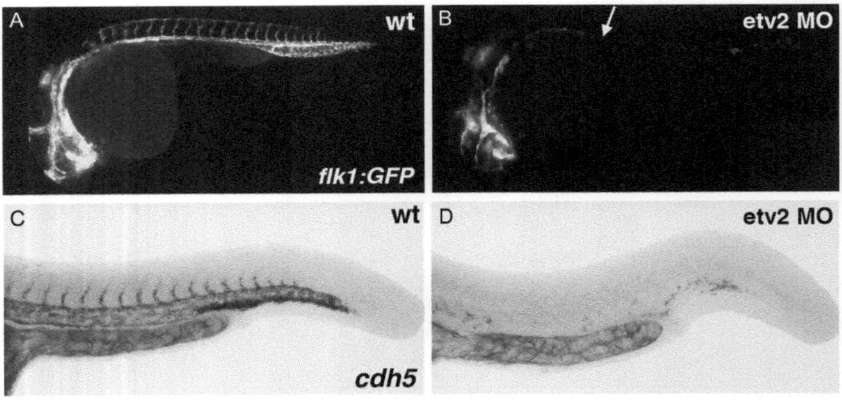

Fig. 2 Etv2 morpholino knockdown results in near complete absence of zebrafish transgenic reporter *flk1:GFP* (A and B) and vascular marker *cdh5/VE-cadherin* (C and D) expression at 24–26 hpf. *Reprinted from Sumanas, S., & Lin, S. (2006). Ets1-related protein is a key regulator of vasculogenesis in zebrafish.* PLoS Biology, 4(1), *e10.* (See the color plate.)

and undergo apoptosis in the absence of Etv2 function (Pham et al., 2007). However, not all vascular endothelial progenitors undergo apoptosis and some of them transdifferentiate into other lineages such as the myocardium, as discussed further in more detail. In addition to vascular defects, *etv2* knockdown embryos show loss of hematopoietic markers in the ALPM region, associated with myelopoiesis, including *scl* and *pu.1* (Sumanas et al., 2008). Consistent with the loss of myeloid marker expression, primitive myeloid lineages, including macrophages and neutrophils are absent in Etv2-deficient embryos. In contrast, primitive erythroid markers are not affected, and *scl* expression is absent only in myeloid and vascular endothelial progenitors and is not affected in erythroid progenitors, consistent with *etv2* being expressed in myeloid and not erythroid progenitors. *Scl* overexpression rescued myeloid but not endothelial differentiation in Etv2-deficient embryos, arguing that Scl functions downstream of Etv2 in myelopoiesis, while Etv2 is directly required for vascular endothelial differentiation (Sumanas et al., 2008). Distinct from zebrafish and mouse embryos, Etv2 MO knockdown in *Xenopus* resulted in a loss of endothelial marker expression, while hematopoietic markers including myeloid specific genes were not affected (Salanga et al., 2010). Also, different from mouse and zebrafish embryos, *Xenopus* Etv2 appears to function downstream of Fli1. *Etv2* expression is initiated later than *fli1*, and *etv2* expression is downregulated in *fli1* morphants (Ciau-Uitz et al., 2013; Liu, Walmsley, Rodaway, & Patient, 2008). Furthermore, *flk1* expression in the DLP region is not affected in *Xenopus* Etv2 morphants, and it is only downregulated at a later stage, when the loss of vascular endothelial cells is observed (Ciau-Uitz et al., 2013). Therefore, while the overall Etv2 function is conserved evolutionarily, it appears that there are notable differences in timing and progenitor cell types that Etv2 functions among different vertebrates.

The finding that *Etv2* expression is downregulated once vessel formation is established suggests that *Etv2* is only transiently required during vessel and blood specification (Kataoka et al., 2011; Koyano-Nakagawa et al., 2012; Lee et al., 2008). Indeed, *Etv2* inactivation after the hemangiogenic program specification leads to normal development, supporting the transient role played by this factor. Specifically, *Etv2* deletion from E8.5 in ROSACreER;*Etv2*$^{f/f}$ mice with tamoxifen resulted in defects in vascular development. However, *Etv2* deletion in such mice from E9.5 did not appear to cause any obvious hematopoietic and vascular defects (Kataoka et al., 2013). *Mesp1-Cre-* or *Hoxb6-Cre*-mediated *Etv2* deletion resulted in embryonic lethality around E9.5–10.5 and E12.5, respectively, due to hematopoietic and vascular defects in the regions including yolk sac, head region, anterior part of the embryo, and endocardium (*Mesp1-Cre*) and in the caudal–lateral parts of the embryo (*Hoxb6-Cre*) (Kataoka et al., 2013). Similarly, *Flk1Cre;Etv2* CKO mice also display hematopoietic and vascular defects (Liu et al., 2015), although Wareing et al. (2012) reported that such mice were born alive with little hematopoietic and vascular defects. Considering transient requirement of this factor in development, deletion efficiency of the floxed *Etv2* alleles at the critical time might have contributed to such different outcomes. Nonetheless, *Etv2* deletion using *Tie2-Cre* (Kataoka et al., 2013; Park et al., 2016), *Vav-Cre* or *VE-cadherin-Cre* (Park et al., 2016) does not lead to any defects in hematopoietic or endothelial cell development. Thus, it has been proposed that ETV2 is required up to the stage until sufficient hematopoietic and vascular progenitors are generated to ensure proper hematopoietic and vascular development (Liu et al., 2015).

Intriguingly, despite severe early defects and near complete absence of endothelial marker expression during early stages of vasculogenesis, zebrafish *etv2* null mutant embryos undergo significant recovery of vascular development during later stages (1–3 dpf) (Craig et al., 2015; Pham et al., 2007). Development of axial vasculature is partially restored, and intersegmental vessel sprouts, although defective, can still form even in complete absence of *etv2* function. It was recently demonstrated that this angiogenic recovery is due to functional redundancy between Etv2 and other related ETS transcription factors. *Fli1b*, a duplicate zebrafish homolog of the mammalian *fli1* gene, has been implicated in the functional redundancy with *etv2* during angiogenesis (Craig et al., 2015). Inducible *etv2* knockdown using photoactivatable morpholinos failed to cause any notable vascular or hematopoietic defects when MO was photoactivated at the 18-somite or later stages, close to or after completion of vasculogenesis. However, while *fli1b* homozygous mutants do not have any apparent defects in vascular development, inducible *Etv2* knockdown in *fli1b−/−* background results in inhibition of angiogenic sprouting without affecting formation of the major axial vessels that form by vasculogenesis (Fig. 3). This argues that Etv2 has an independent role in embryonic angiogenesis, where it functions redundantly with other transcription factors, such as Fli1b. Additional functional redundancy between other ETS transcription factors including *ets1* and *fli1a* has also been suggested (Pham et al., 2007).

While inducible Etv2 inhibition in wild-type zebrafish embryos after completion of vasculogenesis at 24 hpf resulted in no apparent vascular or

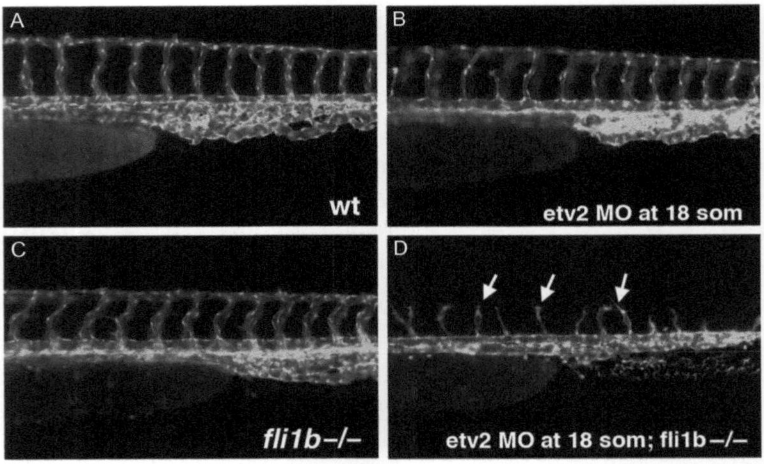

Fig. 3 Etv2 functions redundantly with Fli1b in angiogenesis independent from Etv2 early requirement in vasculogenesis. Inducible Etv2 inhibition was performed at the 18-somite stage using photoactivatable Etv2 morpholino in either wild-type (wt) *fli1a:GFP* transgenic reporter embryos (A and B) or *fli1b−/−* insertional mutants (C and D). Embryos were analyzed at 42 hpf. Note that *fli1b−/−* mutants (C) or Etv2-inhibited embryos in wild-type background (B) are phenotypically normal, while inducible Etv2 inhibition in *fli1b−/−* background (D) results in the inhibition of intersegmental vessel angiogenesis (*arrows*), while vasculogenesis and axial vasculature development are not affected. *Reprinted from Craig, M. P., Grajevskaja, V., Liao, H. K., Balciuniene, J., Ekker, S. C., Park, J. S., et al. (2015). Etv2 and fli1b function together as key regulators of vasculogenesis and angiogenesis.* Arteriosclerosis, Thrombosis, and Vascular Biology, 35(4), *865–876.* (See the color plate.)

hematopoietic defects, lymphangiogenesis was severely inhibited (Davis et al., in review). The thoracic duct, the major lymphatic vessel, and parachordal lymphangioblasts were absent in caged Etv2 MO-injected embryos, which were photoactivated at 24 hpf, after completion of vasculogenesis. This was accompanied by the loss of lymphatic marker expression, and severely reduced expression of *vegfr3/flt4* within the PCV, where lymphangioblasts are known to emerge, while vascular markers were not affected. In the absence of *Etv2* function, lymphatic progenitors failed to respond to VegfC signaling. VegfC/Flt4 signaling has a well-established role in regulating lymphangiogenesis, and *Flt4* is a known direct transcriptional target of Etv2 (De Val et al., 2008; Karkkainen et al., 2004; Yaniv et al., 2006). Consistent with this notion, its expression continues within a subset of PCV at the stages when lymphangioblasts are known to emerge (Davis et al., in review). These results argue that after completion of vasculogenesis, Etv2 function is required to upregulate genes associated with lymphangiogenesis such as *VegfR3*. Future studies are warranted to determine whether *Etv2* is required for mammalian lymphangiogenesis. To this end, ETV2 ChIP-Seq data and transcriptome analyses also support the notion that ETV2 regulates lymphangiogenesis in mice (Liu et al., 2013, 2015). It is also possible that *Etv2* function in lymphangiogenesis in mammalian embryos may be redundant with other ETS factors, or a different *Ets* member has adopted a similar function. Recently, *Ets2* has been implicated in regulating *VegfR3* expression during murine lymphangiogenesis (Yoshimatsu et al., 2011).

5. ETV2, VEGF, AND FLK1 IN HEMATOPOIETIC AND ENDOTHELIAL CELL FORMATION

It is noticeable that *Etv2* inactivation leads to similar hematopoietic and vascular defects to those of *Vegfa* or *Flk1* deficiency (Carmeliet et al., 1996; Ferrara et al., 1996; Shalaby et al., 1995). FLK1 is a receptor tyrosine kinase that binds VEGF (Matthews et al., 1991; Millauer et al., 1993). In primitive streak stage embryos, *Flk1* is first detected in the extraembryonic mesoderm and in the lateral plate mesoderm (Dumont et al., 1995; Ema, Takahashi, & Rossant, 2006; Yamaguchi, Dumont, Conlon, Breitman, & Rossant, 1993). Subsequently, its expression becomes confined to endothelial cells of the yolk sac and embryo proper and to the developing endocardial tube. Besides its functional requirement in hematopoietic and endothelial cell development, FLK1 is also a superb marker for the mesodermal progenitors that will generate the circulatory system. Specifically, in vitro differentiation model of ES cells, which recapitulates the key steps and the same principles governing early embryonic development (Murry & Keller, 2008; Park, Lugus, & Choi, 2005), has been instrumental in determining the developmental potential of FLK1-expressing progenitors. As such, FLK1-expressing progenitors from in vitro differentiated EBs contain the presumptive hemangioblast, a common progenitor of hematopoietic and endothelial cells (Choi, Kennedy, Kazarov, Papadimitriou, & Keller, 1998; Chung et al., 2002; Faloon et al., 2000). ES-derived FLK1+ cells can contribute to endothelial and smooth muscle cells in vivo (Yamashita et al., 2000). Importantly, cell lineage-tracing studies have demonstrated that primitive and definitive blood, and endothelial cells are of the FLK1+mesodermal origin (Lugus, Park, Ma, & Choi, 2009, Motoike, Markham, Rossant, & Sato, 2003).

FLK1 expression can also identify cardiac progenitors. Fate mapping studies have demonstrated that FLK1+ mesoderm also contributes to cardiac and skeletal muscles (Motoike et al., 2003). More recently, Moretti et al. (2006) demonstrated that FLK1+Isl1+Nkx-2.5+ cells from in vitro differentiated ES cells represent a multipotent cardiovascular progenitor, which can give rise to cells of endothelial, smooth muscle, and cardiac cell lineages. Kattman, Huber, and Keller (2006) also reported that Brachyury+FLK1+ cells from in vitro differentiated ES cells could generate cardiomyocyte, endothelial, and vascular smooth muscle cells. Collectively, it can be assumed that FLK1 expression is associated with progenitors that can generate hematopoietic, endothelial, smooth muscle cells, and cardiomyocytes. Recently, it was proposed that PDGFRα expression could further segregate FLK1 progenitors into hemangiogenic vs cardiac, as such while FLK1+PDGFRα− progenitors are enriched for hemangiogenic potential, FLK1+PDGFRα+ progenitors are enriched for cardiac potential (Kattman et al., 2011; Liu et al., 2012). Alternatively, FLK1+PDGFRα− and FLK1+PDGFRα+ progenitors were proposed to represent two distinct developmental stages such that FLK1+PDGFRα− progenitors are derived from more primitive FLK1+PDGFRα+ progenitors (Sakurai et al., 2006). In this scenario, it is possible that FLK1+PDGFRα+ progenitors are heterogeneous and a fraction of these may represent cardiac progenitors.

While *Flk1* is absolutely required for blood and endothelial cell lineage development, *Flk1* seems to be dispensable for cardiomyocyte and skeletal muscle cell development, as morphologically normal splanchnic mesoderm, myocardium, and the dorsolateral part of the somites still develop in *Flk1* null embryos (Ema et al., 2006). By examining LacZ expression from the *Flk1*-LacZ knockin targeted allele, hemangiogenic progenitors were shown to express high expression levels of *Flk1*, while cardiac and somite progenitors exhibit low expression levels of *Flk1* expression. Thus, it was suggested that while FLK1 mesoderm can contribute to blood, endothelial cell, skeletal muscle cells, and cardiomyocytes, FLK1 expression levels, presumably reflecting VEGF signaling strength, might determine the cell lineage outcome of the FLK1+ cells in the embryo (Ema et al., 2006). Thus, while blood and endothelial cells are generated from the hemangiogenic progenitors expressing high levels of FLK1 ($FLK1^{high}$), cardiac, or muscle cells would be generated from the $FLK1^{low}$ progenitors (Ema et al., 2006). Indeed, FLK1+PDGFRα− hemangiogenic progenitors can be already detected around embryonic day (E) 7.5, which progressively expand with time during the course of E7.5–E8.5. Notably, the mean fluorescence intensity of the FLK1 staining within the FLK1+PDGFRα− cell population becomes greater as embryos develop, suggesting an elevated FLK1 signaling activity within FLK1+PDGFRα− cells compared to FLK1+PDGFRα+ cells. Remarkably, $FLK1^{high}$PDGFRα− cell population was exclusively missing in *Etv2−/−* embryos. This cell population represents a small fraction of the embryo, suggesting that hematopoietic and endothelial cells are derived from a small pool of $FLK1^{high}$ progenitors (Liu et al., 2015). Thus, ETV2 function seems to be in the generation of sufficient number of $FLK1^{high}$ hemangiogenic progenitors. Any perturbation in the ETV2, VEGF, and FLK1 balance causing insufficient $FLK1^{high}$ cell generation might lead to defects in hematopoietic and endothelial cell development (Liu et al., 2015).

6. ETV2 REGULATES HEMANGIOGENIC VERSUS CARDIAC MESODERM OUTCOME

There seems to be an antagonistic relationship between hematopoietic and cardiac output. For example, hematopoietic program can be inhibited by enforced *Mesp1* expression in ES cells, which enhanced cardiac differentiation (Lindsley et al., 2008). Enforced *Nkx2-5* expression in ES cells can also enhance cardiac differentiation output by suppressing hematopoietic outcome (Caprioli et al., 2011). Moreover, while enforcing blood and vessel program limits cardiac field, limiting vessel and blood specification result in expansion of the cardiac field in the ALPM in zebrafish (Schoenebeck, Keegan, & Yelon, 2007). In particular, *Etv2* seems to play a key role in the regulation of the hemangiogenic vs cardiogenic output (Fig. 4). Specifically, *scl* and *etsrp* expression in zebrafish embryos can expand hematopoietic and endothelial cell specification, while concurrently reducing the myocardial field in the ALPM. Conversely, knockdown of *scl* and *etsrp* results in an expansion of the boundaries of *hand2* expression, suggesting reciprocal regulation between the hematopoietic/vascular and cardiovascular systems (Schoenebeck et al., 2007). Similarly, enforced *Etv2* expression in ES cells leads to an expansion of FLK1+PDGFRα− hemangiogenic progenitors at the expense of FLK1+PDGFRα+ cardiogenic progenitors (Liu et al., 2012). There was significant increase in PDGFRα single positive and FLK1 and PDGFRα double positive cells in the absence of *Etv2* (Kataoka et al., 2011; Liu et al., 2012, 2015; Rasmussen et al., 2011). In zebrafish embryo, *etv2*-deficient vascular endothelial/endocardial cells upregulate myocardial marker *hand2* and *cmlc2* expression and differentiate as cardiomyocytes (Palencia-Desai et al., 2011). Similarly, presumptive hemangiogenic progenitors differentiated into myocardial cell lineage in the absence of *Etv2* in mice (Rasmussen et al., 2011). Consistently, enforced *Etv2* expression in ES cells results in an increase in genes expressed in the blood and endothelial cell lineages and a decrease in genes expressed in the cardiac lineage. Conversely, genes expressed in hematopoietic as well as endothelial cells were downregulated, but

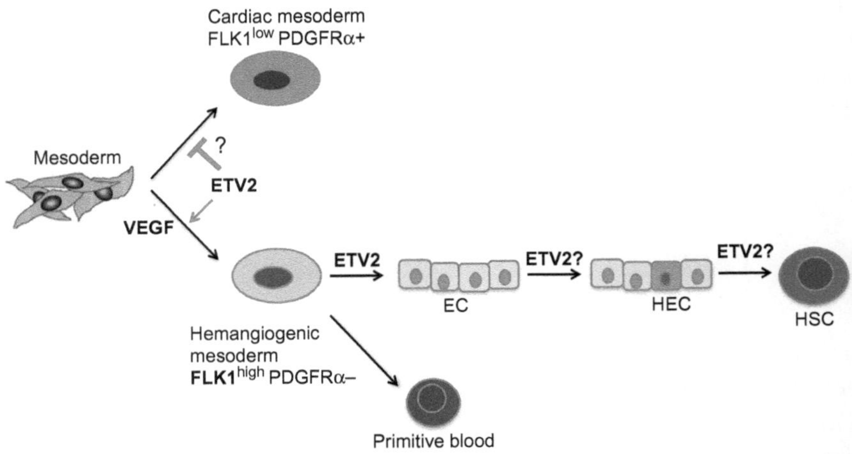

Fig. 4 A model depicting ETV2 function in hematopoietic and endothelial cell differentiation. EC, endothelial cell; HEC, hemogenic endothelial cell; HSC, hematopoietic stem cell. (See the color plate.)

cardiogenic genes were upregulated in *Etv2*-deficient embryos (Lee et al., 2008; Rasmussen et al., 2011). These data are consistent with the idea that ETV2 specifies hemangiogenic fate of the FLK1 mesoderm. Without positive input by ETV2, cardiac differentiation of the FLK1 mesoderm might occur as default.

Previous transcriptional profiling studies between *Etv2*-deficient vs *Etv2*-enforced FLK1 progenitors (Liu et al., 2012), FLK1 + *Etv2*(Venus)+ vs FLK1 + *Etv2*(Venus)− cell populations (Kataoka et al., 2011), *etsrp*-overexpressing late gastrulation/tail bud stage zebrafish embryos vs controls (Gomez et al., 2009; Wong et al., 2009), *Etv2*(EYFP)+ vs *Etv2*(EYFP)− cells from wild-type and *Etv2* mutant littermate embryos (Rasmussen et al., 2011) and *Etv2* enforced (0, 12, and 24 h) *Etv2*−/− EBs vs *Etv2*−/− EBs (Wareing et al., 2012) have identified *Etv2*-dependent genes required for hematopoietic and endothelial cell differentiation. Most recently, ETV2 ChIP-Seq studies have been performed to independently assess ETV2-mediated gene regulatory network and signaling governing hematopoietic and endothelial cell development (Fig. 5; Liu et al., 2015). These studies have revealed that ETV2 direct targets include essential hematopoietic and endothelial cell lineage specifying genes. Specifically, *Flk1*, *Fli1*, *Erg*, *Gata2*, *Scl*, *Meis1*, *Lmo2*, *Tie2*, *VE-cadherin*, *Dll4*, *Flt4*, and *Notch* are among the ETV2 targets,

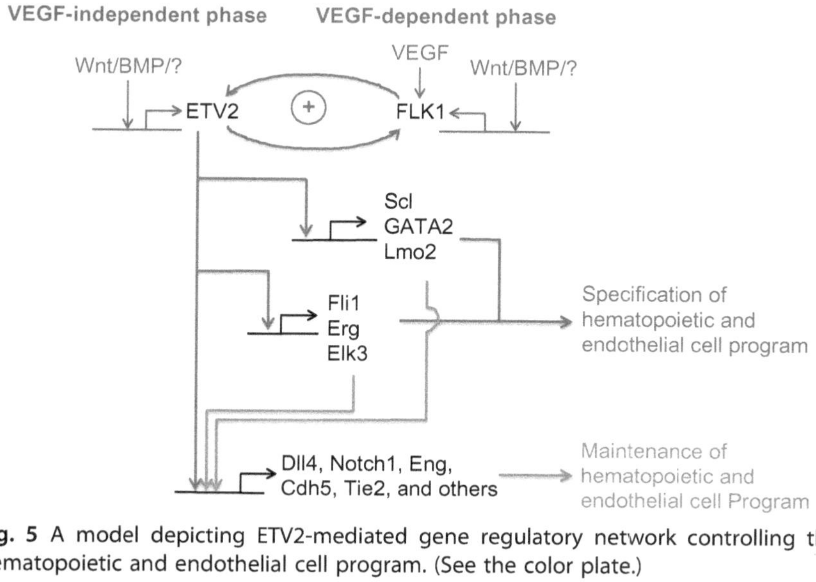

Fig. 5 A model depicting ETV2-mediated gene regulatory network controlling the hematopoietic and endothelial cell program. (See the color plate.)

which play critical roles in hematopoietic and endothelial cell development. ETV2 ChIP-Seq studies have also identified VEGF, Notch, Rho-GTPase, and MAP kinase signaling pathways to be downstream of ETV2 in regulating hematopoietic and endothelial cell development. Importantly, other *Ets* genes are also ETV2 direct targets.

ETV2 binding motif (Liu et al., 2015), which represents ~85% of the peaks, matches perfectly with the known binding specificity of other ETS factors, FLI1 and ERG (Wilson et al., 2010). GATA, SOX, or Ebox motifs are frequently associated with the ETV2 peaks (Liu et al., 2015). About 14% of the binding peaks are enriched in the promoter regions and ~70% in introns or intergenic regions, which suggests that ETV2 functions by interacting with both gene promoters and distal enhancers. ETV2 peak-associated genes are significantly enriched for those that exhibit increased expression in the ETV2 overexpression system, and/or reduced expression in the *Etv2* knockout. On the other hand, there is no significant enrichment of genes with the opposite expression pattern (ie, upregulated in *Etv2* knockout and downregulated in *Etv2* overexpression). This expression pattern suggests that ETV2 primarily functions as a transcriptional activator. Thus, while transcriptome analysis has supported an antagonistic regulatory relationship between the nascent hemangiogenic vs cardiac program, ETV2 ChIP-Seq data did not identify cardiac genes as potential direct targets. We propose that inverse relationship between ETV2 and cardiac genes is probably indirect. As such, expansion of the hemangiogenic program by ETV2 results in reciprocal decrease in cardiac program from the presumptive hemangio-cardiac FLK1 mesoderm. Alternatively, suppression of cardiac genes might be achieved through ER71 target genes, such as *Scl*, which has recently been shown to suppress cardiac genes (Van Handel et al., 2012). Thus, it is possible that the primary role of ETV2 is in the specification of the hemangiogenic outcome from the mesoderm. Once the hemangiogenic fate is specified, ETV2 target genes might maintain hematopoietic and endothelial cell lineage identity by actively suppressing the nonhemangiogenic cell lineage program.

Previous studies have implicated the core gene regulatory network played by the ETS, GATA, and E-Box motifs in hematopoietic and endothelial cell development (Pimanda et al., 2007). Consistent with this idea, previous studies have demonstrated that *Etv2*, *Gata2*, or *Scl* deficiency leads to hematopoietic and vascular defects (Lee et al., 2008; Shivdasani, Mayer, & Orkin, 1995; Tsai et al., 1994). Moreover, *Etv2*, *Gata2*, or *Scl* can independently modulate hemangioblast development (Ismailoglu, Yeamans, Daley,

Perlingeiro, & Kyba, 2008; Liu et al., 2012; Lugus et al., 2007). *Etv2*, *Gata2*, and *Scl* seem to form a core in the hemangiogenic lineage specification, as coexpression of *Etv2*, *Gata2*, and *Scl* during the hemangioblast formation stage can robustly induce hemangioblast cell population (Liu et al., 2013). Notably, comparison of the ETV2 ChIP-Seq data to that of GATA2 and SCL from Wilson et al. (2010) revealed that GATA2 and SCL motifs were significantly enriched within the ETV2 peaks, suggesting that ETV2 and these factors may interact or cobind to some of these sites. Indeed, ETV2 and GATA2 have been recently reported to form a complex to regulate hematopoietic and endothelial cell gene expression (Shi et al., 2014). Thus, the ETS, GATA, and E-Box gene regulatory network is integral to ETV2 mediated hemangiogenic cell fate specification.

Forkhead factors might also be integral to ETV2-mediated hemangiogenic cell fate specification. Specifically, key hematopoietic and endothelial cell genes, including *Flk1*, *Tie2*, *Scl*, *Notch4*, *Mef2c*, and *Cdh5*, contain FOX:ETS motifs in their enhancer and promoter regions (De Val et al., 2008). Coexpression of ETS factors and Forkhead factors, compared to either factor alone, can greatly enhance the expression of these genes. Of the Forkhead factors, *Foxh1*, *Foxf1*, and *Foxc1/c2* might be relevant to ETV2 expression and/or function. For example, *Foxh1* along with *Smad2/3* mediates Nodal-signaling events. Reflecting the dosage effect of Nodal action, *Foxh1* deficiency in mice leads to a varying phenotype, from the gastrulation defects to anterior primitive streak patterning, node, and definitive endoderm formation defects. Intriguingly, *Foxh1* overexpression was shown to inhibit *flk1* expression in zebrafish, although *flk1* expression in *foxh1* mutant embryos has not been measured quantitatively (Choi et al., 2007). *Foxf1* deficiency leads to extraembryonic and lateral plate mesoderm formation defects (Mahlapuu, Ormestad, Enerbäck, & Carlsson, 2001). Recently, *Foxf1* was reported to negatively regulate hematopoietic differentiation from mesodermal progenitors (Fleury, Eliades, Carlsson, Lacaud, & Kouskoff, 2015). Compound *Foxc1* and *Foxc2* inactivation leads to arterial venous fusions, loss of arterial gene expression such as *Dll4* and embryonic lethality. Knockdown of zebrafish *Foxc1a* together with *Etv2* had a synergistic effect in inhibiting vascular endothelial and endocardial differentiation (De Val et al., 2008; Palencia-Desai et al., 2011), supporting the model that *Etv2* and *Foxc1* interact during endothelial differentiation. In the future, it will be important to determine the interplay between ETV2, Forkhead factors, GATA2 and SCL in hemangiogenic mesoderm, hematopoietic, and endothelial cell lineage development.

7. ETS HIERARCHY AND ETS SWITCHING MECHANISMS IN HEMATOPOIETIC AND VASCULAR DEVELOPMENT

Recent ETV2 ChIP-Seq analysis demonstrated that many ETS factors including *Fli1*, *Erg*, *Ets1*, and *Ets2* are potential ETV2 direct targets (Liu et al., 2015). Importantly, *Etv2* was also identified as a target, suggesting that ETV2 positively regulates its own expression and that of other *Ets* genes. Consistent with the idea that ETV2 is upstream of other Ets genes, *Etv2* expression precedes that of other *Ets* genes in the developing mouse embryo and differentiating ES cells. Specifically, *Fli1* expression is detected shortly after *Etv2*, followed by *Erg*, *Elk3*, *Ets1*, and *Ets2*. Moreover, enforced *Etv2* expression in differentiating ES cells prominently induces expression of these *Ets* genes. Conversely, *Fli1*, *Erg*, and *Elk3* expression is greatly reduced in *Etv2* null mouse embryos. *Ets1* and *Ets2* expression is also reduced in *Etv2* null embryos. Importantly, *Etv2*, *Erg*, *Elk3*, *Ets1*, or *Ets2* expression is not impaired in *Fli1* null ES cells differentiated in culture or in embryos. Importantly, *Fli1* null embryos generate similar levels of FLK1^{high}PDGFRα− cells compared to controls, suggesting that FLI1 is dispensable for hemangiogenic progenitor formation. Moreover, enforced *Fli1* expression in differentiating ES cells does not alter *Etv2* expression, although other *Ets* genes are upregulated (Liu et al., 2015). These studies support the notion that ETV2 activates other *Ets* factors including *Fli1* in hematopoietic and endothelial cell development.

One possibility for the nonredundant role of ETV2 in the FLK1^{high}PDGFRα− hemangioblast commitment could be the timing of the ETS factor availability. As such, *Etv2* happens to be expressed before the other *Ets* factors and thus carries out such a robust function. If this is true, it is expected that when other *Ets* genes are expressed during the same time frame when *Etv2* is expressed, they might also be able to enhance hemangiogenic outcome as seen by *Etv2*, ie, enhanced FLK1^{high}PDGFRα− hemangiogenic progenitor formation. However, when tested in the ES system, other *Ets* genes were either less efficient (*Fli1*) or could not (*Ets1* or *Ets2*) affect the hemangiogenic output skewing (Liu et al., 2015). Thus, timing of *Ets* factor expression alone cannot explain the nonredundant ETV2 function, as FLI1, ERG, or ETS1 cannot replace ETV2 function even if they were expressed at the same time frame as ETV2, in the hemangioblast induction from ES cells. Collectively, judging from the hemangiogenic mesoderm skewing potential, it appears that ETS factors are inherently different. ETV2 carries out nonredundant role in the hemangiogenic mesoderm specification.

Functional hierarchy among the ETS factors in the hemangiogenic mesoderm formation has been additionally addressed by coexpressing *Ets* genes in the context of *Gata2* and *Scl*. In particular, FLK1+PDGFRα− hemangiogenic skewing outcome is even more robust if *Etv2* is coexpressed with *Gata2* and *Scl* (*Etv2-Gata2-Scl*) than *Etv2* alone (Liu et al., 2015). In this context, while the FLK1+PDGFRα− hemangiogenic output is increased by *Fli1-Gata2-Scl* or *Erg-Gata2-Scl*, the level of the FLK1+PDGFRα− cells generated by the *Fli1-Gata2-Scl* or *Erg-Gata2-Scl* is not as robust as that by the *Etv2-Gata2-Scl*. Importantly, there is incomplete skewing of the mesoderm toward FLK1+PDGFRα− hemangiogenic progenitors by the *Fli1-Gata2-Scl* or *Erg-Gata2-Scl*, as judged by FLK1+PDGFRα+ cardiac progenitors that are still generated. *Ets1-Gata2-Scl* is the least effective in the skewing of the FLK1+PDGFRα− hemangioblast from ES cells. These studies indicate that ETV2 carries out a uniquely distinct role in the hemangiogenic progenitor generation. No other ETS factors can replace ETV2 in the hemangioblast formation. In the future, structure and function of ETV2 in relationship to other ETS factors may elucidate the unique role of ETV2 in the hematopoietic and endothelial cell development. Additionally, ETV2 may require a unique cofactor(s). Recent finding of OVOL2 as ETV2, not ETS1 or ETS2, cofactor may help in the future to further investigate such nonredundant function played by ETV2 (Kim et al., 2014).

In zebrafish, overexpression of a zebrafish duplicates *fli1* homolog *fli1b*, but not *fli1a*, resulted in a similar precocious and ectopic expansion of vascular marker expression as *etv2* (Craig et al., 2015). *Fli1b* and *etv2* also induced expression of each other in zebrafish embryos. *Fli1b* induced ectopic marker expression even in *etv2* or *cloche* null mutant background, and ChIP analysis suggests that Etv2 can occupy the same promoter sites as Fli1b (Craig et al., 2015). Therefore, these results argue that *Etv2* and *Fli1b* have overlapping functions and may have similar binding targets, and their functional roles are defined largely by the timing of their expression. *Etv2* is expressed earlier than *fli1b* and therefore its function is essential during vasculogenesis, while *Etv2* and *Fli1b* function redundantly during later embryonic stages, when angiogenesis is initiated.

As ETV2 function is transiently required, there must be a mechanism maintaining the hematopoietic and endothelial cell program that was initiated by ETV2. We recently provided an ETS switching mechanism as one such mode. Specifically, when ETV2 target loci *Lmo2*, *Gata2*, *Cdh5*, *Tie2*/

Tek, and *Scl/Tal1* were examined for the ETS factor occupancy during EB development, while the ETS sites of the target genes were occupied by ETV2 when ETV2 expression was robust, the same target sites were occupied by other ETS factors, such as FLI1, when ETV2 was no longer expressed. FLI1 occupancy on these genes was minimal when Etv2 expression was robust. Thus, ETV2 initiated blood and endothelial cell program is maintained by other ETS factors through an ETS switching mechanism. Similar to the GATA-switching mechanism controlling the erythroid cell lineage differentiation (Bresnick, Lee, Fujiwara, Johnson, & Keles, 2010; Grass et al., 2006), ETV2 induces expression of other *Ets* genes, thereby creating an ETS regulatory loop to maintain hematopoietic and endothelial cell program.

8. ETV2 IN DEFINITIVE HEMATOPOIETIC PROGRAM AND FUNCTION

It is well established that HSCs emerge within the hemogenic endothelium of the DA (Bertrand et al., 2010; Boisset et al., 2010; Kissa & Herbomel, 2010; Lam, Hall, Crosier, Crosier, & Flores, 2010). Intriguingly, *Etv2* is expressed within the DA at the stages when HSCs form. Therefore this raises a question if ETV2 is required for HSC development, and if ETV2 is involved in further differentiation of HSCs into different hematopoietic cell lineages. Indeed, several studies support the notion that ETV2 might also be involved in the HSC formation. For example, the +23 Runx1 enhancer region, which marks the hematopoietic stem and progenitor cell population, also marks hemogenic endothelium (Swiers et al., 2013). *Etv2* is expressed within the +23 Runx1 enhancer driven GFP+ hemogenic endothelium, although it is also expressed in GFP− endothelial cells (Swiers et al., 2013). Importantly, *Etv2*-deficient ES cells cannot contribute to definitive hematopoietic system when tested in chimeric mice generated between *Er71−/−* ES cells and wild-type blastocysts (Liu et al., 2012). This suggests a cell-autonomous role for this factor in definitive hematopoietic development. Within the definitive hematopoietic system, ETV2 was shown to be required for HSC maintenance and function (Lee, Kim, & Lim, 2011). Specifically, *Etv2* expression is enriched within HSC and progenitor cell populations. Moreover, *Mx1-Cre*-mediated conditional deletion of the *Etv2* by the poly(I:C) administration leads to rapid loss of HSC and progenitor cell population and long-term repopulation potential. Similarly, HSC markers were absent in zebrafish Etv2 MO knockdown embryos, and Scl-α isoform was sufficient to partially rescue both endothelial and HSC marker expression (Ren, Gomez, Zhang, & Lin, 2010). However, because Etv2 knockdown embryos have a loss of endothelial differentiation, the effect on HSCs may be indirect and is caused by the absence of the DA itself. Supporting this explanation, scl-β partially rescued HSC marker *runx1* expression only when endothelial differentiation was restored in Etv2 knockdown embryos. Therefore, it remains to be established if Etv2 has an independent role in zebrafish HSC development. Whether ETV2 can directly, or indirectly potentially through its downstream ETS factors, control HSC genesis in the hemogenic endothelium needs to be investigated in the future. For example, *Fev*, an ETS factor, has been recently shown to control HSC genesis from hemogenic endothelium (Wang et al., 2013). ETV6 has been shown to control HSC formation through VEGF (Ciau-Uitz et al., 2013). Potentially, ETV2 and VEGF signaling might also be involved in the establishment of the definitive hematopoietic program. To this end, it is also worth mentioning that *Flk1*-deficient ES cells also fail to contribute to definitive hematopoietic system when examined in chimeric mice generated with wild-type blastocysts (Shalaby et al., 1997).

9. REGULATION OF *Etv2* EXPRESSION

Transcriptional regulators and signaling pathways that control *Etv2* expression are beginning to be elucidated. In mice, it appears that ETV2 positively regulates its own expression (Koyano-Nakagawa et al., 2015; Liu et al., 2015). NKX2-5 can activate the 3.9 kb Etv2 promoter/enhancer (Ferdous et al., 2009). PKA/CREB can also activate *Etv2* expression in the ES differentiation system (Yamamizu et al., 2012). MESP1-CREB1 can also activate *Etv2* expression (Shi et al., 2015). However, whether any of these factors is necessary for endogenous *Etv2* expression has not been established yet. Similarly, functional dissection of zebrafish *etv2* promoter identified 2.3 kb minimal region required for *etv2* vascular expression. Foxc1a and Foxc1b were shown to bind to evolutionary conserved regions within this promoter (Veldman & Lin, 2012). However, combined knockdown of *foxc1a* and *foxc1b* resulted in only a small decrease in *etv2* expression, suggesting that *foxc1a* and *foxc1b* may only play a minor role in regulating *etv2* expression. Expression of all anterior hematovascular markers including *etv2* was absent in *gata5/gata6*

MO knockdown embryos, although the effect was only limited to the anterior *etv2* expression in the ALPM, and *etv2* expression seemed to recover at later stages (Peterkin, Gibson, & Patient, 2009). It is not clear if these GATA factors regulate *etv2* expression directly in the ALPM and if there are other GATA homologs involved in regulating *etv2* expression in the PLPM. It has been argued that angiogenic factor AGGF1 regulates *etv2* expression, however only very minor reduction in *etv2* expression has been observed in AGGF1 MO knockdown embryos, and these findings have not been validated by mutant data (Li et al., 2014). Intriguingly, while Nkx2-5 could activate Etv2 promoter driven reporter expression in mice (Ferdous et al., 2009), overexpression of *nkx2.5*, one of the key regulators of myocardial differentiation, resulted in the inhibition of *etv2* expression in zebrafish (Simoes, Peterkin, & Patient, 2011). Direct binding to zebrafish *etv2*-promoter and intron regions by Scl and Nkx2.5 transcription factors has been demonstrated. It has been argued that repression by Scl and Nkx2.5 inhibits *etv2* expression in the myocardial progenitors (Schupp, Waas, Chun, & Ramchandran, 2014). However, it has not been demonstrated if *etv2* expression expands into the myocardium in Scl or Nkx2.5 knockdown embryos; in fact, *etv2* expression is reportedly not affected in Scl or Nkx2.5 MO knockdown embryos (Simoes et al., 2011; Sumanas & Lin, 2006). Furthermore, *scl* expression is not observed in the myocardial progenitors, contrasting with its proposed role in inhibiting *etv2* expression in the myocardium.

As for the upstream signaling pathways, BMP, Wnt, and Notch pathways are required for *Etv2* expression in the ES system (Lee et al., 2008). VEGF signaling can activate *Etv2* expression (Kataoka et al., 2011; Rasmussen et al., 2012). Calcineurin-NFAT and/or p38 MAPK-CREB pathway downstream of VEGF has been suggested to regulate *Etv2* expression (Koyano-Nakagawa et al., 2015; Rasmussen et al., 2012). Additionally, in zebrafish, Shh-Vegf and Notch pathways have also been shown to affect the number of Etv2-expressing cells in the LPM, although this effect appears to be due to effect on the proliferation of Etv2-positive cells in the LPM (Chun et al., 2011). FGF signaling has been implicated in promoting cardiac fate and restricting hemangioblast fate in zebrafish embryos. Inhibition of FGF signaling resulted in the expansion of *etv2* expression into cardiac region. Although VEGF has been implicated to be upstream of *Etv2* in mice, *etv2* expression in zebrafish is not significantly affected in VegfA morpholino knockdown embryos or *vegfa* mutants (Rost et al., in review; Sumanas & Lin, 2006). However, embryos treated with pan-VegfR chemical inhibitors SU5416 or PTK787, show downregulation of all vascular endothelial marker expression including *etv2* at 24 hpf, when circulation is initiated in zebrafish embryos. Furthermore, vascular endothelial *fli1* promoter driven *etv2* expression can restore endothelial differentiation in VegfR-inhibited embryos. However, initiation of *etv2* expression at early somitogenesis stages is not affected in VegfR inhibitor treated embryos (Rost et al., in review). These studies suggest that VEGF signaling is not required to initiate *Etv2* expression. Similarly in mice, *Flk1*-deficient ES cells and embryos still generate hematopoietic progenitors, although at reduced levels (Hidaka, Stanford, & Bernstein, 1999; Schuh, Faloon, Hu, Bhimani, & Choi, 1999). *Etv2* expression is still seen in *Flk1*-deficient EB cells (H. Zhao et al., unpublished data), although *Etv2* expression was reported to be absent in *Flk1*-deficient embryos or EBs (Rasmussen et al., 2012). Future studies on upstream signals regulating *Etv2* and *Flk1* will be critical for further dissecting molecular mechanisms involved in hematopoietic and vascular development.

10. CLOSING THOUGHTS AND FUTURE DIRECTIONS

It appears that interplay between ETV2, VEGF, and FLK1 is critical for hematopoietic and endothelial cell development. Specifically, current studies suggest that *Etv2* and *Flk1* expression is initiated independently of VEGF. Upon VEGF engagement, *Etv2* and *Flk1* positive feedback regulation is formed, which is essential for establishing the hematopoietic and endothelial cell program (Liu et al., 2015; H. Zhao et al., unpublished data). At a cellular level, in vitro-differentiated ES cells or mouse posterior primitive streak regions were shown to harbor VEGF-responsive VEGFR2/FLK1 expressing cells, hemangiogenic progenitor, or hemangioblast, which has a potential generating both blood and endothelial cells in culture at a single cell level (Choi et al., 1998; Faloon et al., 2000; Huber, Kouskoff, Fehling, Palis, & Keller, 2004). Cell lineage tracing studies using *Flk1-Cre* and Rosa-stop-EYFP or Rosa-stop-LacZ reporter mice also supported the notion that blood and endothelial cells originate from the FLK1 mesoderm (Lugus et al., 2009; Motoike et al., 2003). Moreover, tracking single cells of the ventral margin of gastrula-stage zebrafish embryos further supported the idea that blood and endothelial cells develop from the hemangioblast, although this study found that not all blood and endothelial cells arose from the presumptive hemangioblasts (Vogeli et al., 2006). While these studies have supported the notion of the hemangioblast, there are studies that refute the hemangioblast concept. For example, fate mapping studies of early streak, mid streak, or early bud stages of mouse embryos suggested that progenitors giving rise to yolk sac blood islands and vascular endothelium arise independently during early gastrulation (Kinder et al., 1999). Similarly, lineage-mapping studies also suggested a different cell origin of hematopoietic and endothelial cell lineages (Padron-Barthe et al., 2014; Ueno & Weissman, 2006). Clearly, additional studies are still required to clarify the existence of the hemangioblast.

In further probing the subject of the hemangioblast, it has been recently shown that FLK1 expression levels can be used to further identify hemangiogenic progenitors. As such, $FLK1^{high}$ cells contain the presumptive hemangiogenic progenitors giving rise to hematopoietic and endothelial cells (Ema et al., 2006; Liu et al., 2015). Intriguingly, this cell population represents a small fraction of the embryo, indicating that hematopoietic and endothelial cells are derived from a small pool of cells expressing high levels of FLK1. Thus, the future lineage tracing studies need to consider focusing on this cell population. Experimental data also suggest that the hemangiogenic progenitor population is a transient cell population, as such successful marking of this transient cell population at the right time is key to hemangiogenic cell lineage tracing. Specifically, even in differentiating ES cells, the hemangioblast population can only be found during a short period of time when Brachyury expression is detected (Choi et al., 1998). Similarly in the developing embryo, while a great portion of the hematopoietic and endothelial cells was shown to derive from a common progenitor when cells were labeled early during the gastrulation stage (Padron-Barthe et al., 2014; Vogeli et al., 2006), the frequency of such cells giving rise to both hematopoietic and endothelial cells greatly decreases when cells were labeled at a later stage (Padron-Barthe et al., 2014). Furthermore, in the developing embryo, while the frequency of such cell population was most abundant from the primitive streak regions, yolk sac was found to have greatly reduced levels of the hemangioblast cell population (Huber et al., 2004). This suggested that hemangioblast commitment occurs within the posterior primitive streak region and that the hemangioblast potential to generate both hematopoietic and endothelial cells is rapidly lost by the time when they reach yolk sac. In this case, committed hemangioblasts would generate two daughter cells, hematopoietic, and endothelial cells, en route before reaching the yolk sac. The hematopoietic and endothelial committed progenitors then migrate to different sites in the yolk sac. Thus, each cell lineage would appear to have derived from already restricted independent progenitors within the primitive streak, as seen in several studies (Padron-Barthe et al., 2014; Ueno & Weissman, 2006). Vogeli et al. (2006) also found that blood and endothelial cells originating from single labeled cells were found at spatially distinct sites. Alternatively, it is also possible that hemangioblasts upon arriving yolk sac may encounter different microenvironments of the yolk sac, guided by available growth factor and/or morphogen gradients, such as VEGF, and become committed immediately to either cell fates. Indeed, recent *Xenopus* studies support this notion that microenvironments influence the hemangioblast differentiation outcome (Myers & Krieg, 2013).

Since its discovery in 1992, ETV2 has emerged as an obligatory factor, whose function is required at the earliest stage in hematopoietic and vascular development. A great deal of upstream regulators and signaling pathways controlling *Etv2* expression has been identified so far. Moreover, ETV2 downstream effectors regulating hematopoietic and vascular development have also been elucidated. However, our knowledge of the hemangiogenic lineage commitment from mesoderm is still incomplete. Thus, future studies will still require further delineating its function in the hemangiogenic specification within mesoderm. Such information will be important for further understanding molecular and cellular mechanisms involved in hematopoietic and vascular development. Such knowledge will additionally aid our efforts generating hematopoietic and vascular progenitors from pluripotent stem cells.

Although investigations on ETV2 function in the developmental context have been instrumental for understanding the cell lineage commitment mechanisms involved in hematopoietic and vascular systems, its function in adult is vastly lacking. Currently, there are limited studies on this gene in adult hematopoietic and vascular systems. For example, ETV2 seems to be required within the adult hematopoietic system, as its deletion leads to seemingly rapid loss of HSCs and function (Lee et al., 2011). Moreover, ETV2 is reactivated and required for vascular regeneration in injury, as endothelial *Etv2* deletion leads to impaired vascular regeneration and tissue recovery (Park et al., 2016). ETV2–VEGFR2 interaction seems to be also required for optimal vascular regeneration (Park et al., 2016). Detailed molecular mechanisms involved in ETV2 and vascular regeneration and repair need to be elucidated in the future. Whether ETV2 is also required for hematopoietic regeneration would also be important and need to be elucidated. Finally, as genetic programs regulating developmental processes are often activated and drive pathogenic conditions, it will also be important to determine if ETV2 is involved in pathologic conditions, such as tumor angiogenesis or hematologic malignancies. Such comprehensive understanding could provide a new research platform for the development of novel therapeutic strategies utilizing this gene in regenerative medicine and pathologic conditions.

ACKNOWLEDGMENTS

We would like to thank members of both Choi and Sumanas laboratories for constructive criticism throughout the work and sharing unpublished studies. We made every effort to include all relevant studies, and we apologize to those whose work was not referenced either due to space limitations or our oversight. This study was supported by NIH Grants HL55337 and HL63736 (to K.C.) and HL107369 (to S.S.).

REFERENCES

Abedin, M. J., Nguyen, A., Jiang, N., Perry, C. E., Shelton, J. M., Watson, D. K., et al. (2014). Fli1 acts downstream of Etv2 to govern cell survival and vascular homeostasis via positive autoregulation. *Circulation Research*, *114*, 1690–1699.

Adamo, P., & Ladomery, M. R. (2015). The oncogene ERG: A key factor in prostate cancer. *Oncogene*. http://dx.doi.org/10.1038/onc.2015.109. Apr 27 [Epub ahead of print].

Bartel, F. O., Higuchi, T., & Spyropoulos, D. D. (2000). Mouse models in the study of the Ets family of transcription factors. *Oncogene*, *19*, 6443–6454.

Bertrand, J. Y., Chi, N. C., Santoso, B., Teng, S., Stainier, D. Y., & Traver, D. (2010). Haematopoietic stem cells derive directly from aortic endothelium during development. *Nature*, *464*, 108–111.

Birdsey, G. M., Dryden, N. H., Amsellem, V., Gebhardt, F., Sahnan, K., Haskard, D. O., et al. (2008). Transcription factor ERG regulates angiogenesis and endothelial apoptosis through VE-cadherin. *Blood*, *111*, 3498–3506.

Boisset, J. C., van Cappellen, W., Andrieu-Soler, C., Galjart, N., Dzierzak, E., & Robin, C. (2010). In vivo imaging of haematopoietic cells emerging from the mouse aortic endothelium. *Nature*, *464*, 116–120.

Bresnick, E. H., Lee, H. Y., Fujiwara, T., Johnson, K. D., & Keles, S. (2010). GATA switches as developmental drivers. *The Journal of Biological Chemistry*, *285*, 31087–31093.

Brown, T. A., & McKnight, S. L. (1992). Specificities of protein-protein and protein-DNA interaction of GABP alpha and two newly defined ets-related proteins. *Genes & Development*, *6*, 2502–2512.

Caprioli, A., Koyano-Nakagawa, N., Iacovino, M., Shi, X., Ferdous, A., Harvey, R. P., et al. (2011). Nkx2-5 represses Gata1 gene expression and modulates the cellular fate of cardiac progenitors during embryogenesis. *Circulation*, *123*, 1633–1641.

Carmeliet, P., Ferreira, V., Breier, G., Pollefeyt, S., Kieckens, L., Gertsenstein, M., et al. (1996). Abnormal blood vessel development and lethality in embryos lacking a single VEGF allele. *Nature*, *380*, 435–439.

Choi, J., Dong, L., Ahn, J., Dao, D., Hammerschmidt, M., & Chen, J. N. (2007). FoxH1 negatively modulates flk1 gene expression and vascular formation in zebrafish. *Developmental Biology*, *304*, 735–744.

Choi, K., Kennedy, M., Kazarov, A., Papadimitriou, J. C., & Keller, G. (1998). A common precursor for hematopoietic and endothelial cells. *Development*, *125*, 725–732.

Chun, C. Z., Remadevi, I., Schupp, M. O., Samant, G. V., Pramanik, K., Wilkinson, G. A., et al. (2011). Fli+ etsrp+ hemato-vascular progenitor cells proliferate at the lateral plate mesoderm during vasculogenesis in zebrafish. *PLoS One*, *6*, e14732.

Chung, Y. S., Zhang, W. J., Arentso, E., Kingsley, P. D., Palis, J., & Choi, K. (2002). Lineage analysis of the hemangioblast as defined by Flk-1 and SCL expression. *Development, 129*, 5511–5520.

Ciau-Uitz, A., Pinheiro, P., Kirmizitas, A., Zuo, J., & Patient, R. (2013). VEGFA-dependent and -independent pathways synergise to drive Scl expression and initiate programming of the blood stem cell lineage in Xenopus. *Development, 140*, 2632–2642.

Craig, M. P., Grajevskaja, V., Liao, H. K., Balciuniene, J., Ekker, S. C., Park, J. S., et al. (2015). Etv2 and fli1b function together as key regulators of vasculogenesis and angiogenesis. *Arteriosclerosis, Thrombosis, and Vascular Biology, 35*(4), 865–876.

Davidson, A. J., & Zon, L. I. (2004). The "definitive" (and "primitive") guide to zebrafish hematopoiesis. *Oncogene, 23*, 7233–7246.

Davis, J. A., Lubert, A., Palencia Desai, S., Winkler, T., & Sumanas, S. (in review). ETS transcription factor Etsrp/Etv2 is required for the initiation of lymphangiogenesis.

De Val, S., Chi, N. C., Meadows, S. M., Minovitsky, S., Anderson, J. P., Harris, I. S., et al. (2008). Combinatorial regulation of endothelial gene expression by ets and forkhead transcription factors. *Cell, 135*, 1053–1064.

Dejana, E., Taddei, A., & Randi, A. M. (2007). Foxs and Ets in the transcriptional regulation of endothelial cell differentiation and angiogenesis. *Biochimica et Biophysica Acta, 1775*, 298–312.

Dumont, D. J., Fon, G. H., Puri, M. C., Gradwohl, G., Alitalo, K., & Breitman, M. L. (1995). Vascularization of the mouse embryo: A study of flk-1, tek, tie, and vascular endothelial growth factor expression during development. *Developmental Dynamics, 203*, 80–92.

Ema, M., Takahashi, S., & Rossant, J. (2006). Deletion of the selection cassette, but not cis-acting elements, in targeted Flk1-lacZ allele reveals Flk1 expression in multipotent mesodermal progenitors. *Blood, 107*, 111–117.

Faloon, P., Arentson, E., Kazarov, A., Deng, C. X., Porcher, C., Orkin, S., et al. (2000). Basic fibroblast growth factor positively regulates hematopoietic development. *Development, 127*, 1931–1941.

Ferdous, A., Caprioli, A., Iacovino, M., Martin, C. M., Morris, J., Richardson, J. A., et al. (2009). Nkx2-5 transactivates the Ets-related protein 71 gene and specifies an endothelial/endocardial fate in the developing embryo. *Proceedings of the National Academy of Sciences of the United States of America, 106*, 814–819.

Ferrara, N., Carver-Moore, K., Chen, H., Dowd, M., Lu, L., O'Shea, K. S., et al. (1996). Heterozygous embryonic lethality induced by targeted inactivation of the VEGF gene. *Nature, 380*, 439–442.

Fleury, M., Eliades, A., Carlsson, P., Lacaud, G., & Kouskoff, V. (2015). FOXF1 inhibits hematopoietic lineage commitment during early mesoderm specification. *Development, 142*, 3307–3320.

Ginsberg, M., James, D., Ding, B. S., Nolan, D., Geng, F., Butler, J. M., et al. (2012). Efficient direct reprogramming of mature amniotic cells into endothelial cells by ETS factors and TGFβ suppression. *Cell, 151*, 559–575.

Glenn, N. O., Schumacher, J. A., Kim, H. J., Zhao, E. J., Skerniskyte, J., & Sumanas, S. (2014). Distinct regulation of the anterior and posterior myeloperoxidase expression by Etv2 and Gata1 during primitive Granulopoiesis in zebrafish. *Developmental Biology, 393*, 149–159.

Gomez, G., Lee, J. H., Veldman, M. B., Lu, J., Xiao, X., & Lin, S. (2012). Identification of vascular and hematopoietic genes downstream of etsrp by deep sequencing in zebrafish. *PLoS One*, 7, e31658.

Gomez, G. A., Veldman, M. B., Zhao, Y., Burgess, S., & Lin, S. (2009). Discovery and characterization of novel vascular and hematopoietic genes downstream of etsrp in zebrafish. *PLoS One*, 4, e4994.

Grass, J. A., Jing, H., Kim, S. I., Martowicz, M. L., Pal, S., Blobel, G. A., et al. (2006). Distinct functions of dispersed GATA factor complexes at an endogenous gene locus. *Molecular and Cellular Biology, 26*, 7056–7067.
Han, J. K., Chang, S. H., Cho, H. J., Choi, S. B., Ahn, H. S., Lee, J., et al. (2014). Direct conversion of adult skin fibroblasts to endothelial cells by defined factors. *Circulation, 130*, 1168–1178.
Hayashi, M., Pluchinotta, M., Momiyama, A., Tanaka, Y., Nishikawa, S., & Kataoka, H. (2012). Endothelialization and altered hematopoiesis by persistent Etv2 expression in mice. *Experimental Hematology, 40*, 738–750.
Hidaka, M., Stanford, W. L., & Bernstein, A. (1999). Conditional requirement for the Flk-1 receptor in the in vitro generation of early hematopoietic cells. *Proceedings of the National Academy of Sciences of the United States of America, 96*, 7370–7375.
Hollenhorst, P. C., McIntosh, L. P., & Graves, B. J. (2011). Genomic and biochemical insights into the specificity of ETS transcription factors. *Annual Review of Biochemistry, 80*, 437–471.
Huber, T. L., Kouskoff, V., Fehling, H. J., Palis, J., & Keller, G. (2004). Haemangioblast commitment is initiated in the primitive streak of the mouse embryo. *Nature, 432*, 625–630.
Ismailoglu, I., Yeamans, G., Daley, G. Q., Perlingeiro, R. C., & Kyba, M. (2008). Mesodermal patterning activity of SCL. *Experimental Hematology, 36*, 1593–1603.
Jin, H., Li, L., Xu, J., Zhen, F., Zhu, L., Liu, P. P., et al. (2012). Runx1 regulates embryonic myeloid fate choice in zebrafish through a negative feedback loop inhibiting Pu.1 expression. *Blood, 119*, 5239–5249.
Karim, F. D., Urness, L. D., Thummel, C. S., Klemsz, M. J., McKercher, S. R., Celada, A., et al. (1990). The ETS-domain: A new DNA-binding motif that recognizes a purine-rich core DNA sequence. *Genes & Development, 4*, 1451–1453.
Karkkainen, M. J., Haiko, P., Sainio, K., Partanen, J., Taipale, J., Petrova, T. V., et al. (2004). Vascular endothelial growth factor C is required for sprouting of the first lymphatic vessels from embryonic veins. *Nature Immunology, 5*, 74–80.
Kataoka, H., Hayashi, M., Kobayashi, K., Ding, G., Tanaka, Y., & Nishikawa, S. (2013). Region-specific Etv2 ablation revealed the critical origin of hemogenic capacity from Hox6-positive caudal-lateral primitive mesoderm. *Experimental Hematology, 41*, 567–581.
Kataoka, H., Hayashi, M., Nakagawa, R., Tanaka, Y., Izumi, N., Nishikawa, S., et al. (2011). Etv2/ER71 induces vascular mesoderm from Flk1+PDGFRα+ primitive mesoderm. *Blood, 118*, 6975–6986.
Kattman, S. J., Huber, T. L., & Keller, G. M. (2006). Multipotent FLK-1+cardiovascular progenitor cells give rise to the cardiomyocyte, endothelial, and vascular smooth muscle lineages. *Developmental Cell, 11*, 723–732.
Kattman, S. J., Witty, A. D., Gagliardi, M., Dubois, N. C., Niapour, M., Hotta, A., et al. (2011). Stage-specific optimization of activin/nodal and BMP signaling promotes cardiac differentiation of mouse and human pluripotent stem cell lines. *Cell Stem Cell, 8*, 228–240.
Kim, J. Y., Lee, R. H., Kim, T. M., Kim, D. W., Jeon, Y. J., Huh, S. H., et al. (2014). OVOL2 is a critical regulator of ER71/ETV2 in generating FLK1+, hematopoietic, and endothelial cells from embryonic stem cells. *Blood, 124*, 2948–2952.
Kinder, S. J., Tsang, T. E., Quinlan, G. A., Hadjantonakis, A. K., Nagy, A., & Tam, P. P. (1999). The orderly allocation of mesodermal cells to the extraembryonic structures and the anteroposterior axis during gastrulation of the mouse embryo. *Development, 126*, 4691–4701.
Kissa, K., & Herbomel, P. (2010). Blood stem cells emerge from aortic endothelium by a novel type of cell transition. *Nature, 464*, 112–115.

Kohli, V., Schumacher, J. A., Desai, S. P., Rehn, K., & Sumanas, S. (2013). Arterial and venous progenitors of the major axial vessels originate at distinct locations. *Developmental Cell, 25*, 196–206.

Koyano-Nakagawa, N., Kweon, J., Iacovino, M., Shi, X., Rasmussen, T. L., Borges, L., et al. (2012). Etv2 is expressed in the yolk sac hematopoietic and endothelial progenitors and regulates Lmo2 gene expression. *Stem Cells, 30*, 1611–1623.

Koyano-Nakagawa, N., Shi, X., Rasmussen, T. L., Das, S., Walter, C. A., & Garry, D. J. (2015). Feedback mechanisms regulate Ets variant 2 (Etv2) gene expression and hematoendothelial lineages. *The Journal of Biological Chemistry, 290*, 28107–28119. Sep 22 [Epub ahead of print].

Lam, E. Y., Hall, C. J., Crosier, P. S., Crosier, K. E., & Flores, M. V. (2010). Live imaging of Runx1 expression in the dorsal aorta tracks the emergence of blood progenitors from endothelial cells. *Blood, 116*, 909–914.

Lee, D., Kim, T., & Lim, D. S. (2011). The Er71 is an important regulator of hematopoietic stem cells in adult mice. *Stem Cells, 29*, 539–548.

Lee, D. J., Park, C., Lee, H., Lugus, J. J., Kim, S. H., Arentson, E., et al. (2008). ER71 acts downstream of BMP, Notch, and Wnt signaling in blood and vessel progenitor specification. *Cell Stem Cell, 2*, 497–507.

Leprince, D., Gegonne, A., Coll, J., de Taisne, C., Schneeberger, A., Lagrou, C., et al. (1983). A putative second cell-derived oncogene of the avian leukaemia retrovirus E26. *Nature, 306*, 395–397.

Li, L., Chen, D., Li, J., Wang, X., Wang, N., Xu, C., et al. (2014). Aggf1 acts at the top of the genetic regulatory hierarchy in specification of hemangioblasts in zebrafish. *Blood, 123*, 501–508.

Lindsley, R. C., Gill, J. G., Murphy, T. L., Langer, E. M., Cai, M., Mashayekhi, M., et al. (2008). Mesp1 coordinately regulates cardiovascular fate restriction and epithelial-mesenchymal transition in differentiating ESCs. *Cell Stem Cell, 3*, 55–68.

Liu, F., Bhang, S. H., Arentson, E., Sawada, A., Kim, C. K., Kang, I., et al. (2013). Enhanced hemangioblast generation and improved vascular repair and regeneration from embryonic stem cells by defined transcription factors. *Stem Cell Reports, 1*, 166–182.

Liu, F., Kang, I., Park, C., Chang, L. W., Wang, W., Lee, D., et al. (2012). ER71 specifies Flk-1+ hemangiogenic mesoderm by inhibiting cardiac mesoderm and Wnt signaling. *Blood, 119*, 3295–3305.

Liu, F., Li, D., Yu, Y. Y., Kang, I., Cha, M. J., Kim, J. Y., et al. (2015). Induction of hematopoietic and endothelial cell program orchestrated by ETS transcription factor ER71/ETV2. *EMBO Reports, 16*, 654–669.

Liu, F., Walmsley, M., Rodaway, A., & Patient, R. (2008). Fli1 acts at the top of the transcriptional network driving blood and endothelial development. *Current Biology, 18*, 1234–1240.

Loughran, S. J., Kruse, E. A., Hacking, D. F., de Graaf, C. A., Hyland, C. D., Willson, T. A., et al. (2008). The transcription factor ERG is essential for definitive hematopoiesis and the function of adult hematopoietic stem cells. *Nature Immunology, 9*, 810–819.

Lugus, J. J., Chung, Y. S., Mills, J. C., Kim, S. I., Grass, J., Kyba, M., et al. (2007). GATA2 functions at multiple steps in hemangioblast development and differentiation. *Development, 134*, 393–405.

Lugus, J. J., Park, C., Ma, Y. D., & Choi, K. (2009). Both primitive and definitive blood cells are derived from FLK-1+mesoderm. *Blood, 113*, 563–566.

Mahlapuu, M., Ormestad, M., Enerbäck, S., & Carlsson, P. (2001). The forkhead transcription factor Foxf1 is required for differentiation of extra-embryonic and lateral plate mesoderm. *Development, 128*, 155–166.

Matthews, W., Jordan, C. T., Gavin, M., Jenkins, N. A., Copeland, N. G., & Lemischka, I. R. (1991). A receptor tyrosine kinase cDNA isolated from a population of enriched primitive hematopoietic cells and exhibiting close genetic linkage to c-kit. *Proceedings of the National Academy of Sciences of the United States of America*, *88*, 9026–9030.

McLaughlin, F., Ludbrook, V. J., Cox, J., von Carlowitz, I., Brown, S., & Randi, A. M. (2001). Combined genomic and antisense analysis reveals that the transcription factor ERG is implicated in endothelial cell differentiation. *Blood*, *98*, 3332–3339.

Millauer, B., Wizigmann-Voos, S., Schnurch, H., Martinez, R., Moller, N. P., Risau, W., et al. (1993). High affinity VEGF binding and developmental expression suggest Flk-1 as a major regulator of vasculogenesis and angiogenesis. *Cell*, *72*, 835–846.

Moore, J. C., Sheppard-Tindell, S., Shestopalov, I. A., Yamazoe, S., Chen, J. K., & Lawson, N. D. (2013). Post-transcriptional mechanisms contribute to Etv2 repression during vascular development. *Developmental Biology*, *384*, 128–140.

Moretti, A., Caron, L., Nakano, A., Lam, J. T., Bernshausen, A., Chen, Y., et al. (2006). Multipotent embryonic isl1 + progenitor cells lead to cardiac, smooth muscle, and endothelial cell diversification. *Cell*, *127*, 1151–1165.

Morita, R., Suzuki, M., Kasahara, H., Shimizu, N., Shichita, T., Sekiya, T., et al. (2015). ETS transcription factor ETV2 directly converts human fibroblasts into functional endothelial cells. *Proceedings of the National Academy of Sciences of the United States of America*, *112*, 160–165.

Motoike, T., Markham, D. W., Rossant, J., & Sato, T. N. (2003). Evidence for novel fate of Flk-1^+ progenitor: Contribution to muscle lineage. *Genesis*, *35*, 153–159.

Murray, P. D. F. (1932). The development in vitro of the blood of the early chick embryo. *Proceedings of the Royal Society of London*, *11*, 497–521.

Murry, C. E., & Keller, G. (2008). Differentiation of embryonic stem cells to clinically relevant populations: Lessons from embryonic development. *Cell*, *132*, 661–680.

Myers, C. T., & Krieg, P. A. (2013). BMP-mediated specification of the erythroid lineage suppresses endothelial development in blood island precursors. *Blood*, *122*, 3929–3939.

Neuhaus, H., Muller, F., & Hollemann, T. (2010). Xenopus er71 is involved in vascular development. *Developmental Dynamics*, *239*, 3436–3445.

Padron-Barthe, L., Temino, S., Villa del Campo, C., Carramolino, L., Isern, J., & Torres, M. (2014). Clonal analysis identifies hemogenic endothelium as the source of the blood-endothelial common lineage in the mouse embryo. *Blood*, *124*, 2523–2532.

Palencia-Desai, S., Kohli, V., Kang, J., Chi, N. C., Black, B. L., & Sumanas, S. (2011). Vascular endothelial and endocardial progenitors differentiate as cardiomyocytes in the absence of Etsrp/Etv2 function. *Development*, *138*, 4721–4732.

Park, C., Lee, T., Bhang, S. H., Liu, F., Nakamura, R., Oladipupo, S. S., et al. (2016). Injury mediated vascular regeneration requires endothelial ER71/ETV2. *Arteriosclerosis, Thrombosis, and Vascular Biology*, *36*, 86–96. pii: ATVBAHA.115.306430 [Epub ahead of print].

Park, C., Lugus, J. J., & Choi, K. (2005). Stepwise commitment from embryonic stem to hematopoietic and endothelial cells. *Current Topics in Developmental Biology*, *66*, 1–36.

Peterkin, T., Gibson, A., & Patient, R. (2009). Common genetic control of haemangioblast and cardiac development in zebrafish. *Development*, *136*, 1465–1474.

Pham, V. N., Lawson, N. D., Mugford, J. W., Dye, L., Castranova, D., Lo, B., et al. (2007). Combinatorial function of ETS transcription factors in the developing vasculature. *Developmental Biology*, *303*, 772–783.

Pimanda, J. E., Ottersbach, K., Knezevic, K., Kinston, S., Chan, W. Y., Wilson, N. K., et al. (2007). Gata2, Fli1, and Scl form a recursively wired gene-regulatory circuit during early hematopoietic development. *Proceedings of the National Academy of Sciences of the United States of America*, *104*, 17692–17697.

Proulx, K., Lu, A., & Sumanas, S. (2010). Cranial vasculature in zebrafish forms by angioblast cluster-derived angiogenesis. *Developmental Biology*, *348*, 34–46.

Rasmussen, T. L., Kweon, J., Diekmann, M. A., Belema-Bedada, F., Song, Q., Bowlin, K., et al. (2011). ER71 directs mesodermal fate decisions during embryogenesis. *Development*, *138*, 4801–4812.
Rasmussen, T. L., Shi, X., Wallis, A., Kweon, J., Zirbes, K. M., Koyano-Nakagawa, N., et al. (2012). VEGF/Flk1 signaling cascade transactivates Etv2 gene expression. *PLoS One*, *7*, e50103.
Ren, X., Gomez, G. A., Zhang, B., & Lin, S. (2010). Scl isoforms act downstream of etsrp to specify angioblasts and definitive hematopoietic stem cells. *Blood*, *115*, 5338–5346.
Rost, M. S., Enriquez, J. R., Schumacher, J. A., Baltrunaite, K., Rossi, A., & Stainier, D. Y. (in review). Vegf signaling promotes endothelial differentiation of arterial and venous progenitors by modulating etv2 expression.
Sabin, F. R. (1920). Studies on the origin of blood vessels and of red corpuscles as seen in the living blastoderm of the chick during the second day of incubation. *Contributions to Embryology*, *9*, 213–262.
Sakurai, H., Era, T., Jakt, L. M., Okada, M., Nakai, S., Nishikawa, S., et al. (2006). In vitro modeling of paraxial and lateral mesoderm differentiation reveals early reversibility. *Stem Cells*, *24*, 575–586.
Salanga, M. C., Meadows, S. M., Myers, C. T., & Krieg, P. A. (2010). ETS family protein ETV2 is required for initiation of the endothelial lineage but not the hematopoietic lineage in the Xenopus embryo. *Developmental Dynamics*, *239*, 1178–1187.
Schoenebeck, J. J., Keegan, B. R., & Yelon, D. (2007). Vessel and blood specification override cardiac potential in anterior mesoderm. *Developmental Cell*, *13*, 254–267.
Schuh, A. C., Faloon, P., Hu, Q. L., Bhimani, M., & Choi, K. (1999). In vitro hematopoietic and endothelial potential of flk-1(-/-) embryonic stem cells and embryos. *Proceedings of the National Academy of Sciences of the United States of America*, *96*, 2159–2164.
Schupp, M. O., Waas, M., Chun, C. Z., & Ramchandran, R. (2014). Transcriptional inhibition of etv2 expression is essential for embryonic cardiac development. *Developmental Biology*, *393*, 71–78.
Shalaby, F., Ho, J., Stanford, W. L., Fischer, K. D., Schuh, A. C., Schwartz, L., et al. (1997). A requirement for Flk1 in primitive and definitive hematopoiesis and vasculogenesis. *Cell*, *89*, 981–990.
Shalaby, F., Rossant, J., Yamaguchi, T. P., Gertsenstein, M., Wu, X. F., Breitman, M. L., et al. (1995). Failure of blood-island formation and vasculogenesis in Flk-1-deficient mice. *Nature*, *376*, 62–66.
Sharrocks, A. D. (2001). The ETS-domain transcription factor family. *Nature Reviews. Molecular Cell Biology*, *2*, 827–837.
Shi, X., Richard, J., Zirbes, K. M., Gong, W., Lin, G., Kyba, M., et al. (2014). Cooperative interaction of Etv2 and Gata2 regulates the development of endothelial and hematopoietic lineages. *Developmental Biology*, *389*, 208–218.
Shi, X., Zirbes, K. M., Rasmussen, T. L., Ferdous, A., Garry, M. G., Koyano-Nakagawa, N., et al. (2015). The transcription factor Mesp1 interacts with cAMP-responsive element binding protein 1 (Creb1) and coactivates Ets variant 2 (Etv2) gene expression. *The Journal of Biological Chemistry*, *290*, 9614–9625.
Shivdasani, R. A., Mayer, E. L., & Orkin, S. H. (1995). Absence of blood formation in mice lacking the T-cell leukaemia oncoprotein tal-1/SCL. *Nature*, *373*, 432–434.
Simoes, F. C., Peterkin, T., & Patient, R. (2011). Fgf differentially controls cross-antagonism between cardiac and haemangioblast regulators. *Development*, *138*, 3235–3245.
Spyropoulos, D. D., Pharr, P. N., Lavenburg, K. R., Jackers, P., Papas, T. S., Ogawa, M., et al. (2000). Hemorrhage, impaired hematopoiesis, and lethality in mouse embryos carrying a targeted disruption of the Fli1 transcription factor. *Molecular and Cellular Biology*, *20*, 5643–5652.

Sumanas, S., Gomez, G., Zhao, Y., Park, C., Choi, K., & Lin, S. (2008). Interplay among Etsrp/ER71, Scl, and Alk8 signaling controls endothelial and myeloid cell formation. *Blood*, *111*, 4500–4510.

Sumanas, S., Jorniak, T., & Lin, S. (2005). Identification of novel vascular endothelial-specific genes by the microarray analysis of the zebrafish cloche mutants. *Blood*, *106*, 534–541.

Sumanas, S., & Lin, S. (2006). Ets1-related protein is a key regulator of vasculogenesis in zebrafish. *PLoS Biology*, *4*(1), e10.

Swiers, G., Baumann, C., O'Rourke, J., Giannoulatou, E., Taylor, S., Joshi, A., et al. (2013). Early dynamic fate changes in haemogenic endothelium characterized at the single-cell level. *Nature Communications*, *4*, 2924.

Tsai, F. Y., Keller, G., Kuo, F. C., Weiss, M., Chen, J., Rosenblatt, M., et al. (1994). An early haematopoietic defect in mice lacking the transcription factor GATA-2. *Nature*, *371*, 221–226.

Ueno, H., & Weissman, I. L. (2006). Clonal analysis of mouse development reveals a polyclonal origin for yolk sac blood islands. *Developmental Cell*, *11*, 519–533.

Van Handel, B., Montel-Hagen, A., Sasidharan, R., Nakano, H., Ferrari, R., Boogerd, C. J., et al. (2012). Scl represses cardiomyogenesis in prospective hemogenic endothelium and endocardium. *Cell*, *150*, 590–605.

Veldman, M. B., & Lin, S. (2012). Etsrp/Etv2 is directly regulated by Foxc1a/b in the zebrafish angioblast. *Circulation Research*, *110*, 220–229.

Veldman, M. B., Zhao, C., Gomez, G. A., Lindgren, A. G., Huang, H., Yang, H., et al. (2013). Transdifferentiation of fast skeletal muscle into functional endothelium in vivo by transcription factor Etv2. *PLoS Biology*, *11*, e1001590.

Vogeli, K. M., Jin, S. W., Martin, G. R., & Stainier, D. Y. (2006). A common progenitor for haematopoietic and endothelial lineages in the zebrafish gastrula. *Nature*, *443*, 337–339.

Wang, L., Liu, T., Xu, L., Gao, Y., Wei, Y., Duan, C., et al. (2013). Fev regulates hematopoietic stem cell development via ERK signaling. *Blood*, *122*, 367–375.

Wareing, S., Mazan, A., Pearson, S., Göttgens, B., Lacaud, G., & Kouskoff, V. (2012). The Flk1-Cre-mediated deletion of ETV2 defines its narrow temporal requirement during embryonic hematopoietic development. *Stem Cells*, *30*, 1521–1531.

Warga, R. M., Kane, D. A., & Ho, R. K. (2009). Fate mapping embryonic blood in zebrafish: Multi- and unipotential lineages are segregated at gastrulation. *Developmental Cell*, *16*, 744–755.

Wei, G., Srinivasan, R., Cantemir-Stone, C. Z., Sharma, S. M., Santhanam, R., Weinstein, M., et al. (2009). Ets1 and Ets2 are required for endothelial cell survival during embryonic angiogenesis. *Blood*, *114*, 1123–1130.

Wilson, N. K., Foster, S. D., Wang, X., Knezevic, K., Schutte, J., Kaimakis, P., et al. (2010). Combinatorial transcriptional control in blood stem/progenitor cells: Genome-wide analysis of ten major transcriptional regulators. *Cell Stem Cell*, *7*, 532–544.

Wong, K. S., Proulx, K., Rost, M. S., & Sumanas, S. (2009). Identification of vasculature-specific genes by microarray analysis of etsrp/etv2 overexpressing zebrafish embryos. *Developmental Dynamics*, *238*, 1836–1850.

Wythe, J. D., Dang, L. T., Devine, W. P., Boudreau, E., Artap, S. T., He, D., et al. (2013). ETS factors regulate Vegf-dependent arterial specification. *Developmental Cell*, *26*, 45–58.

Yamaguchi, T. P., Dumont, D. J., Conlon, R. A., Breitman, M. L., & Rossant, J. (1993). Flk-1, an flt-related receptor tyrosine kinase is an early marker for endothelial cell precursors. *Development*, *118*, 489–498.

Yamamizu, K., Matsunaga, T., Katayama, S., Kataoka, H., Takayama, N., Eto, K., et al. (2012). PKA/CREB signaling triggers initiation of endothelial and hematopoietic cell differentiation via Etv2 induction. *Stem Cells*, *30*, 687–696.

Yamashita, J., Itoh, H., Hirashima, M., Ogawa, M., Nishikawa, S., Yurugi, T., et al. (2000). Flk-1-positive cells derived from embryonic stem cells serve as vascular progenitors. *Nature*, *408*, 92–96.

Yaniv, K., Isogai, S., Castranova, D., Dye, L., Hitomi, J., & Weinstein, B. M. (2006). Live imaging of lymphatic development in the zebrafish. *Nature Medicine*, *12*, 711–716.

Yoshimatsu, Y., Yamazaki, T., Mihira, H., Itoh, T., Suehiro, J., Yuki, K., et al. (2011). Ets family members induce lymphangiogenesis through physical and functional interaction with Prox1. *Journal of Cell Science*, *124*, 2753–2762.

CHAPTER FIVE

Taking the Leap: Runx1 in the Formation of Blood from Endothelium

J. Tober*[,2], M.W. Maijenburg*[,2], N.A. Speck*[,1]
*University of Pennsylvania, Philadelphia, PA, United States
[1]Corresponding author: e-mail address: nancyas@exchange.upenn.edu

Contents

Abstract

Blood cell formation in the embryo occurs from multiple anatomic sites and results in the production of hematopoietic stem and progenitor cells that appear in overlapping waves. The transcription factor Runx1 is involved in a dramatic step of this process, for the transition from an endothelial cell that is integrated in a monolayer to a non-adherent circulating blood cell, a process conceptually similar to the epithelial to mesenchymal cell transition. Here we will review the role of Runx1 in the so-called hemogenic endothelium. We will describe the blood cell progenitors for which Runx1 is required, the proximal upstream transcription factors and signaling events that regulate its expression, and some of its important downstream targets in the hemogenic endothelium.

[2] Equal contribution.

Current Topics in Developmental Biology, Volume 118
ISSN 0070-2153
http://dx.doi.org/10.1016/bs.ctdb.2016.01.008

Embryonic development involves complex alterations of cell fate, shape, migration, and behavior. A dramatic example of all of these, repeated many times over in development, is the transition that a cell that is highly connected to other cells in a monolayer undergoes to become a mobile cell. The classic, most well-understood examples of this phenomenon are the epithelial to mesenchymal transitions that occur in the formation of mesendodermal cells from the epiblast during gastrulation, the cardiac cushion from the atrioventricular canal and outflow tract, and the delamination of neural crest cells from the dorsal neural tube. A more recently identified example, similar in concept to the epithelial to mesenchymal transition but less well understood at a molecular level is the formation of blood from endothelial cells, through what is referred to as an endothelial to hematopoietic cell transition (Kissa & Herbomel, 2010). One of the most crucial proteins for executing that transition, and that is required for the formation of all blood cells in the adult organism is the transcription factor Runx1.

Runx1 is a sequence-specific DNA-binding transcription factor that is expressed in essentially all blood cells with the exception of erythrocytes (Lorsbach et al., 2004; North, Stacy, Matheny, Speck, & de Bruijn, 2004). A non-DNA binding subunit, core binding factor β (CBFβ) increases Runx1's affinity for DNA, and is essential for most of Runx1's in vivo functions (Gu, Goetz, Graves, & Speck, 2000; Kim et al., 1999; Niki et al., 1997; Ogawa et al., 1993; Sasaki et al., 1996; Wang, Stacy, Miller, et al., 1996; Wang et al., 1993). Runx1 is one of three RUNX proteins that, along with CBFβ are collectively known as core binding factors (CBF). Mutations in both the *RUNX1* and *CBFB* genes are common in leukemia and in myelodysplastic syndrome, a preleukemic condition. Mutations include chromosomal translocations and loss-of-function mutations in *RUNX1*, and chromosomal translocations in *CBFB*. Many reviews have been written over the last 5 years on the contributions of *RUNX1* and *CBFB* mutations to hematopoietic diseases (De Braekeleer et al., 2011; Duployez, Willekens, Marceau-Renaut, Boudry-Labis, & Preudhomme, 2015; Engel & Hiebert, 2010; Harada & Harada, 2009; Ichikawa et al., 2013; Lam & Zhang, 2012; Mangan & Speck, 2011; Sangle & Perkins, 2011). This review will focus on the first and most essential role of Runx1 in hematopoiesis and embryonic development, in the formation of blood cells from endothelium.

1. THE THREE WAVES OF BLOOD CELL FORMATION

Blood is derived from the mesoderm, one of the three germ layers along with ectoderm and endoderm. Mesoderm forms from epiblast cells that have migrated through the primitive streak during gastrulation. When and where the epiblast cell transits through the streak determines the final location of the mesoderm derived from it, and to which future tissues and organs that mesoderm will contribute (Kinder et al., 1999; Lawson, Meneses, & Pedersen, 1991; Parameswaran & Tam, 1995; Tam & Beddington, 1987). The first epiblast cells that enter the primitive streak (at embryonic day (E) 6.5 in the mouse) contribute primarily to extraembryonic mesoderm that forms the yolk sac blood islands. Slightly later in development (E7.0), many epiblast cells transiting through the streak will contribute to the yolk sac vitelline vasculature. At E7.5 epiblast cells entering the posterior end of the streak will contribute mostly to the allantois, an extraembryonic structure that is a precursor to the umbilical artery and the fetal vasculature in the labyrinth region of the placenta. Cells entering the medial portion of the streak at E7.5 will make the greatest contribution to lateral plate mesoderm, and cells in the most anterior portion of the streak will contribute mostly to somites (Kinder et al., 1999). Blood is derived from several of these mesoderm populations, including mesoderm in the yolk sac blood islands, mesoderm that produces the vitelline vasculature, the allantoic mesoderm, and lateral plate mesoderm.

The first wave of blood formation, often referred to as the primitive wave, produces a large number of unipotent progenitors for primitive erythrocytes, a smaller number of bipotent primitive erythrocyte/megakaryocyte progenitors, and primitive macrophages, all of which differentiate from mesoderm in the yolk sac blood islands beginning at E7.25 (Bertrand et al., 2005; Palis, Robertson, Kennedy, Wall, & Keller, 1999; Tober et al., 2007; Fig. 1). The primitive erythrocytes are morphologically distinct from their adult counterparts; they are relatively large cells that enucleate in the circulation and express embryonic hemoglobin, in contrast to adult definitive erythrocytes that are smaller, enucleate in avascular spaces, and do not express embryonic globin (Kingsley, Malik, Fantauzzo, & Palis, 2004; Palis, 2014). Primitive megakaryocytes are also morphologically distinct from their adult counterparts, as they are diploid, as opposed to the highly polyploid megakaryocytes found in the bone marrow (Potts et al., 2014). Primitive macrophages in fish embryos are derived from the anterior most lateral mesoderm of the head, and macrophage progenitors migrate from there to the yolk sac where they differentiate into macrophages (Herbomel, Thisse, & Thisse, 1999) (it is unknown whether macrophage precursors in the mouse embryo also originate in the head). With the onset of circulation, primitive macrophages are distributed throughout the embryo. Primitive macrophages are the only cells that emerge during this

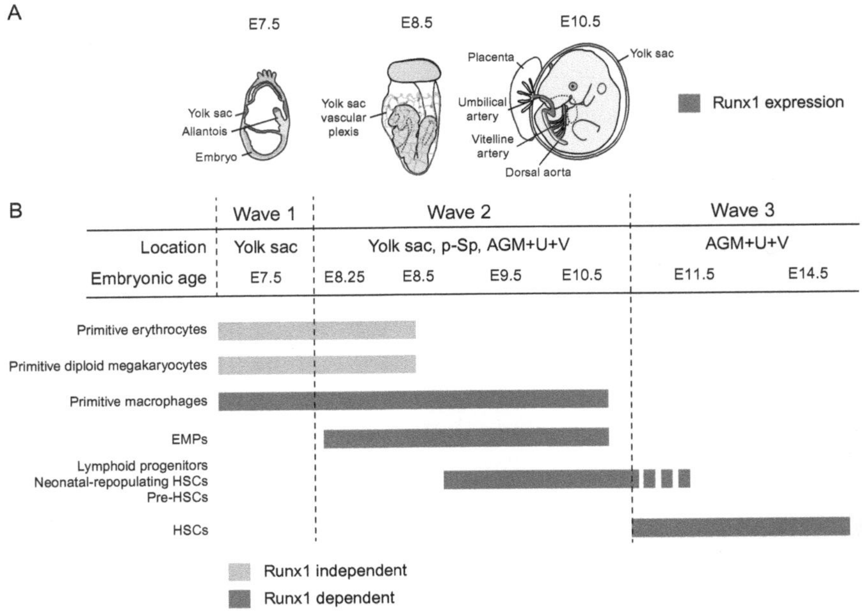

Fig. 1 Waves of hematopoietic cell emergence, sites of Runx1 expression, and progenitors/HSCs dependent on Runx1 activity. (A) Schematic diagrams of (*left* to *right*) E7.5 (neural plate), E8.5 (1–4 sp stage), and E10.5 mouse embryos. Sites of Runx1 expression relevant to hematopoietic progenitor/HSC emergence are colored in red. These include the yolk sac blood islands and the distal end of the allantois in the E7.5 embryo; in hemogenic endothelial cells and small clusters of EMPs in the yolk sac vascular plexus of the E8.5 embryo (not shown is Runx1 expression at the juncture of the dorsal aorta, vitelline and umbilical vasculature); and in the ventral aspect of the dorsal aorta, the umbilical artery, vitelline artery, and in small hematopoietic clusters in the yolk sac of an E10.5 embryo. (B) The three waves of blood formation as defined by Yoder (2014). The progenitors that emerge are indicated on the *left*. Progenitors/HSCs that require Runx1 for emergence are indicated with red bars. Lymphoid progenitors were shown to be dependent on functional CBFβ (Chen et al., 2011; Kobayashi et al., 2014). Only the times and sites of progenitor/HSC emergence are shown, and the persistence of the cells in the adult mouse is not illustrated. AGM+U+V, AGM region, umbilical, and vitelline arteries. (See the color plate.)

first wave of primitive hematopoiesis that are thought to persist in the adult, where they contribute to tissue resident macrophages in epithelia of adult tissues and microglia in the brain (Ginhoux et al., 2010; Gomez Perdiguero et al., 2015; Haldar & Murphy, 2014; Schulz et al., 2012).

Wave 2 in the mouse embryo refers to the appearance of the first "definitive" progenitors. The term "definitive" originally referred to adult (as opposed to primitive) erythrocytes (Palis, 2014). Nowadays the term definitive, as it refers to embryonic hematopoiesis, is used in two different ways; some use it to describe all committed embryonic progenitors except for those that appear in wave 1, while others restrict their use of the word "definitive" to describe adult-repopulating HSCs. Yoder has suggested that the term "wave 2" be used to describe the emergence of all committed progenitors prior to the HSC (excluding those that emerge in wave 1), and "wave 3" to refer only to adult-repopulating HSCs and their progeny (Yoder, 2014). We have adopted Yoder's terminology in this review.

The first progenitors that appear in wave 2 produce definitive erythroid and myeloid lineage cells, and are referred to as erythroid–myeloid progenitors or EMPs. EMPs emerge at E8.25 in the yolk sac of the mouse embryo, and small numbers have also been detected in the heart (McGrath et al., 2015; Nakano et al., 2013; Palis et al., 1999). Progenitors for more adult like, highly polyploid megakaryocytes also emerge in wave 2. Wave 2 megakaryocytes generate proplatelets after a lag of several days, unlike the primitive diploid megakaryocytes from wave 1 that rapidly produce proplatelets and platelets (Potts et al., 2014). Wave 2 also includes lymphoid progenitors that appear at E9.5 both in the yolk sac and in the para-aortic splanchnopleura (P-Sp) in the caudal part of the embryo, which contains lateral plate mesoderm (Boiers et al., 2013; Godin, Garcia-Porrero, Coutinho, Dieterlen-Lièvre, & Marcos, 1993; Lin, Yoder, & Yoshimoto, 2014; Yokota et al., 2006; Yoshimoto et al., 2011, 2012). Erythroid/myeloid and lymphoid progenitors are also found in the placenta, and in two major arteries connecting the dorsal aorta to the yolk sac and placenta, the vitelline (also known as omphalomesenteric) and umbilical arteries, respectively (de Bruijn, Speck, Peeters, & Dzierzak, 2000; Gekas, Dieterlen-Lievre, Orkin, & Mikkola, 2005; Gordon-Keylock, Sobiesiak, Rybtsov, Moore, & Medvinsky, 2013; North et al., 1999; Rhodes et al., 2008).

Although none of the progenitors that emerge in wave 2 can provide high level, multilineage engraftment when transplanted directly into adult mice, various forms of wave 2 progenitors possess limited repopulating activity. These include neonatal repopulating hematopoietic stem cells (hematopoietic stem cells able to engraft neonates but not adults) that appear at E9.0 (Yoder, Hiatt, & Mukherjee, 1997), and colony-forming unit-spleen (CFU-S) progenitors at E9.5 (Medvinsky, Samoylina, Müller, & Dzierzak, 1993). Pre-HSCs, unable to engraft adult mice directly, but able to mature into adult-repopulating HSCs following explant culture, or following coculture with stromal or endothelial cells, also emerge in the P-Sp, vitelline, and umbilical arteries beginning at E9.0 (Gordon-Keylock et al., 2013; Hadland et al., 2015; Medvinsky & Dzierzak, 1996; Rybtsov et al., 2014, 2011; Taoudi et al., 2008). Although these multilineage progenitors cannot directly engraft adult mice, many may mature into adult-repopulating HSCs in the fetal liver and are likely to be a major in vivo source of adult HSCs (Kieusseian, Brunet de la Grange, Burlen-Defranoux, Godin, & Cumano, 2012).

The final wave of blood formation, or wave 3, is defined by the emergence of long-term repopulating HSCs that can provide long-term, high-level, multilineage engraftment when transplanted directly into adult mice, and can self renew (Yoder, 2014). HSC emergence occurs at E10.5–E11.5 in mice, and at Carnegie stages 14–15 (days 32–33) in human embryos in the aorta/gonad/mesonephros (AGM) region, which consists of the dorsal aorta where it is flanked by the developing gonads and mesonephros (kidney) (AGM is the more commonly used term for the P-Sp at this stage) (Ivanovs et al., 2011; Medvinsky & Dzierzak, 1996; Müller, Medvinsky, Strouboulis, Grosveld, & Dzierzak, 1994). HSCs are also found at approximately the same time in the vitelline and umbilical arteries, the placenta, and the head (de Bruijn et al., 2000; Gekas et al., 2005; Gordon-Keylock et al., 2013; Li et al., 2012; Ottersbach & Dzierzak, 2005). The wave 2 multilineage progenitors and wave 3 HSCs will colonize the fetal liver and subsequently the bone marrow, and they or their progeny are, with the exception of a subset of macrophages, the source of all adult blood.

2. THE ROLE OF RUNX1 IN THE FORMATION OF EMBRYONIC PROGENITORS AND HSCs

Deletion of Runx1 in the germline eliminated almost all hematopoietic progenitors and HSCs, sparing only a subset of wave 1 progenitors, the primitive erythrocytes and primitive diploid megakaryocytes (Cai et al., 2000; Okuda, van Deursen, Hiebert, Grosveld, & Downing, 1996; Potts et al., 2014; Wang, Stacy, Binder, et al., 1996; Fig. 1). The majority of Runx1-deficient embryos hemorrhaged in specific locations (sensory nerve ganglia, mid brain, hindbrain, dorsal root ganglia) (Okuda et al., 1996; Wang, Stacy, Binder, et al., 1996). Embryos that hemorrhaged were anemic due to blood leakage into the brain ventricles and tissues, but the remainder had grossly normal numbers of primitive erythrocytes in the circulation (Okuda et al., 1996; Wang, Stacy, Binder, et al., 1996). The primitive erythrocytes and megakaryocytes did not escape unscathed; a portion of primitive erythrocytes (fewer than half) had abnormal, deformed morphology, reduced expression of the erythroid maker Ter119, and lower levels of the erythroid transcription factors Klf1 and Gata1 (Yokomizo et al., 2008). Primitive diploid megakaryocytes were fewer in number in Runx1-deficient embryos and produced fewer platelets (Potts et al., 2014). Aside from the hemorrhaging, which was associated with apoptosis of endothelial and neural crest cells (Wang, Stacy, Binder, et al., 1996), the vasculature of Runx1-deficient embryos was mostly normal, with subtle deficiencies in the number of small capillaries and vascular branching (Takakura et al., 2000). The most dramatic blood phenotypes in Runx1-deficient embryos, however, were the complete absence of macrophages, definitive EMPs, CFU-S, and HSCs (Cai et al., 2000; Lacaud, Kouskoff, Trumble, Schwantz, & Keller, 2004; Li, Chen, Stacy, & Speck, 2006; Okuda et al., 1996; Wang, Stacy, Binder, et al., 1996; Yokomizo et al., 2000). Knockout of CBFβ, the non-DNA binding partner of Runx1 caused a similar phenotype (Niki et al., 1997; Sasaki et al., 1996; Wang, Stacy, Miller, et al., 1996), although EMPs were not completely eliminated as they were with Runx1 deficiency, indicating that some residual Runx1 function is maintained in the absence of CBFβ (Wang, Stacy, Miller, et al., 1996).

The complete loss of all definitive blood lineages in the absence of pronounced vascular defects suggested that Runx1 was acting after formation of the vasculature, at an early stage in the generation of definitive blood produced in waves 2 and 3 (and also in primitive macrophages in wave 1). To gain insights into where the block in blood formation was occurring, North et al. created a Runx1 reporter allele in which β-galactosidase was expressed from the endogenous Runx1 locus (North et al., 1999). From examining the expression of Runx1-lacZ in embryos they made a quite unexpected finding, which was that Runx1 was expressed in a small population of endothelial cells prior to the appearance of definitive blood progenitors. These endothelial cells were located in the sites from which definitive blood progenitors are first detected, including the yolk sac, the dorsal aorta within the P-Sp, and the vitelline and umbilical arteries. Shortly after Runx1-expressing endothelial cells appeared, clusters of round Runx1-lacZ+ hematopoietic cells were found attached to endothelial cells in the lumens of the vessels in the same locations, suggesting a precursor/progeny relationship between the endothelial cells and hematopoietic cluster cells (North et al., 1999). This observation dovetailed beautifully with two discoveries, including one made by Dieterlen-Lievre and colleagues a year prior, who demonstrated through lineage-tracing studies that endothelial cells in the dorsal aorta of chick embryos gave rise to the hematopoietic cluster cells (Jaffredo, Gautier, Eichmann, & Dieterlen-Lièvre, 1998). The second study by Nishikawa et al. showed that endothelial cells, defined by presence of cell surface vascular endothelial cadherin (VE-cad) and the absence of hematopoietic markers (CD45 and Ter119), purified from E9.5 yolk sac or P-Sp could produce lymphoid cells when cultured ex vivo (Nishikawa et al., 1998). The discovery that Runx1, a transcription factor associated with hematopoiesis, was expressed in the endothelium in blood-producing sites prior to the appearance of blood cells lent credence to the hypothesis put forth by Jaffredo and Nishikawa that endothelial cells were the immediate precursors of blood in the major arteries.

The concept of endothelial precursors for blood, a so-called hemogenic endothelium was first articulated in the early 1900s (Emmel, 1916; Jordon, 1916; Sabin, 1920), but the idea fell into disfavor, and at the time of the Jaffredo, North, and Nishikawa studies (Jaffredo et al., 1998; Nishikawa et al., 1998; North et al., 1999) the more widely accepted hypothesis was that all blood formed from a hemangioblast, a bipotent cell could give rise to either a blood cell or an endothelial cell (Murray, 1932). Bolstering that hypothesis was that cells fitting the description of a hemangioblast had been identified in the primitive streak of E7.5 mouse embryos and in embryonic stem (ES) cell cultures (Choi, Kennedy, Kazarov, Papadimitriou, & Keller, 1998; Huber, Kouskoff, Fehling, Palis, & Keller, 2004; Kennedy et al., 1997). Hemogenic endothelium, however, posited a linear relationship between blood and endothelium, whereby the former differentiates directly from latter. The debate between hemangioblasts vs hemogenic endothelium as the direct precursor for definitive blood was finally settled 10 years later by a series of live imaging studies that showed blood forming from endothelium in real time (Bertrand, Chi, et al., 2010; Boisset et al., 2010; Eilken, Nishikawa, & Schroeder, 2009; Kissa & Herbomel, 2010; Lam, Hall, Crosier, Crosier, & Flores, 2010). It is now widely believed that definitive EMPs, lymphoid progenitors, and HSCs differentiate from hemogenic endothelium. In fact, lineage-tracing studies using a Cre recombinase driven by the VE-cad (encoded by *Cdh5*) regulatory sequences to activate a Rosa26 reporter allele demonstrated that most, if not all HSCs in the adult mouse are derived from hemogenic endothelium in the embryo (Chen, Yokomizo, Zeigler, Dzierzak, & Speck, 2009; Zovein et al., 2008).

Runx1 expression anticipates the appearance of blood in all of the anatomic sites in which blood forms (Fig. 1). Runx1 is expressed in the yolk sac mesoderm prior to the appearance of primitive erythrocytes, and Runx1 reporter alleles can be used to isolate cells with hemangioblast activity from embryonic stem cell cultures, and primitive erythroid progenitors from the yolk sac (Lacaud et al., 2002, 2004; North et al., 1999; Zeigler et al., 2006). Runx1 is expressed in cells at the distal end of the allantois, the precursor of the umbilical and placental vasculature, and in endothelial cells at the base of the allantois, at the point where the vitelline, umbilical, and dorsal aortic arteries amalgamate, beginning at 4sp (Daane & Downs, 2011), all of which are later sites of blood formation. Runx1 is widely expressed in endothelial cells in the vitelline artery, the first intraembryonic site of hematopoietic cluster formation in the mouse (Garcia-Porrero, Godin, & Dieterlen-Lièvre, 1995), in endothelial cells in the dorsal aorta and umbilical arteries (North et al., 1999), and in fetal vessels of the placental labyrinth (Ottersbach & Dzierzak, 2005; Rhodes et al., 2008; Fig. 1). By E14.5 Runx1 expression in endothelial cells is no longer visible, hematopoietic cluster cells in the major arteries have almost disappeared, and the cells that were born in the arteries have been released into the circulation and taken up residency in the fetal liver and thymus (North et al., 1999; Yokomizo et al., 2012).

Runx1 expression in endothelium not only anticipated the appearance of blood progenitors, germline mutation of Runx1 blocked the formation of hematopoietic cells from the hemogenic endothelium both in vivo, and from purified endothelial cells cultured ex vivo (North et al., 1999; Yokomizo et al., 2001). Morpholino knockdown or mutations in Runx1 had the same phenotype in zebrafish (Gering & Patient, 2005; Kalev-Zhylinska et al., 2002; Sood et al., 2010). It was thus proposed that Runx1 acts as a "molecular switch specifying the conversion from an endothelial to hematopoietic cell fate" (North et al., 1999). Runx1 is not required for the formation of either the primitive erythrocytes or the diploid platelet-forming megakaryocytes (Okuda et al., 1996; Potts et al., 2014; Wang, Stacy, Binder, et al., 1996), both of which are believed to differentiate directly from yolk sac mesoderm. Although the yolk sac mesoderm that gives rise to primitive erythrocytes and diploid megakaryocytes briefly expresses endothelial markers such as VE-cad, CD31, Tie2, CD34, and endoglin at E7.5 the expression of endothelial markers declines substantially by E8.5, and the cells do not adopt an endothelial-like morphology before differentiating into blood cells (Ema et al., 2006; Yokomizo et al., 2007). EMPs and HSCs, on the other hand, form from a morphological endothelial layer and remain VE-cad^{+} until they after they reach the fetal liver (Goldie, Lucitti, Dickinson, & Hirschi, 2008; Li, Ferkowicz, Johnson, Shelley, &

Yoder, 2005; Nadin, Goodell, & Hirschi, 2003; North et al., 2002; Padron-Barthe et al., 2014; Taoudi et al., 2005; Yokomizo et al., 2001). Collectively these data suggest that Runx1 is necessary for the differentiation of progenitors and HSCs derived from VE-cad^{+} endothelium, and not from mesoderm that only transiently expresses VE-cad.

Important questions were whether Runx1 was required only in the hemogenic endothelium for blood formation, and if so, whether Runx1 expression in endothelium was sufficient for blood formation. The former question was addressed by deleting *Runx1* in the endothelium using a Cre recombinase driven by the VE-cad regulatory sequences. This produced a phenotype essentially identical to germline deletion, indicating that Runx1 is required for blood cell formation in the hemogenic endothelium per se (or more precisely, in a cell in which the VE-cad-driven Cre was active) (Chen et al., 2009). With respect to sufficiency, recent efforts to reprogram endothelial cells into blood cells identified four transcription factors that together could promote an endothelial to hematopoietic cell conversion, FOSB, GFI1, RUNX1, and SPI1 (Sandler et al., 2014). The most effective single factor of the four was Runx1, suggesting that not only is Runx1 required in the endothelium for blood formation, it may also be sufficient in at least a subset of endothelial cells.

3. THE TEMPORAL REQUIREMENT FOR RUNX1 IN HEMOGENIC ENDOTHELIUM

Runx1 is expressed in the hemogenic endothelium in the major arteries at least 1 d before hematopoietic clusters appear, suggesting that specification of the endothelium by Runx1 is occurring over time. In an elegant study, Swiers et al. examined the hematopoietic and endothelial potential of single endothelial cells purified from the major arteries at different times of development to determine when hematopoietic specification occurred (Swiers et al., 2013). The endothelial cells were purified based on expression of a green fluorescent reporter transgene driven from a enhancer located 23.5 kb downstream from exon 1 in the *Runx1* gene (referred to as the +23 enhancer and the transgene as 23GFP), and then cultured under conditions permissive for the differentiation of either blood or endothelium (Nottingham et al., 2007; Swiers et al., 2013). Swiers et al. showed that at E8.5, single 23GFP^{+} endothelial cells produced only endothelial tubules in culture, and no blood cells, thus were not yet specified as hemogenic. 23GFP^{+} endothelial cells purified from E9.5 embryos could form either endothelial tubules or blood cells when cultured under conditions permissive for both fates, but the frequency of 23GFP$^+$ endothelial cells that differentiated into endothelial tubules was eightfold higher than the frequency of 23GFP$^+$ endothelial cells that differentiated into blood. Therefore at E9.5, some 23GFP$^+$ endothelial cells are specified as hemogenic and can produce blood cells in a permissive environment, but the vast majority maintained an endothelial fate (Swiers et al., 2013). At E10.5 this relationship was inverted, and the frequency of 23GFP$^+$ endothelial cells that produced blood was twice that of 23GFP$^+$ endothelial cells that formed endothelial tubules, thus the majority of E10.5 23GFP$^+$ endothelial cells were specified as hemogenic (Swiers et al., 2013). A caveat was that 23GFP was not an entirely faithful reporter of endogenous Runx1 expression, as fewer than 50% of 23GFP$^+$ endothelial cells expressed endogenous Runx1. Thus it is plausible that all Runx1-expressing endothelial cells at E10.5 are specified as hemogenic and that the 23GFP$^+$ cells that retained an endothelial fate at E10.5 may have been the same subset of 23GFP$^+$ cells that did not express endogenous Runx1. In summary, hematopoietic specification of hemogenic endothelium (defined from the onset of 23GFP expression to when the majority of 23GFP$^+$ endothelial cells formed blood when cultured under permissive conditions) occurs between E8.5 and E10.5. Although the precise time when commitment of hemogenic endothelial cells to the blood lineage has not been determined, high levels of cell surface CD41 is a reliable marker of commitment (Hashimoto et al., 2007; Li et al., 2005). CD41 is expressed on megakaryocytes and platelets, but in the embryo it also becomes upregulated on hemogenic endothelial cells early during the endothelial to hematopoietic cell transition, before the blood cell is released from the endothelium (Eilken et al., 2009; Kissa et al., 2008). Once CD41 is upregulated, the hemogenic endothelial cell has passed the point of no return and is committed to differentiating into a blood cell, and endothelial potential has been lost (Hashimoto et al., 2007; Li et al., 2005).

The temporal requirement for Runx1 in hemogenic endothelium for blood cell formation has been examined using several different strategies. Tober et al. assessed when Runx1 was required in hemogenic endothelium by deleting *Runx1* in endothelium in 24 h intervals spanning E7.5 to E14.5 using a tamoxifen-regulated form of Cre driven from the VE-cad regulatory sequences. They found no blood cells in which both alleles of *Runx1* had been deleted with VE-cad-Cre^{ERT} at any time point, indicating that Runx1 is absolutely required in VE-cad-Cre^{ERT} expressing cells for blood cell formation (Tober, Yzaguirre, Piwarzyk, & Speck, 2013). They did not, however, determine whether Runx1 was required in hematopoietic cluster cells. VE-cad mRNA is downregulated in hematopoietic cluster cells, but VE-cad protein remains on the cell surface until sometime after HSCs colonize the fetal liver (Bollerot, Pouget, & Jaffredo, 2005; Kim, Yilmaz, & Morrison, 2005; North et al., 2002; Taoudi et al., 2005). By analogy, whether Cre^{ERT} remains active in the clusters that are no longer transcribing the VE-cad-Cre^{ERT} transgene will depend on how quickly the Cre^{ERT} mRNA and protein are degraded. Therefore, although Runx1 was clearly required at all times in endothelial cells for blood cell formation, whether it is still required once an endothelial cell has transitioned into a hematopoietic cluster cell could not be determined (Tober et al., 2013). On the other hand, once EMPs and HSCs had colonized the fetal liver, the absolute requirement for Runx1 has clearly ended, as deletion with Vav1-Cre, which is active in fetal liver EMPs and HSCs, but does not appear to be active in the endothelium or in hematopoietic cluster cells, did not eliminate either EMPs or HSCs (Cai et al., 2015, 2011; Chen et al., 2009).

Liakhovitskaia et al. (2014) asked a different question, which was in a Runx1-deficient embryo, in what cells could Runx1 expression be restored and rescue hematopoiesis? Specifically, they asked whether if they activated Runx1 in a CD41 (*Itga2b*) expressing cell, using a conditionally activated *Runx1* allele in conjunction with CD41-Cre, could they rescue blood cell formation in a Runx1-deficient embryo. Restoring Runx1 expression in CD41-Cre expressing cells rescued blood formation, including HSCs, indicating that hemogenic endothelium can progress in its differentiation all the way to a $CD41^+$ cell before Runx1 is required. A question raised by this experiment is at exactly what time of development CD41-Cre was active. Swiers et al. showed that at E8.5 all nascent endothelial cells (hemogenic and nonhemogenic) express low levels of CD41 (Swiers et al., 2013). CD41 is then downregulated in endothelial cells at E9.5–E10.5, and its expression again increases in hemogenic endothelial cells as they initiate the endothelial to hematopoietic cell transition (Eilken et al., 2009; Kissa et al., 2008; Swiers et al., 2013). Thus, it is not clear whether Runx1 expression was rescued in all endothelial cells expressing low levels of CD41 at E8.5, or in the small subset of hemogenic endothelial cells at E9.5–E10.5 that have again upregulated CD41 and are poised to transition into blood cells.

The dosage of *Runx1* also affects the kinetics of blood cell formation. Reductions in *Runx1* dosage due to haploinsufficiency of *Runx1*, or in embryos with hypomorphic *Runx1* alleles resulted in decreased numbers of erythroid/myeloid progenitors in the yolk sac, AGM region, and fetal liver (Matheny et al., 2007; Wang, Stacy, Binder, et al., 1996; Wang, Stacy, Miller, et al., 1996). More interesting, though, was that the temporal appearance of HSCs was altered in *Runx1*$^{+/-}$ embryos. Specifically, HSCs were detectable in the yolk sac, AGM region, and fetal livers at E10.5 in *Runx1*$^{+/-}$ embryos, whereas they are extremely rare (at a frequency of 0.03 HSC/embryo) in wild-type embryos (Cai et al., 2011; Kumaravelu et al., 2002; Müller et al., 1994). There was also an early appearance of blast-colony-forming units and primitive erythroid colonies in cultures of *Runx1*$^{+/-}$ embryonic stem cells, and the expression of mesodermal markers was advanced by 12 h, suggesting that Runx1 may negatively regulate mesoderm commitment and hemangioblast specification (Lacaud et al., 2004). The mechanisms underlying the observed temporal changes associated with reduced Runx1 dosage are unclear.

4. DOWNSTREAM RUNX1 TARGETS THAT EXECUTE THE ENDOTHELIAL TO HEMATOPOIETIC CELL TRANSITION

The process of blood formation from endothelium involves dramatic changes in cell morphology and gene expression. In the mouse, hemogenic endothelial cells were shown to have distinguishing morphological features such as a more oblong and rounded shape than their endothelial neighbors, and filopodia-like protrusions of the membrane (Bos, Hawkins, & Zovein, 2015). In the zebrafish embryo, where the endothelial to hematopoietic cell transition has been best visualized, hemogenic endothelial cells are first seen to contract, and then bend toward the direction of cluster formation, forming a crescent-like structure (Kissa & Herbomel, 2010). The hemogenic endothelial cell remains in its bent conformation for 1–2 h, then bends further in the medial to lateral direction, bringing the two lateral endothelial cells together to seal the gap in the endothelial layer that will be made by the departing cell. The hemogenic endothelial cell releases its contacts with its lateral neighbors while maintaining contact with its rostral and caudal neighbors, then rounds up and detaches from the endothelium. In the absence of Runx1 (achieved through injection of a morpholino) most endothelial cells did not attempt to undergo the endothelial to hematopoietic transition, and those that began to contract, apoptosed (Kissa & Herbomel, 2010).

Two of Runx1's important direct targets in the hemogenic endothelium are *Gfi1* and *Gfi1b*, which encode zinc finger transcriptional repressors that play an important role in repressing the expression of endothelial genes. Overexpression of Gfi1 in Runx1-deficient endothelial cells downregulated the expression of endothelial genes including VE-cad, claudin 5 (*Cldn5*), *Pecam1*, and endoglin (*Eng*). It did not downregulate all endothelial genes, as *Cd34*, *Kdr*, and *Tie2* (*Tek*), were still expressed (Lancrin et al., 2012). mRNAs for several hematopoietic genes including *Gata1*, *Tal*, *Myb*, and *Lmo2* were upregulated upon Gfi1 overexpression in Runx1-deficient cells, and morphological changes that accompany the endothelial to hematopoietic cell transition occurred, even to the point of generating round non-adherent cells (Lancrin et al., 2012). However the round cells generated by overexpression of Gfi1 in the absence of Runx1 had no clonogenic activity, and hence were not functional hematopoietic progenitors. In embryos with functional Runx1 that lacked both Gfi1 and Gfi1b, blood cells formed but remained in the yolk sac and failed to disseminate throughout the embryo. Nevertheless yolk sacs from Gfi1/Gfi1b-deficient embryos contained hematopoietic progenitors that were revealed once the cells were disaggregated and plated as single cells in clonogenic assays, although cells within the Gfi1/Gfi1b-deficient hematopoietic colonies failed to properly downregulate endothelial markers such as *Sox7*, *Eng*, *Tek*, *Cdh5*, and *Flk1* (Lancrin et al., 2012). Therefore at later stages during the transition of hemogenic endothelium to hematopoietic cells Gfi1/Gfi1b represses the endothelial fate, while Runx1 activates the hematopoietic program.

Runx1 presumably regulates the expression of genes in addition to *Gfi1* and *Gfi1b* that are involved in executing the endothelial to hematopoietic cell transition. Lie et al. (2014) used DNA adenine methyltransferase identification (DamID) to identify Runx1 targets in hemogenic endothelium differentiated from ES cell cultures. The DamID technique involves creating a fusion protein containing the DNA-binding protein of interest (in this case Runx1) linked to the *E. coli* enzyme adenine methyltransferase (Dam) (Aughey & Southall, 2015). The Runx1–Dam fusion protein will introduce methyl groups at GATC sites in DNA adjacent to the Runx1 occupied sites, which can be detected by cleavage with a methylation-sensitive restriction enzyme combined with high-throughput sequencing. The advantage of DamID is that it requires relatively few cells compared to that required for transcription factor ChIP-Seq, and it can detect weakly associated DNA-binding proteins (Hoogenkamp et al., 2009).

Lie-A-Ling generated Runx1-deficient ES cells in which Runx1 expression could be restored via a doxycycline inducible transgene encoding the Runx1–Dam fusion protein (Lie et al., 2014). In the absence of doxycycline,

Runx1-deficient ES cells can be differentiated into $Flk1^+$ Kit^+ $Tie2^+$ $CD41^-$ cells that are arrested at an early hemogenic endothelial stage (Lancrin et al., 2009; Lichtinger et al., 2012). Administration of doxycycline induces functional Runx1 (Runx1–Dam) expression, allowing the hemogenic endothelial cells to differentiate into committed $Flk1^{+/-}$ Kit^+ $Tie2^-$ $CD41^+$ hematopoietic progenitors (EMPs). However, conveniently, Runx1–Dam expression in uninduced cells was somewhat leaky, such that in the absence of doxycycline the cells expressed a low level of Runx1–Dam that was insufficient to overcome the differentiation arrest at the $Flk1^+$ Kit^+ $Tie2^+$ $CD41^-$ hemogenic endothelial stage, but was adequate for detecting Runx1 occupancy by performing DamID. They found >200 genes that were both bound by Runx1 and differentially expressed upon Runx1–Dam induction. The target genes were primarily enriched for genes associated with integrin signaling, cell adhesion, cellular movement, endothelial cell migration, and interaction with the extracellular matrix. Interestingly, very few hematopoietic-specific genes were occupied by Runx1 at this stage, consistent with data from Swiers et al. showing that the majority of Runx1-expressing (23GFP+) endothelial cells isolated from E8.5 and E9.5 embryos had maintained their endothelial fate (Swiers et al., 2013). Thus an early function of Runx1 in hemogenic endothelium appears to be the activation of adhesion and migration genes involved in executing the endothelial to hematopoietic cell transition, and the activation of hematopoietic genes may be a later activity.

5. EPIGENETIC REGULATION BY RUNX1

Due to the rarity of hemogenic endothelial cells in the embryo, the epigenetic regulation by Runx1 has primarily been examined in hemogenic endothelial cells differentiated from ES cell cultures under conditions that recapitulate yolk sac hematopoiesis (Hoogenkamp et al., 2009; Lichtinger et al., 2012; Tanaka et al., 2012). Several studies utilized a system similar to that employed by Lie et al. (2014), consisting of a Runx1-deficient ES cell line and a doxycyline-inducible Runx1 cDNA that can be used to correct Runx1 deficiency in a temporally controlled manner (Lancrin et al., 2009). Lichtinger et al. (2012) used these ES cells to generate Runx1-deficient $Flk1^+$ Kit^+ $Tie2^+$ $CD41^-$ cells that are arrested at an early hemogenic endothelial stage (Lancrin et al., 2009; Lichtinger et al., 2012). They performed a genome wide analysis of Tal1, Fli1, C/EBPβ, and RNA Pol II occupancy, and H3K9 acetylation in the $Flk1^+$ Kit^+ $Tie2^+$ $CD41^-$ early hemogenic endothelial cells, in Flk1$^+$ Kit$^+$ Tie2$^+$ CD41$^+$ later hemogenic endothelial cells following a 24-h Runx1 induction, and in Flk1$^{+/-}$ Kit$^+$ Tie2$^-$ CD41$^+$ hematopoietic progenitors differentiated from hemogenic endothelial cells (Runx1 occupancy was also analyzed in the latter two populations). They found that Tal1, Fli1, and C/EBPβ occupied a large number of genomic sites in Flk1$^+$ Kit$^+$ Tie2$^+$ CD41$^-$ early hemogenic endothelial cells in the absence of Runx1. Hematopoietic genes were highly represented in genes cooccupied by Tal1 and Fli1, and a higher enrichment score for HSC-related genes was associated with genes occupied by all three transcription factors. The authors suggested that occupancy by Tal1, Fli1, and C/EBPβ may be part of a priming mechanism that prepares hematopoietic genes for rapid upregulation by Runx1 in the hemogenic endothelium. Twenty-four hours after Runx1 induction by doxycycline, the Runx1-bound sites in later hemogenic endothelial cells (Flk1$^+$ Kit$^+$ Tie2$^+$ CD41$^+$) that had differentiated from the previously arrested cells (Flk1$^+$ Kit$^+$ Tie2$^+$ CD41$^-$) could be clustered into two groups. One group consisted mostly of promoter sequences previously marked by H3K9Ac, indicating that Runx1 bound to preexisting transcription factor complexes flanked by acetylated histones. Runx1 strongly increased and extended H3K9 acetylation at these sites. The second group of Runx1 occupied sites consisted primarily of promoter distal sites previously marked by little or no H3K9Ac, and Runx1 also strongly increased H3K9Ac at these sites. Tal1 and Fli1 underwent a rapid genomic redistribution following Runx1 induction, forming transcription complexes at new *cis*-elements that were associated with increased H3K9 acetylation. Runx1 appeared to recruit Tal1 and Fli1 to its bound sites, as Tal1 and Fli1 peaks located within 400 bp of the Runx1 peaks significantly increased. The new Tal1 and Fli1 occupancy induced by Runx1 was shifted toward that found in committed hematopoietic progenitors. Runx1 directly binds the Tal1 and Fli1 proteins (Huang et al., 2009; Lichtinger et al., 2012; Palii et al., 2011), and by this mechanism may recruit these transcription factors to its bound sites. Runx1 also mediates long-range chromatin interactions that may facilitate the recruitment of other transcription factors to its occupied sites (Collins et al., 2011; Levantini et al., 2011).

6. *CIS*-ACTING SEQUENCES REGULATING RUNX1 TRANSCRIPTION

Determining how transcription of the *Runx1* gene is regulated is key to understanding the transcription factors and signaling pathways that control blood cell formation from hemogenic endothelium. Prior to the development of high-throughput sequencing approaches, characterizing the transcriptional regulation of the *Runx1* gene was a laborious undertaking due to the fact it is large (murine *Runx1* spans 224,684 bp) and contains two promoters, P1 (distal) and P2 (proximal) that are separated by 160 kb (Bee, Liddiard, et al., 2009; Bee et al., 2010; Fujita, Nishimura, Taniwaki, Abe, & Okuda, 2001; Ghozi, Bernstein, Negreanu, Levanon, & Groner, 1996; Pozner et al., 2007; Telfer & Rothenberg, 2001). All vertebrate Runx genes have P1 and P2 promoters, thus this configuration is highly conserved throughout evolution. Transcripts from the two promoters differ in their 5′ untranslated sequences, whether they are translated from a 5′ cap or from an internal ribosome entry site, the efficiency at which they are translated, and in coding sequences for the N-terminal 19 amino acids of the Runx1 protein (Levanon et al., 1996; Levanon & Groner, 2004; Pozner et al., 2000). No compelling data indicates that the different N-termini that result from P1 vs P2 usage confer distinct properties to the Runx1 protein (Challen & Goodell, 2010; Telfer, Hedblom, Anderson, Laurent, & Rothenberg, 2004). Promoter usage does, however, control the timing and levels of Runx1 expression at different times of development.

Sroczynska, Lancrin, Pearson, Kouskoff, and Lacaud (2009) analyzed transcription from P1 and P2 in ES cell differentiation cultures that recapitulate yolk sac hematopoiesis, and in mouse embryos, by introducing distinguishable markers downstream from each promoter in the endogenous *Runx1* locus. P2 was the first to become active, at E7.25 (neural plate stage), in a subset of mesodermal cells marked by expression of Flk1, the receptor for vascular endothelial growth factor (VEGF). P2 was also active at a slightly later (headfold) stage in a population of hemogenic endothelial cells that expressed the endothelial marker Tie2 (encoded by *Tek*) and low levels of Kit. In contrast, the distal P1 was not utilized in Flk1$^+$ mesoderm or in Tie2$^+$ Kit$^+$ hemogenic endothelial cells, and only became active once the cells upregulated CD41. EMPs and lymphoid progenitors were more highly represented in cells utilizing P1 (with or without transcription from P2) as compared to P2 alone, while most primitive erythroid progenitors utilized P2 only. In summary, P2 (and not P1) is active in yolk sac mesoderm and hemogenic endothelium, and initiation of transcription from P1 occurs in committed CD41$^+$ progenitors (Hashimoto et al., 2007; Li et al., 2005; Sroczynska et al., 2009).

Consistent with its activity in hemogenic endothelium, transcription from P2 was essential for the formation of definitive progenitors, as a neomycin resistance gene cassette inserted 1 kb upstream of P2 that severely attenuated transcription greatly decreased the number of EMPs and hematopoietic clusters in the aorta, and caused neonatal lethality (Bee et al., 2010; Pozner et al., 2007). Deletion of P1 also decreased EMPs and hematopoietic clusters, but to a much lesser extent than attenuation of P2 activity, and the absence of transcripts from P1 had no effect on fetal or adult viability (Bee et al., 2010; Sroczynska et al., 2009). Therefore P2 is essential for initiation of hematopoiesis in the embryo and fetal viability, most likely due to its activity in hemogenic endothelium (Bee et al., 2010; Pozner et al., 2007; Sroczynska et al., 2009). P1 contributes to the overall levels of Runx1, as its deletion decreased the number of white blood cells in the adult and increased the percentage of bone marrow lineage negative, Sca-1^+, Kit^+ (LSK) cells, a phenotype reminiscent of that associated with reduced Runx1 dosage in the adult mouse (Growney et al., 2005; Ichikawa et al., 2004; Matheny et al., 2007). Conserved consensus sites in P1 include a CCAAT box, a cAMP-responsive element (CRE), GATA, RUNX1, and SMAD motifs. P2 contains several conserved CCAAT boxes, initiator sequences, GC-box, OCT, and ETS motifs (Ghozi et al., 1996; Pimanda, Donaldson, et al., 2007). Neither promoter conferred consistent hematopoietic expression in the embryo in the absence of the Runx1 +23 hematopoietic enhancer, but both promoters were appropriately activated in their normal embryonic hematopoietic sites by the +23 enhancer (Bee, Ashley, et al., 2009).

The aforementioned *Runx1* +23 hematopoietic enhancer is located between P1 and P2, and is conserved in tetrapods such as human (*Homo sapiens*), dog (*Canis familiaris*), horse (*Equus caballus*), opossum (*Monodelphis domestica*), chicken (*Gallus gallus*), frog (*Xenopus tropicalis*), and lizard (*Anolis carolinensis*) (Ng et al., 2010; Nottingham et al., 2007). Zebrafish lack a conserved +23 hematopoietic enhancer, but zebrafish P2 has more specific activity in sites of hematopoietic cell emergence such as the dorsal aorta compared to murine P2, and is likely to function as the +23 enhancer equivalent (Lam et al., 2009). Although expression of a reporter driven by the +23 enhancer recapitulated most of Runx1's normal expression in hematopoietic sites in the mouse embryo there were some subtle differences, as the +23 enhancer was active in fewer endothelial cells than endogenous Runx1 (Ng et al., 2010; North et al., 2002; Nottingham et al., 2007). Also, Runx1 is expressed in the subaortic mesenchyme in the AGM region, and the +23 enhancer lacked this activity (North et al., 1999; Nottingham et al., 2007). Nevertheless the +23 enhancer is a very useful reagent that has been used to study hematopoiesis in mouse and zebrafish embryos, and in human ES cell cultures (Ferrell, Xi, Ma, Adlakha, & Kaufman, 2015; Koh et al., 2015; Swiers et al., 2013; Tamplin et al., 2015).

7. THE GENE REGULATORY NETWORK UPSTREAM OF RUNX1

Cis-acting regulatory modules (CRMs) determine the expression of key tissue-specific genes and can be used as a starting point to dissect the gene regulatory network (GRN) that controls the specification of a particular tissue. The general approach for assembling a GRN is to identify a CRM in a gene encoding a key regulatory transcription factor, then determine which transcription factor motifs in the CRM are essential for its activity in the tissue of interest. Next, the transcription factors present in the tissue of interest able to bind to those motifs are identified. Finally, perturbations of the network through inhibition or ectopic activation of the GRN components are used to test the validity of the GRN model (Davidson et al., 2002).

The *Runx1* +23 hematopoietic enhancer, which is expressed in the hemogenic endothelium and hematopoietic clusters, contains conserved ETS, GATA, MYB, E-BOX, and RUNX motifs (Ng et al., 2010; Nottingham et al., 2007). Mutation of the GATA and ETS sites in the *Runx1* +23 enhancer had the most deleterious effects on its activity in the yolk sac blood islands, endothelial cells, and hematopoietic cluster cells, whereas mutation of the RUNX motif had very little effect on embryonic expression. Therefore the GATA and ETS sites (and potentially the E-BOX site, which was not tested) constitute a CRM in the *Runx1* gene that drives its expression in the sites of hematopoietic cell emergence in the embryo (Nottingham et al., 2007).

The GATA, ETS, and E-BOX sites constitute a CRM that is present in and regulate the expression of three transcription factors that bind the CRM. Specifically, the *Gata2* (encoding a GATA factor), *Fli1* (an ETS factor), and *Tal1* (encoding an E-BOX binding factor) genes all contain a GATA/ETS/E-BOX CRM. All three genes are required for the formation of blood in the embryo (Pimanda, Ottersbach, et al., 2007; Robb et al., 1995; Spyropoulos et al., 2000; Tsai et al., 1994; Table 1). The three transcription factors regulate the expression of their own and the other two genes, with the GRN circuitry wired as a positive feedback loop that locks the expression of all three genes into an "ON" state (Pimanda, Ottersbach, et al., 2007; Fig. 2). Several lines of evidence indicate that Gata2 and Fli1 regulate *Runx1* expression and/or activity in the hemogenic endothelium. *Gata2*, for example, is expressed in the lateral plate mesoderm adjacent to the primitive streak, and in all endothelial cells, thus is present in cells (or the precursors of cells) in which *Runx1* is later expressed (Khandekar et al., 2007; Minegishi et al., 1999; North et al., 1999). Endothelial cell-specific expression of *Gata2*

Table 1 Regulation of Runx1 Expression in the Dorsal Aorta

Gene	Species	Sites of Expression	Mutation	Phenotype	Proposed Mechanism	References
Runx1	Mouse	Endothelial cells in the vitelline and umbilical arteries and ventral aspect of the dorsal aorta, mesenchyme ventral to dorsal aorta	Germline	No primitive macrophages, EMPs, or HSCs, no endothelial to hematopoietic cell transitions, hemorrhaging		1–5
			Tek-Cre, Cdh5-Cre	Same as germline		6,7
			Mx1-Cre, Vav1-Cre	HSCs present at three- to fourfold decreased frequency, expansion of LSK cells in bone marrow, defects in lymphopoiesis and megakaryopoiesis		7–12
runx1	Zebrafish	Ventral dorsal aorta	Morpholino, germline mutation	Loss of definitive blood (*myb* and *ikaros* expression) in dorsal aorta, arterial specification normal. Overexpression of Notch intracellular domain could not rescue *myb* expression in *runx1* morphant	Runx1 required for Notch activity in HSC formation	13–16

Gata2	Mouse	Lateral plate mesoderm, all endothelial cells	Germline	Decreased number of primitive erythrocytes, endothelial cells normal. Embryonic lethality prior to aortic hematopoiesis		17, 18
			Gata2 endothelial enhancer deletion or *Cdh5-Cre*-mediated deletion	Severely decreased EMPs, loss of HSCs, no endothelial to hematopoietic cell transitions, hemorrhaging. Similar to germline *Runx1* mutation	Gata2 is part of the Gata2/Tal1/Fli1 GRN; binds Runx1 +23 enhancer; loss of Gata2 decreases *Runx1* expression	19–23
gata2b	Zebrafish	Endothelial cells in dorsal aorta	Morpholino	Decreased *runx1* and *myb* expression. Phenocopies reduction of *runx1* expression. *myb* expression rescued by restoring *runx1* in morphant	*gata2b* functions downstream of *notch1a and notch1b* and upstream of *runx1* in endothelium	24
Fli1	Mouse	Lateral plate mesoderm, all endothelial cells	Germline	Decreased EMPs (fetal liver), hemorrhaging. Similar to germline *Runx1* mutation	Fli1 is part of the Gata2/Tal1/Fli1 GRN, binds the Runx1 +23 enhancer	25–29

Continued

Table 1 Regulation of Runx1 Expression in the Dorsal Aorta—cont'd

Gene	Species	Sites of Expression	Mutation	Phenotype	Proposed Mechanism	References
fli1	Zebrafish	Lateral plate mesoderm, endothelial cells	Morpholino, constitutively active allele	Morphant lost blood and endothelial programs; *runx1* and *gata2* expression reduced. Constitutively active Fli1 caused increased and ectopic expression of *gata2*	Fli1 is an upstream regulator of *gata2* and *runx1*	30
BMP4	Mouse	Mesenchyme ventral to dorsal aorta	No mutants; AGM explant treatment with BMP agonists and antagonists	Inhibition of BMP4 abolishes HSC activity in AGM explant cultures. Progenitors reduced threefold	BMP4 activates *Runx1* expression in ventral endothelial cells, likely through upregulating Gata2	31, 32
	Zebrafish	Mesenchyme ventral to dorsal aorta	Morpholino, inducible dominant negative allele	Loss of hematopoietic clusters, runx1 expression reduced in dorsal aorta	BMP required for the initiation and maintenance of *Runx1* expression in ventral wall of the dorsal aorta	33

Notch1	Mouse	Endothelium of dorsal aorta	Germline	Failure to generate subset of wave 2 and all wave 3 hematopoietic cells (EMPs normal), abnormal angiogenesis and arterial specification. Notch1-deficient ES cells failed to contribute to adult (wave 3) hematopoiesis		34, 35
				Transduction of *Runx1* in *Notch1*$^{-/-}$ P-Sp explants rescues erythroid/myeloid progenitor formation	Notch1 required upstream of Runx1	36
notch1a *notch1b*	Zebrafish	Endothelium	Morpholino	Loss of *runx1* and *myb* expression. Activation of NICD both early (14 hpf) and later (20 hpf) rescued *runx1* and *myb* expression. Expression of NICD in endothelial cells could rescue *notch1a* and *notch1b* morphants while expression in somites could not	*notch1a/b* required upstream of *runx1* in endothelial cells	37

Continued

Table 1 Regulation of Runx1 Expression in the Dorsal Aorta—cont'd

Gene	Species	Sites of Expression	Mutation	Phenotype	Proposed Mechanism	References
notch3	Zebrafish	Schlerotome of somites, dorsal aorta endothelium	Morpholino	Loss of *runx1* and *myb* expression, no defect in arterial specification. Activation of NICD early (14 hpf) rescued *runx1* and *myb* expression, while activation later (20 hpf) did not. Expression of NICD in somites could rescue *runx1* and *myb* expression	*notch3* is required in somites to relay a signal to hemogenic endothelial cells. This requirement is temporally distinct from the requirement for *notch1a/b* in endothelial cells	37
Rbpj	Mouse		Germline	Loss of definitive hematopoietic compartments, loss of *Runx1*, *Tal1*, and *Gata2* expression	Notch1 directly binds *Gata2* promoter	38, 39
Jag1	Mouse	Endothelial and hematopoietic cluster cells in dorsal aorta	Germline	Decreased erythroid/myeloid progenitors (AGM region), decreased *runx1* and *gata2* expression. Arterial specification normal. Transduction of Gata2 rescued $Jag1^{-/-}$ defect	Jag1 mediates Notch1 activation of *Gata2* expression in hemogenic endothelium. Notch1 requirements for arterial specification and blood formation are distinct	40

Mib1	Mouse	Endothelium and clusters	Germline	Similar to *Notch1* mutant	Mindbomb is an E3 ubiquitin ligase required in a signal-sending cell for Notch activation in a signal-receiving cell	41
mib	Zebrafish	Ubiquitous	Germline	Absence of *runx1* and *myb* expression. Expression of NICD could rescue *runx1* and *myb* expression	Phenotype in blood formation due to loss of Notch signaling	14, 42, 43
fgfr1, *fgfr2*, *fgfr3*	Zebrafish	*fgfr2* in pronephric ducts, hypochord, and neural tube	Inducible dominant negative *fgfr1* (FGF receptor)	Dominant negative *fgfr1* (potentially inhibits signaling from FGFR1, FGFR2, and FGFR3) increased *bmp4* expression and extended *runx1* expression to more dorsal endothelial cells. Expression of the BMP antagonists *gremlin1a* and *noggin2* expression increased. Active *fgfr1* had the opposite effect	FGF signaling represses *bmp4* expression and also represses BMP4 activity by inducing expression of the BMP antagonists *noggin2* and *gremlin1a*	44

Continued

Table 1 Regulation of Runx1 Expression in the Dorsal Aorta—cont'd

Gene	Species	Sites of Expression	Mutation	Phenotype	Proposed Mechanism	References
		fgfr3 in somites and throughout trunk	Inducible constitutively activated *fgfr1*	*runx1* expression in embryos expressing dominant negative fgfr1 was rescued by a constitutively active *bmpr1* transgene		
fgf10a	Zebrafish	Throughout trunk	Morpholino	Overexpression of *fgf10a* decreased *runx1* expression		44

Abbreviations: *HSC*, hematopoietic stem cell; *EMPs*, yolk sac-derived erythroid–myeloid progenitors; *BMP*, bone morphogenetic protein; *AGM*, aorta gonad mesonephros region; *FGF*, fibroblast growth factor; *FGFr*, FGF receptor; *hpf*, hours postfertilization; *NICD*, Notch intracellular domain; *P-Sp*, para-aortic splanchnopleura; *LSK*, lineage negative Sca-1$^+$ Kit$^+$; *GRN*, gene regulatory network.

1. Okuda et al. (1996); 2. Wang, Stacy, Binder, et al. (1996); 3. Cai et al. (2000); 4. North et al. (1999); 5. Yokomizo et al. (2001); 6. Li et al. (2006); 7. Chen et al. (2009); 8. Ichikawa et al. (2004); 9. Growney et al. (2005); 10. Putz, Rosner, Nuesslein, Schmitz, and Buchholz (2006); 11. Cai et al. (2011); 12. Cai et al. (2015); 13. Gering and Patient (2005); 14. Burns, Traver, Mayhall, Shepard, and Zon (2005); 15. Sood et al. (2010); 16. Kalev-Zhylinska et al. (2002); 17. Tsai et al. (1994); 18. Minegishi et al. (1999); 19. Khandekar et al. (2007); 20. de Pater et al. (2013); 21. Gao et al. (2013); 22. Johnson et al. (2012); 23. Lim et al. (2012); 24. Butko et al. (2015); 25. Spyropoulos et al. (2000); 26. Hart et al. (2000); 27. Brown et al. (2000); 28. Pimanda, Ottersbach, et al. (2007); 29. Nottingham et al. (2007); 30. Liu, Walmsley, Rodaway, and Patient (2008); 31. Durand et al. (2007); 32. Oren, Torregroza, and Evans (2005); 33. Wilkinson et al. (2009); 34. Kumano et al. (2003); 35. Hadland et al. (2004); 36. Nakagawa et al. (2006); 37. Kim, Melick, et al. (2014); 38. Robert-Moreno, Espinosa, de la Pompa, and Bigas (2005); 39. Yokoyama et al. (2009); 40. Robert-Moreno et al. (2008); 41. Yoon et al. (2008); 42. Bertrand, Cisson, Stachura, and Traver (2010); 43. Zhang, Li, Lim, Qiu, and Jiang (2007); 44. Pouget et al. (2014); 45. Bigas, Guiu, and Gama-Norton (2013).

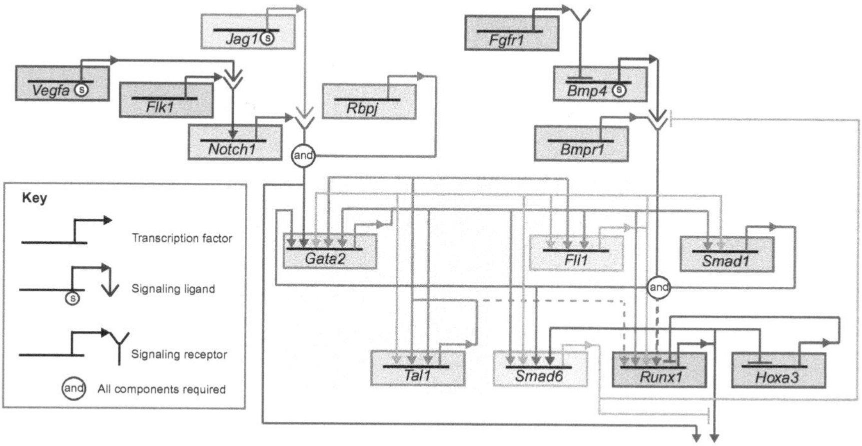

Fig. 2 The gene regulatory network in the hemogenic endothelium. References for each of the interactions illustrated are indicated in the main text. The key for the GRN symbols is indicated in the *box*. Shaded in pale green is the interconnected GRN kernel consisting of *Gata1*, *Fli1*, and *Tal1* (Pimanda, Ottersbach, et al., 2007). *Dotted lines* indicate interactions assumed but not yet definitively demonstrated. (See the color plate.)

is driven by an enhancer located 9.5 kb downstream of the promoter (Gao et al., 2013; Johnson et al., 2012). Mutation of the *Gata2* endothelial enhancer resulted in decreased endothelial *Gata2* and *Runx1* expression, and the absence of hematopoietic clusters (Gao et al., 2013). The phenotype of embryos lacking the endothelial cell-specific *Gata2* enhancer was strikingly similar to that of Runx1 deficiency, in that they had relatively normal primitive erythropoiesis, severely impaired definitive hematopoiesis, a lack of hematopoietic clusters and HSCs, and had hemorrhaged into the neural tube, suggesting that loss of Runx1 expression or function in hemogenic endothelium may be a major downstream consequence of losing *Gata2* expression in endothelium (de Pater et al., 2013; Johnson et al., 2012; Lim et al., 2012; Okuda et al., 1996; Wang, Stacy, Binder, et al., 1996; Table 1). Recent experiments in zebrafish support the hypothesis that a major function of endothelial Gata2 is to upregulate *Runx1* expression. Zebrafish has two *gata2* paralogues, *gata2a* and *gata2b*; *gata2a* is expressed in lateral plate mesoderm and in all endothelial cells, while *gata2b* is specifically expressed in hemogenic endothelium (Butko et al., 2015). Morpholino knockdown of *gata2b* decreased *runx1* expression in the hemogenic endothelium and also decreased expression of the *runx1* downstream target *myb*. Importantly, ectopic expression of *runx1* could partially restore *myb* expression in the hemogenic endothelium in *gata2b* morphants, indicating that *gata2b* is required upstream of *runx1* in hemogenic endothelium, and that *runx1* is an important downstream target of *gata2b* (Butko et al., 2015).

Fli1 is one of the earliest blood and endothelial cell markers expressed in the dorsal lateral plate mesoderm, where it functions upstream of *Tal1* and *Gata2* (Liu et al., 2008; Walmsley, Ciau-Uitz, & Patient, 2002). *Fli1* is also expressed in endothelial and hematopoietic cluster cells, thus is present in cells in which Runx1 is expressed (Brown et al., 2000; Hart et al., 2000; Pimanda, Donaldson, et al., 2007). Mutation of *Fli1* in the mouse caused hemorrhaging in the brain with blood spilling into the neural tube, almost identical to the hemorrhagic phenotypes caused by Runx1 deficiency, and by the ablation of *Gata2* expression in endothelial cells (Gao et al., 2013; Hart et al., 2000; Johnson et al., 2012; Okuda et al., 1996; Spyropoulos et al., 2000; Wang, Stacy, Binder, et al., 1996; Table 1). *Fli1*-deficient embryos had many fewer erythroid/myeloid progenitors in the fetal liver, although progenitors were not completely absent as was observed with *Runx1* deficiency (Okuda et al., 1996; Spyropoulos et al., 2000; Wang, Stacy, Binder, et al., 1996).

The striking hemorrhaging phenotype shared by the *Runx1*, *Fli1*, and endothelial-specific *Gata2* mutations is not well understood. One study that addressed this phenomenon in Runx1-deficient embryos identified a defect in angiogenesis in explant cultures of AGM regions that could be rescued by the addition of adult bone marrow cells or by soluble angiopoietin 1 (Takakura et al., 2000). The authors hypothesized that in Runx1-deficient embryos, a lack of HSCs (or in hindsight more likely a more primitive population of wave 1 or wave 2 blood cells) that would normally secrete angiopoietin in the vicinity of the capillaries was responsible for the defects in angiogenic sprouting and hemorrhaging. More recent studies have revealed important roles for wave 1 or wave 2 blood platelets in development of the lymphatic vasculature (Bertozzi et al., 2010; Finney et al., 2012; Fu et al., 2008; Suzuki-Inoue et al., 2010), thus the lack of a common critical blood cell population may explain the shared hemorrhaging phenotype of *Runx1*, *Fli1*, and endothelial cell-specific *Gata2* mutant fetuses.

The role of Tal1 in regulating *Runx1* expression in endothelium is less clear, complicated by the functional redundancy of E-BOX binding proteins. Germline disruption of *Tal1* has profound effects on hematopoiesis; embryos die by E9.5 with no blood cells, including primitive erythrocytes (Porcher et al., 1996; Robb et al., 1995; Visvader, Fujiwara, & Orkin, 1998; Table 1). Tal1-deficient ES cells generate hemangioblasts, but these cells fail to differentiate into the Flk1^{+} Kit^{+} Tie2^{+} CD41^{-} early hemogenic endothelial cells, the stage at which Runx1 is required (Lancrin et al., 2009). Thus Tal1 deficiency blocks hematopoietic development at least one step earlier than Runx1 deficiency (Lancrin et al., 2009; Mikkola, Fujiwara, Schlaeger, Traver, & Orkin, 2003; Robertson, Kennedy, Shannon, & Keller, 2000). However, deletion of Tal1 in endothelial cells with Tie2-Cre had very modest effects on the number of erythroid/myeloid progenitors in the E12.5 fetal liver or on fetal liver HSCs, which suggests that Tal1 activity in endothelium is not important for *Runx1* expression or activity (Schlaeger, Mikkola, Gekas, Helgadottir, & Orkin, 2005). The closely related Lyl1 protein is expressed in the endothelium of the embryo proper, and likely compensates for Tal1 deficiency in endothelium (Curtis, Salmon, & Pimanda, 2012; Giroux et al., 2007).

As *Gata2*, *Fli1*, and *Lyl1* are expressed throughout the embryonic endothelium, they cannot be solely responsible for Runx1 expression, which is restricted to only a small number of hemogenic endothelial cells. Other transcription factors and signaling pathways must function to restrict Runx1 expression to the hemogenic endothelial cells that will ultimately form blood. One of these transcription factors is the homeobox protein, HoxA3. HoxA3 is a proendothelial factor that suppresses hematopoietic fate (Iacovino et al., 2011). HoxA3 and Runx1 are expressed in a mutually exclusive pattern in murine embryos; at E8.5 Runx1 is expressed in the yolk sac and endothelial cells of the vitelline artery, while HoxA3 is expressed in endothelial cells in the dorsal aortae. This relationship flips at E9.5-E10.5, at which time endothelial cells in the dorsal aorta acquire Runx1 and lose HoxA3 expression. Forced expression of HoxA3 in ES cell-derived blast colonies reduced the expression of hematopoietic markers and the number of hematopoietic progenitors, while transduction of Runx1 into HoxA3-induced cells rescued the expression of hematopoietic genes. HoxA3 occupied sites near P2 in the *Runx1* gene, demonstrating that *Runx1* is a direct HoxA3 target. Thus, through mutual inhibition, HoxA3 and Runx1 influence the choice between following a hematopoietic or endothelial program (Iacovino et al., 2011; Fig. 2).

8. SIGNALING PATHWAYS REGULATING RUNX1 EXPRESSION IN THE AORTA

8.1 Notch Signaling

Runx1 expression in, and blood cell formation from endothelium is most well understood for the dorsal aorta, a structure that is conserved in all vertebrate organisms. Runx1 expression in the dorsal aorta is primarily localized to its ventral aspect, by signals originating from the adjacent tissues. The signals involved in patterning the dorsal/ventral, anterior/posterior, and lateral/medial axes in the embryo, and that instruct formation of the dorsal aorta and blood have been extensively reviewed elsewhere (Ciau-Uitz, Monteiro, Kirmizitas, & Patient, 2014; Clements & Traver, 2013; Kaimakis, Crisan, & Dzierzak, 2013; Kim, Stachura, & Traver, 2014; Marks-Bluth & Pimanda, 2012; Mirshekar-Syahkal, Fitch, & Ottersbach, 2014; Zhang, Patient, & Liu, 2013). Some of the major signaling pathways that are most proximal to Runx1 and that directly or indirectly regulate Runx1 expression and activity in the aorta are briefly described here.

The landscape of tissues surrounding the dorsal aorta, from the perspective of a transverse section through an E8.0 mouse embryo is illustrated in Fig. 3. Ectoderm is the dorsal most tissue layer; ventral to ectoderm is the

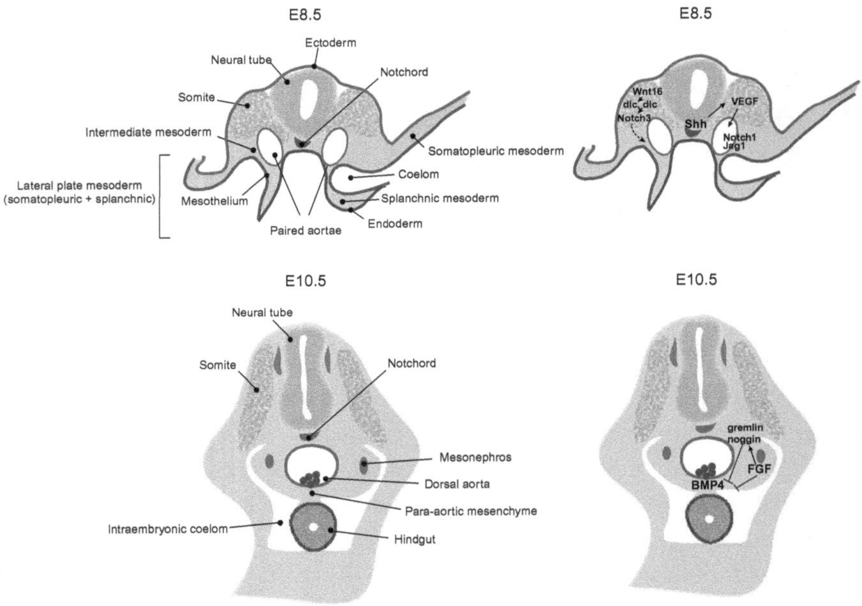

Fig. 3 Signaling at the level of the P-Sp/AGM region. On top are schematic diagrams of a transverse section through an E8.5 mouse embryo with the tissues mentioned in the text indicated with labels (*left*), and the relevant signaling pathways involved in migration of the paired aortae to the midline and activation of Notch signaling in aortic endothelial cells on the *right*. Also shown is the Notch3 pathway downstream of Wnt16, which is a separate, earlier requirement for Notch signaling (Clements et al., 2011). Below are schematic diagrams of E10.5 embryos to illustrate the location of the ventral para-aortic mesenchyme (*left*), which is a source of BMP4 (*right*). In zebrafish BMP4 is negatively regulated by FGF signaling (Pouget et al., 2014). (See the color plate.)

neural tube, and ventral to that is the notochord. Flanking the neural tube are the somites, which are derived from paraxial mesoderm, and bordering the somites on either side is lateral plate mesoderm. Lateral plate mesoderm is separated into two layers by a cavity called the coelom; the layer of mesoderm dorsal to the coelom is referred to as the somatopleuric mesoderm, and the layer ventral to the coelom is the splanchnic mesoderm.

The dorsal aorta forms from angioblasts in the lateral plate mesoderm that, in mouse and avian embryos, first appear as parallel cords of cells flanking the midline axis. These cords will hollow out to form paired lumenized aortic vessels, which then migrate to the midline and fuse into a single vessel (Sato, 2013). Zebrafish is somewhat different as individual angioblasts rather than lumenized vessels migrate to the midline then coalesce into a single cord structure that subsequently forms a lumen. Runx1 expression in mouse and chicken embryos initiates in endothelial cells in the ventral aspects of the paired dorsal aortae (North et al., 1999; Nottingham et al., 2007; Richard et al., 2013), however blood cell formation occurs only after the paired aortae have fused into a single vessel.

Specification of arterial angioblasts and their migration toward the midline requires hedgehog and VEGF signaling. Sonic hedgehog (Shh), which is secreted from the floorplate of the neural tube, the notochord, and hypochord of zebrafish embryos, and from the neural tube, notochord, and endoderm in frog, chick, and mouse embryos, induces the expression and activity of VEGF in the somites (Byrd et al., 2002; Coultas et al., 2010; Lawson et al., 2001; Lawson, Vogel, & Weinstein, 2002; Vokes & Krieg, 2002; Vokes et al., 2004). VEGF, in turn, drives the proliferation and migration of endothelial cells or their angioblast precursors toward the midline through its activation of the VEGF receptor (encoded by *Flk1* in mice and *kdrl* in zebrafish) on angioblasts and endothelial cells. VEGF also activates the expression of Notch signaling pathway components in arterial endothelial cells, and the Notch pathway is a key driver of both arterial fate and of blood cell formation.

Notch receptors and ligands are transmembrane proteins that signal laterally between neighboring cells by activating Notch signaling in the signal-receiving cell, and suppressing it in the signal-producing cell (Kopan & Ilagan, 2009; Lai, 2004). Activation of the Notch receptor results in its cleavage by two proteases, the ADAM TACE proteases and γ-secretase, which release the Notch intracellular domain (NICD) from its extracellular and transmembrane domains. NICD translocates to the nucleus where it forms a complex with the transcription factor CBF1, Suppressor of Hairless, Lag-1 (CSL, otherwise known as RBPJκ) and a coregulator named mastermind, and the assembled complex activates transcription. Mice have four Notch receptors (Notch 1–4) and five ligands, including three Delta ligands and two Jagged ligands (Dll1, Dll3, Dll4, and Jag1, Jag2). Zebrafish also have four Notch receptors, Notch1a, Notch1b, Notch2, and Notch3. *notch1a* and *notch1b* resulted from the duplication of the *Notch1* gene, and together they replicate the expression pattern of mammalian *Notch1* (Kim, Melick, et al., 2014). Zebrafish has five Delta ligands (Dla, Dlb, Dlc, Dld, Dll4) and three Jagged ligands (Jagged1a, Jagged1b, Jagged2).

Several Notch receptors and ligands are required for arterial specification in mice, including Notch1, Notch4, and Dll4 (Bigas et al., 2013; Duarte et al., 2004; Krebs et al., 2004; Lawson et al., 2001; Quillien et al., 2014). A partially overlapping set of receptors and ligands, Notch1 and Jag1, are required for hematopoietic cell formation from arterial endothelium (Hadland et al., 2004; Kumano et al., 2003; Nakagawa et al., 2006; Robert-Moreno et al., 2005, 2008). The requirement for Notch signaling in hematopoietic cell formation can be separated genetically from the requirement for arterial specification. For example, the Notch ligand Jag1 is required for hematopoietic cell formation but not for arterial specification (Robert-Moreno et al., 2008).

In zebrafish, two temporally distinct Notch signaling events are required for blood cell formation from aortic endothelium (Kim, Melick, et al., 2014). The first Notch signaling event originates in the somites, where Wnt16 activates the expression of two delta-like ligands, *dlc* and *dld*, which in turn activate signaling from the Notch3 receptor (Clements et al., 2011). Activation of Notch3 in the somites delivers a necessary relay signal to endothelial cells in the dorsal aorta by a mechanism that is not yet understood (Kim, Melick, et al., 2014). This Notch3-mediated signaling event is not required for arterial identity, but it is required for blood cell formation from arterial endothelium. The requirement for Notch3 signaling in somites is followed by a later requirement for Notch1 signaling in aortic endothelial cells, which in zebrafish is mediated through *notch1a* and *notch1b* (Kim, Melick, et al., 2014).

The requirement for Notch signaling is restricted to hematopoietic cell development from arteries, as yolk sac hematopoiesis (both wave 1 and wave 2) is intact in both *Notch1* mutants and mutants for the Notch-effector E3-ubiquitin ligase *mindbomb* (Bertrand, Cisson, et al., 2010; Burns et al., 2005; Hadland et al., 2004; Kumano et al., 2003; Yoon et al., 2008). Runx1 expression is lost in the arterial endothelium of both Notch1 and Notch-effector mutants (Burns et al., 2005; Gering & Patient, 2005; Robert-Moreno et al., 2008), and forced expression of the NICD in zebrafish induced ectopic expression of *runx1* and *c-myb* in the dorsal aorta, indicating that Notch signaling is situated upstream of *runx1* (Burns et al., 2005). Runx1 could also rescue the formation of erythroid and myeloid lineage cells in Notch1 mutants, as retroviral delivery of Runx1 in *Notch1*-null p-Sp rescued the formation of $CD45^+$ hematopoietic cells, lineage positive cells, and colony-forming units (whether it rescued lymphopoiesis was not examined) (Nakagawa et al., 2006). However there is currently no evidence that *Runx1* is a direct Notch target. The +23 hematopoietic enhancer in murine *Runx1* lacks binding sites for CSL/RBPJκ, the downstream transcriptional effector of Notch signaling (Nottingham et al., 2007). Chromatin immunoprecipitation experiments in mammalian hematopoietic cell lines expressing activated Notch1 determined that Notch1 bound the *Gata2*, but not the *Runx1* promoter, suggesting that Notch signaling may regulate Runx1 expression indirectly through Gata2 (Robert-Moreno et al., 2005). This is supported by genetic data in zebrafish demonstrating that Notch1 signaling was required for *gata2b* expression, and *gata2b* is required for *runx1* expression (Butko et al., 2015).

In more differentiated blood cells Notch and Runx proteins appear to function in parallel. ChIP-seq for Notch1 and CSL/RBPJκ in T-lymphoblastic leukemia cells revealed that Notch1 and CSL/RBPJκ were commonly bound near RUNX consensus sites in distal enhancers, suggesting functional cooperativity (Wang et al., 2014). In *Drosophila* both Notch and the Runx homologue Lozenge are expressed in and required for the formation of a lineage of blood cells called crystal cells (Lebestky, Chang, Hartenstein, & Banerjee, 2000; Lebestky, Jung, & Banerjee, 2003). ChIP-Seq for Notch in a precursor of crystal cells revealed that Notch-occupied sites were frequently adjacent to RUNX consensus sites, and functional analyses indicated that together Notch and Lozenge regulated their common targets (Terriente-Felix et al., 2013). Notch did not, however, occupy the *Lozenge* locus. In mice, Notch1 and Runx1 act in parallel to specify the T cell lineage, as both are required for T cell specification, but not for the expression or activity of each other (Guo, Maillard, Chakraborti, Rothenberg, & Speck, 2008). In zebrafish embryos, expression of the NICD could not rescue *myb* expression in the dorsal aorta of a *runx1* morphant (Burns et al., 2005), indicating that Notch signaling and *runx1* are required in parallel for blood cell formation.

8.2 Fibroblast Growth Factor and Bone Morphogenic Protein Signaling

At E8.5 in the mouse, the paired aortae are sandwiched beneath the somites and somatopleuric mesoderm on the dorsal side, and endoderm on the ventral side (Fig. 3). After the paired aortae fuse into a single vessel at the midline, the ventral aspect of the fused aorta is adjacent to mesenchyme. This subaortic mesenchyme is derived from splanchnic mesoderm that migrates under the aortae toward the midline from more lateral positions, similar to the migration of the paired aortae (Richard et al., 2013; Wasteson et al., 2008; Wiegreffe, Christ, Huang, & Scaal, 2009). The presence of subaortic mesenchyme is essential for hematopoiesis, as physically blocking its migration will ablate Runx1 expression in the endothelium and the endothelial to hematopoietic cell transition (Richard et al., 2013).

The dorsal aorta, Runx1 expression, and the distribution of hematopoietic clusters are polarized along the dorsal/ventral axis. In avian embryos hematopoietic clusters and Runx1 expression are localized to the ventral aspect of the dorsal aorta (Jaffredo et al., 1998; Marshall et al., 1999; Tavian et al., 1996). In the mouse embryo, hematopoietic clusters are found in both the ventral and dorsal aspects of the dorsal aorta, but careful dissection and transplantation experiments showed that HSCs are confined to the ventral half (Taoudi & Medvinsky, 2007; Yokomizo & Dzierzak, 2010). The polarization of the aorta in the dorsal–ventral axis is mediated by growth factors produced by the surrounding tissues, and at least one of these growth factors, bone morphogenic protein 4 (BMP4), is thought to regulate *Runx1*. BMP4 is a ventralizing factor and a major signaling pathway required for the formation of HSCs (Pimanda, Donaldson, et al., 2007; Wilkinson et al., 2009), and like Runx1 is expressed in a ventral to dorsal gradient in the mesenchyme surrounding the dorsal aorta (Durand et al., 2007; Pimanda, Donaldson, et al., 2007; Suonpaa et al., 2005; Wilkinson et al., 2009). Inhibition of BMP4 either prior to or during the temporal window that Runx1 is expressed did not affect arterial specification or the integrity of the endothelium, but dramatically downregulated Runx1 expression, indicating that BMP4 in the subaortic mesenchyme is required for both the initiation and maintenance of Runx1 expression in the endothelium (Wilkinson et al., 2009).

Smad1, which encodes a transcription factor effector of BMP4 signaling, is also expressed in the dorsal aorta in a ventral to dorsal gradient, similar to the expression pattern of *Runx1*, and like *Runx1* is activated by

Gata2 and Fli1 (Marks-Bluth et al., 2015). Smad1 binds to a consensus site in the distal *Runx1* promoter (P1), and therefore can activate *Runx1* transcription directly (Pimanda, Donaldson, et al., 2007; Fig. 2). Smad1 also activates the expression of *Gata2*, thus can indirectly activate Runx1 expression through the *Runx1* +23 enhancer (Oren et al., 2005). Indirect regulation through *Gata2* is the more likely explanation for the activation of *Runx1* in hemogenic endothelium, as there is no apparent binding site for Smad1 in the *Runx1* +23 enhancer, and the P1 promoter, although directly bound by Smad1, is not active in hemogenic endothelium (Sroczynska et al., 2009).

Multiple molecules negatively regulate BMP signaling, and thereby determine the location and level of *Runx1* expression (Knezevic et al., 2011; Pimanda, Donaldson, et al., 2007; Pouget et al., 2014). The Smad proteins fall into three categories: the receptor-regulated Smads (eg, Smad1), the common-partner Smads, and the inhibitory Smads (eg, Smad6). BMP signaling phosphorylates and activates Smad1 and the expression of Runx1. Smad1 and Runx1 in turn occupy and activate transcription of the *Smad6* gene. Smad6 downregulates Runx1 levels by directly inhibiting BMP signaling, and additionally by facilitating Runx1 protein degradation. These negative feedback loops act as a rheostat to fine tune BMP signaling and Runx1 levels (Ishida et al., 2000; Knezevic et al., 2011; Pimanda, Donaldson, et al., 2007; Fig. 2). Fibroblast growth factor (FGF) also negatively regulates Runx1 expression in the dorsal aorta through the modulation of BMP signaling (Pouget et al., 2014; Fig. 2). Expression of a dominant negative form of the FGF receptor 1 (*fgfr1*) just prior to when *runx1* is normally expressed in the dorsal aorta in zebrafish increased both endogenous *bpm4* and *runx1* expression, while expression of a constitutively active *fgfr1* decreased *bmp4* and *runx1* expression. Bmp4 is genetically downstream of FGF signaling and upstream of *runx1*, as an activated form of the BMP receptor (*bmpr1b*) could rescue *runx1* expression in embryos expressing constitutively active *fgfr1*, while a dominant negative form of *bmpr1b* repressed *runx1* expression when FGF signaling was repressed. In addition to affecting *bmp4* expression, FGF signaling also modulates BMP signaling activity by activating the expression of two negative regulators of BMP signaling in the somites, *noggin2* and *gremlin1a*. Thus FGF signaling determines both the amount and location of BMP signaling in the dorsal aorta by directly affecting *bpm4* expression and by activating negative regulators of BMP signaling in the more dorsal aspects of the aorta (Pouget et al., 2014).

9. CONCLUSIONS AND FUTURE DIRECTIONS

Morphological, functional, and genetic studies in multiple experimental organisms have provided detailed information about the formation of blood in the embryo. However our current state of knowledge is akin to a partially assembled jigsaw puzzle, with parts of the puzzle well assembled, the overall picture discernable, but large gaps remaining. One such gap is in our understanding of the precise mechanism by which Runx1 expression and blood cell formation is localized to specific embryonic sites. Blood cells form in several different anatomic locations. This review has focused on blood cell formation in the dorsal aorta, as it is a highly evolutionarily conserved structure. But amniotes have additional blood-producing structures, for which the powerful genetic studies in zebrafish can provide no insights. What local signals are responsible for Runx1 expression in the other arteries in the mouse embryo, such as the vitelline and umbilical arteries? And the placenta, which was not discussed in this chapter, harbors many HSCs and autonomously produces erythroid/myeloid and lymphoid progenitors (Corbel, Salaun, Belo-Diabangouaya, & Dieterlen-Lievre, 2007; Gekas et al., 2005; Ottersbach & Dzierzak, 2005; Rhodes et al., 2008; Zeigler et al., 2006). What is the local signaling environment of the placenta that activates Runx1 expression and promotes blood cell formation in that tissue?

Epigenetic regulation by Runx1 has been studied in embryonic stem cell cultures, which approximate yolk sac hematopoiesis. Will the principles learned there also apply to Runx1 activity in the major arteries, and will direct Runx1 targets in the major arteries be identical to those in the yolk sac endothelium? Notch signaling is required for blood cell formation from the dorsal aorta, but not from the yolk sac. Why is that the case, and how does Runx1 integrate with Notch signaling to produce blood from arterial endothelium?

The endothelial to hematopoietic cell transition is a fascinating process to behold. However almost nothing is known about the cell biological mechanisms responsible for converting a flat endothelial cell tightly integrated in a monolayer into a round, nonadherent hematopoietic cell. How does this occur, and how similar or different is that process from epithelial to mesenchymal transitions, or from the vascular sprouting that occurs during angiogenesis? And, what exactly does Runx1 contribute to this process?

Inherited and in utero somatically acquired mutations in the human *RUNX1* gene greatly increase the risk of leukemia (Ford et al., 1998;

Song et al., 1999; Wiemels, Cazzaniga, et al., 1999; Wiemels, Ford, van Wering, Postma, & Greaves, 1999). For example, the most frequent chromosomal translocation in childhood acute lymphocytic leukemia is the t (12;21), which fuses the *RUNX1* and *ETV6* genes, and it occurs in utero. Is there some contribution of the embryonic environment that makes this translocation especially potent in this setting?

Finally, despite much effort, the production of HSCs capable of robust, long-term multilineage engraftment from embryonic stem cells, or induced pluripotent stem cells has thus far failed (Vo & Daley, 2015). Blood cell formation in ES cell cultures also proceeds through a hemogenic endothelial intermediate (Eilken et al., 2009; Lancrin et al., 2009). Can we use our knowledge of how Runx1 expression and activity is regulated to improve this process? Or, will the approach taken by Sandler et al. (2014) introducing Runx1 alone, or in combination with other transcription factors into human endothelial cells produce HSCs capable of self-renewal and of contributing to all blood cell lineages? In the next decade we should have the answers to most if not all of these questions.

REFERENCES

Aughey, G. N., & Southall, T. D. (2015). Dam it's good! DamID profiling of protein-DNA interactions. *Wiley Interdisciplinary Reviews. Developmental Biology*, *5*, 25–37.

Bee, T., Ashley, E. L., Bickley, S. R., Jarratt, A., Li, P. S., Sloane-Stanley, J., et al. (2009). The mouse Runx1 +23 hematopoietic stem cell enhancer confers hematopoietic specificity to both Runx1 promoters. *Blood*, *113*, 5121–5124.

Bee, T., Liddiard, K., Swiers, G., Bickley, S. R., Vink, C. S., Jarratt, A., et al. (2009). Alternative Runx1 promoter usage in mouse developmental hematopoiesis. *Blood Cells, Molecules & Diseases*, *43*, 35–42.

Bee, T., Swiers, G., Muroi, S., Pozner, A., Nottingham, W., Santos, A. C., et al. (2010). Nonredundant roles for Runx1 alternative promoters reflect their activity at discrete stages of developmental hematopoiesis. *Blood*, *115*, 3042–3050.

Bertozzi, C. C., Schmaier, A. A., Mericko, P., Hess, P. R., Zou, Z., Chen, M., et al. (2010). Platelets regulate lymphatic vascular development through CLEC-2-SLP-76 signaling. *Blood*, *116*, 661–670.

Bertrand, J. Y., Chi, N. C., Santoso, B., Teng, S., Stainier, D. Y., & Traver, D. (2010). Haematopoietic stem cells derive directly from aortic endothelium during development. *Nature*, *464*, 108–111.

Bertrand, J. Y., Cisson, J. L., Stachura, D. L., & Traver, D. (2010). Notch signaling distinguishes 2 waves of definitive hematopoiesis in the zebrafish embryo. *Blood*, *115*, 2777–2783.

Bertrand, J. Y., Jalil, A., Klaine, M., Jung, S., Cumano, A., & Godin, I. (2005). Three pathways to mature macrophages in the early mouse yolk sac. *Blood*, *106*, 3004–3011.

Bigas, A., Guiu, J., & Gama-Norton, L. (2013). Notch and Wnt signaling in the emergence of hematopoietic stem cells. *Blood Cells, Molecules & Diseases*, *51*, 264–270.

Boiers, C., Carrelha, J., Lutteropp, M., Luc, S., Green, J. C., Azzoni, E., et al. (2013). Lymphomyeloid contribution of an immune-restricted progenitor emerging prior to definitive hematopoietic stem cells. *Cell Stem Cell, 13*, 535–548.
Boisset, J. C., van Cappellen, W., Andrieu-Soler, C., Galjart, N., Dzierzak, E., & Robin, C. (2010). In vivo imaging of haematopoietic cells emerging from the mouse aortic endothelium. *Nature, 464*, 116–120.
Bollerot, K., Pouget, C., & Jaffredo, T. (2005). The embryonic origins of hematopoietic stem cells: A tale of hemangioblast and hemogenic endothelium. *APMIS, 113*, 790–803.
Bos, F. L., Hawkins, J. S., & Zovein, A. C. (2015). Single-cell resolution of morphological changes in hemogenic endothelium. *Development, 142*, 2719–2724.
Brown, L. A., Rodaway, A. R., Schilling, T. F., Jowett, T., Ingham, P. W., Patient, R. K., et al. (2000). Insights into early vasculogenesis revealed by expression of the ETS-domain transcription factor Fli-1 in wild-type and mutant zebrafish embryos. *Mechanisms of Development, 90*, 237–252.
Burns, C. E., Traver, D., Mayhall, E., Shepard, J. L., & Zon, L. I. (2005). Hematopoietic stem cell fate is established by the Notch-Runx pathway. *Genes & Development, 19*, 2331–2342.
Butko, E., Distel, M., Pouget, C., Weijts, B., Kobayashi, I., Ng, K., et al. (2015). Gata2b is a restricted early regulator of hemogenic endothelium in the zebrafish embryo. *Development, 142*, 1050–1061.
Byrd, N., Becker, S., Maye, P., Narasimhaiah, R., St-Jacques, B., Zhang, X., et al. (2002). Hedgehog is required for murine yolk sac angiogenesis. *Development, 129*, 361–372.
Cai, Z., de Bruijn, M. F. T. R., Ma, X., Dortland, B., Luteijn, T., Downing, J. R., et al. (2000). Haploinsufficiency of AML1/CBFA2 affects the embryonic generation of mouse hematopoietic stem cells. *Immunity, 13*, 423–431.
Cai, X., Gao, L., Teng, L., Ge, J., Oo, Z., Kumar, A. R., et al. (2015). Runx1 deficiency decreases ribosome biogenesis and confers stress resistance to hematopoietic stem and progenitor cells. *Cell Stem Cell, 17*, 165–177. http://dx.doi.org/10.1016/j.stem.2015.06.002.
Cai, X., Gaudet, J. J., Mangan, J. K., Chen, M. J., De Obaldia, M. E., Oo, Z., et al. (2011). Runx1 loss minimally impacts long-term hematopoietic stem cells. *PLoS One, 6*, e28430.
Challen, G. A., & Goodell, M. A. (2010). Runx1 isoforms show differential expression patterns during hematopoietic development but have similar functional effects in adult hematopoietic stem cells. *Experimental Hematology, 38*, 403–416.
Chen, M. J., Li, Y., De Obaldia, M. E., Yang, Q., Yzaguirre, A. D., Yamada-Inagawa, T., et al. (2011). Erythroid/myeloid progenitors and hematopoietic stem cells originate from distinct populations of endothelial cells. *Cell Stem Cell, 9*, 541–552.
Chen, M. J., Yokomizo, T., Zeigler, B. M., Dzierzak, E., & Speck, N. A. (2009). Runx1 is required for the endothelial to haematopoietic cell transition but not thereafter. *Nature, 457*, 889–891.
Choi, K., Kennedy, M., Kazarov, A., Papadimitriou, J. C., & Keller, G. (1998). A common precursor for hematopoietic and endothelial cells. *Development, 125*, 725–732.
Ciau-Uitz, A., Monteiro, R., Kirmizitas, A., & Patient, R. (2014). Developmental hematopoiesis: Ontogeny, genetic programming and conservation. *Experimental Hematology, 42*, 669–683.
Clements, W. K., Kim, A. D., Ong, K. G., Moore, J. C., Lawson, N. D., & Traver, D. (2011). A somitic Wnt16/Notch pathway specifies haematopoietic stem cells. *Nature, 474*, 220–224.
Clements, W. K., & Traver, D. (2013). Signalling pathways that control vertebrate haematopoietic stem cell specification. *Nature Reviews. Immunology, 13*, 336–348.
Collins, A., Hewitt, S. L., Chaumeil, J., Sellars, M., Micsinai, M., Allinne, J., et al. (2011). RUNX transcription factor-mediated association of Cd4 and Cd8 enables coordinate gene regulation. *Immunity, 34*, 303–314.

Corbel, C., Salaun, J., Belo-Diabangouaya, P., & Dieterlen-Lievre, F. (2007). Hematopoietic potential of the pre-fusion allantois. *Developmental Biology, 301*, 478–488.
Coultas, L., Nieuwenhuis, E., Anderson, G. A., Cabezas, J., Nagy, A., Henkelman, R. M., et al. (2010). Hedgehog regulates distinct vascular patterning events through VEGF-dependent and -independent mechanisms. *Blood, 116*, 653–660.
Curtis, D. J., Salmon, J. M., & Pimanda, J. E. (2012). Concise review: Blood relatives: Formation and regulation of hematopoietic stem cells by the basic helix-loop-helix transcription factors stem cell leukemia and lymphoblastic leukemia-derived sequence 1. *Stem Cells, 30*, 1053–1058.
Daane, J. M., & Downs, K. M. (2011). Hedgehog signaling in the posterior region of the mouse gastrula suggests manifold roles in the fetal-umbilical connection and posterior morphogenesis. *Developmental Dynamics, 240*, 2175–2193.
Davidson, E. H., Rast, J. P., Oliveri, P., Ransick, A., Calestani, C., Yuh, C. H., et al. (2002). A provisional regulatory gene network for specification of endomesoderm in the sea urchin embryo. *Developmental Biology, 246*, 162–190.
De Braekeleer, E., Douet-Guilbert, N., Morel, F., Le Bris, M. J., Ferec, C., & De Braekeleer, M. (2011). RUNX1 translocations and fusion genes in malignant hemopathies. *Future Oncology, 7*, 77–91.
de Bruijn, M. F., Speck, N. A., Peeters, M. C., & Dzierzak, E. (2000). Definitive hematopoietic stem cells first develop within the major arterial regions of the mouse embryo. *The EMBO Journal, 19*, 2465–2474.
de Pater, E., Kaimakis, P., Vink, C. S., Yokomizo, T., Yamada-Inagawa, T., van der Linden, R., et al. (2013). Gata2 is required for HSC generation and survival. *The Journal of Experimental Medicine, 210*, 2843–2850.
Duarte, A., Hirashima, M., Benedito, R., Trindade, A., Diniz, P., Bekman, E., et al. (2004). Dosage-sensitive requirement for mouse Dll4 in artery development. *Genes & Development, 18*, 2474–2478.
Duployez, N., Willekens, C., Marceau-Renaut, A., Boudry-Labis, E., & Preudhomme, C. (2015). Prognosis and monitoring of core-binding factor acute myeloid leukemia: Current and emerging factors. *Expert Review of Hematology, 8*, 43–56.
Durand, C., Robin, C., Bollerot, K., Baron, M. H., Ottersbach, K., & Dzierzak, E. (2007). Embryonic stromal clones reveal developmental regulators of definitive hematopoietic stem cells. *Proceedings of the National Academy of Sciences of the United States of America, 104*, 20838–20843.
Eilken, H. M., Nishikawa, S., & Schroeder, T. (2009). Continuous single-cell imaging of blood generation from haemogenic endothelium. *Nature, 457*, 896–900.
Ema, M., Yokomizo, T., Wakamatsu, A., Terunuma, T., Yamamoto, M., & Takahashi, S. (2006). Primitive erythropoiesis from mesodermal precursors expressing VE-cadherin, PECAM-1, Tie2, endoglin, and CD34 in the mouse embryo. *Blood, 108*, 4018–4024.
Emmel, V. E. (1916). The cell clusters in the dorsal aorta of mammalian embryos. *The American Journal of Anatomy, 19*, 401–421.
Engel, M. E., & Hiebert, S. W. (2010). Proleukemic RUNX1 and CBFbeta mutations in the pathogenesis of acute leukemia. *Cancer Treatment and Research, 145*, 127–147.
Ferrell, P. I., Xi, J., Ma, C., Adlakha, M., & Kaufman, D. S. (2015). The RUNX1 +24 enhancer and P1 promoter identify a unique subpopulation of hematopoietic progenitor cells derived from human pluripotent stem cells. *Stem Cells, 33*, 1130–1141.
Finney, B. A., Schweighoffer, E., Navarro-Nunez, L., Benezech, C., Barone, F., Hughes, C. E., et al. (2012). CLEC-2 and Syk in the megakaryocytic/platelet lineage are essential for development. *Blood, 119*, 1747–1756.
Ford, A. M., Bennett, C. A., Price, C. M., Bruin, M. C., van Wering, E. R., & Greaves, M. (1998). Fetal origins of the TEL-AML1 fusion gene in identical twins with leukemia. *Proceedings of the National Academy of Sciences of the United States of America, 95*, 4584–4588.

Fu, J., Gerhardt, H., McDaniel, J. M., Xia, B., Liu, X., Ivanciu, L., et al. (2008). Endothelial cell O-glycan deficiency causes blood/lymphatic misconnections and consequent fatty liver disease in mice. *The Journal of Clinical Investigation*, *118*, 3725–3737.

Fujita, Y., Nishimura, M., Taniwaki, M., Abe, T., & Okuda, T. (2001). Identification of an alternatively spliced form of the mouse AML1/RUNX1 gene transcript AML1c and its expression in early hematopoietic development. *Biochemical and Biophysical Research Communications*, *281*, 1248–1255.

Gao, X., Johnson, K. D., Chang, Y. I., Boyer, M. E., Dewey, C. N., Zhang, J., et al. (2013). Gata2 cis-element is required for hematopoietic stem cell generation in the mammalian embryo. *The Journal of Experimental Medicine*, *210*, 2833–2842.

Garcia-Porrero, J. A., Godin, I. E., & Dieterlen-Lièvre, F. (1995). Potential intraembryonic hemogenic sites at pre-liver stages in the mouse. *Anatomy and Embryology*, *192*, 425–435.

Gekas, C., Dieterlen-Lievre, F., Orkin, S. H., & Mikkola, H. K. (2005). The placenta is a niche for hematopoietic stem cells. *Developmental Cell*, *8*, 365–375.

Gering, M., & Patient, R. (2005). Hedgehog signaling is required for adult blood stem cell formation in zebrafish embryos. *Developmental Cell*, *8*, 389–400.

Ghozi, M. C., Bernstein, Y., Negreanu, V., Levanon, D., & Groner, Y. (1996). Expression of the human acute myeloid leukemia gene *AML1* is regulated by two promoter regions. *Proceedings of the National Academy of Sciences of the United States of America*, *93*, 1935–1940.

Ginhoux, F., Greter, M., Leboeuf, M., Nandi, S., See, P., Gokhan, S., et al. (2010). Fate mapping analysis reveals that adult microglia derive from primitive macrophages. *Science*, *330*, 841–845.

Giroux, S., Kaushik, A. L., Capron, C., Jalil, A., Kelaidi, C., Sablitzky, F., et al. (2007). lyl-1 and tal-1/scl, two genes encoding closely related bHLH transcription factors, display highly overlapping expression patterns during cardiovascular and hematopoietic ontogeny. *Gene Expression Patterns*, 7, 215–226.

Godin, I. E., Garcia-Porrero, J. A., Coutinho, A., Dieterlen-Lièvre, F., & Marcos, M. A. R. (1993). Para-aortic splanchnopleura from early mouse embryos contain B1a cell progenitors. *Nature*, *364*, 67–70.

Goldie, L. C., Lucitti, J. L., Dickinson, M. E., & Hirschi, K. K. (2008). Cell signaling directing the formation and function of hemogenic endothelium during murine embryogenesis. *Blood*, *112*, 3194–3204.

Gomez Perdiguero, E., Klapproth, K., Schulz, C., Busch, K., Azzoni, E., Crozet, L., et al. (2015). Tissue-resident macrophages originate from yolk-sac-derived erythro-myeloid progenitors. *Nature*, *518*, 547–551.

Gordon-Keylock, S., Sobiesiak, M., Rybtsov, S., Moore, K., & Medvinsky, A. (2013). Mouse extraembryonic arterial vessels harbor precursors capable of maturing into definitive HSCs. *Blood*, *122*, 2338–2345.

Growney, J. D., Shigematsu, H., Li, Z., Lee, B. H., Adelsperger, J., Rowan, R., et al. (2005). Loss of Runx1 perturbs adult hematopoiesis and is associated with a myeloproliferative phenotype. *Blood*, *106*, 494–504.

Gu, T.-L., Goetz, T. L., Graves, B. J., & Speck, N. A. (2000). Autoinhibition and partner proteins, CBFb and Ets-1, modulate DNA binding by CBFa2(AML1). *Molecular and Cellular Biology*, *20*, 91–103.

Guo, Y., Maillard, I., Chakraborti, S., Rothenberg, E. V., & Speck, N. A. (2008). Core binding factors are necessary for natural killer cell development, and cooperate with Notch signaling during T cell specification. *Blood*, *112*, 480–492.

Hadland, B. K., Huppert, S. S., Kanungo, J., Xue, Y., Jiang, R., Gridley, T., et al. (2004). A requirement for Notch1 distinguishes 2 phases of definitive hematopoiesis during development. *Blood*, *104*, 3097–3105.

Hadland, B. K., Varnum-Finney, B., Poulos, M. G., Moon, R. T., Butler, J. M., Rafii, S., et al. (2015). Endothelium and NOTCH specify and amplify aorta-gonad-mesonephros-derived hematopoietic stem cells. *The Journal of Clinical Investigation*, *125*, 2032–2045.

Haldar, M., & Murphy, K. M. (2014). Origin, development, and homeostasis of tissue-resident macrophages. *Immunological Reviews*, *262*, 25–35.
Harada, Y., & Harada, H. (2009). Molecular pathways mediating MDS/AML with focus on AML1/RUNX1 point mutations. *Journal of Cellular Physiology*, *220*, 16–20.
Hart, A., Melet, F., Grossfeld, P., Chien, K., Jones, C., Tunnacliffe, A., et al. (2000). Fli-1 is required for murine vascular and megakaryocytic development and is hemizygously deleted in patients with thrombocytopenia. *Immunity*, *13*, 167–177.
Hashimoto, K., Fujimoto, T., Shimoda, Y., Huang, X., Sakamoto, H., & Ogawa, M. (2007). Distinct hemogenic potential of endothelial cells and CD41+ cells in mouse embryos. *Development, Growth & Differentiation*, *49*, 287–300.
Herbomel, P., Thisse, B., & Thisse, C. (1999). Ontogeny and behaviour of early macrophages in the zebrafish embryo. *Development*, *126*, 3735–3745.
Hoogenkamp, M., Lichtinger, M., Krysinska, H., Lancrin, C., Clarke, D., Williamson, A., et al. (2009). *Early chromatin unfolding by RUNX1—A molecular explanation for differential requirements during specification versus maintenance of the hematopoietic gene expression program.* Blood.
Huang, H., Yu, M., Akie, T. E., Moran, T. B., Woo, A. J., Tu, N., et al. (2009). Differentiation-dependent interactions between RUNX-1 and FLI-1 during megakaryocyte development. *Molecular and Cellular Biology*, *29*, 4103–4115.
Huber, T. L., Kouskoff, V., Fehling, H. J., Palis, J., & Keller, G. (2004). Haemangioblast commitment is initiated in the primitive streak of the mouse embryo. *Nature*, *432*, 625–630.
Iacovino, M., Chong, D., Szatmari, I., Hartweck, L., Rux, D., Caprioli, A., et al. (2011). HoxA3 is an apical regulator of haemogenic endothelium. *Nature Cell Biology*, *13*, 72–78.
Ichikawa, M., Asai, T., Saito, T., Yamamoto, G., Seo, S., Yamazaki, I., et al. (2004). AML-1 is required for megakaryocytic maturation and lymphocytic differentiation, but not for maintenance of hematopoietic stem cells in adult hematopoiesis. *Nature Medicine*, *10*, 299–304.
Ichikawa, M., Yoshimi, A., Nakagawa, M., Nishimoto, N., Watanabe-Okochi, N., & Kurokawa, M. (2013). A role for RUNX1 in hematopoiesis and myeloid leukemia. *International Journal of Hematology*, *97*, 726–734.
Ishida, W., Hamamoto, T., Kusanagi, K., Yagi, K., Kawabata, M., Takehara, K., et al. (2000). Smad6 is a Smad1/5-induced smad inhibitor. Characterization of bone morphogenetic protein-responsive element in the mouse Smad6 promoter. *The Journal of Biological Chemistry*, *275*, 6075–6079.
Ivanovs, A., Rybtsov, S., Welch, L., Anderson, R. A., Turner, M. L., & Medvinsky, A. (2011). Highly potent human hematopoietic stem cells first emerge in the intraembryonic aorta-gonad-mesonephros region. *The Journal of Experimental Medicine*, *208*, 2417–2427.
Jaffredo, T., Gautier, R., Eichmann, A., & Dieterlen-Lièvre, F. (1998). Intraaortic hemopoietic cells are derived from endothelial cells during ontogeny. *Development*, *125*, 4575–4583.
Johnson, K. D., Hsu, A. P., Ryu, M. J., Wang, J., Gao, X., Boyer, M. E., et al. (2012). Cis-element mutated in GATA2-dependent immunodeficiency governs hematopoiesis and vascular integrity. *The Journal of Clinical Investigation*, *122*, 3692–3704.
Jordon, H. E. (1916). Evidence of hemogenic capacity of endothelium. *The Anatomical Record*, *10*, 417–420.
Kaimakis, P., Crisan, M., & Dzierzak, E. (2013). The biochemistry of hematopoietic stem cell development. *Biochimica et Biophysica Acta*, *1830*, 2395–2403.
Kalev-Zhylinska, M., Horsfield, J. A., Flores, M. V. C., Postlethwait, J. H., Vitas, M. R., Baas, A. M., et al. (2002). Runx1 is required for zebrafish blood and vessel development and expression of a human RUNX1-CBF2T1 transgene advances a model for studies of leukemogenesis. *Development*, *129*, 2015–2030.

Kennedy, M., Firpo, M., Choi, K., Roberston, W. C., Kabrun, N., & Keller, G. (1997). A common precursor for primitive erythropoiesis and definitive hematopoiesis. *Nature, 386*, 488–493.

Khandekar, M., Brandt, W., Zhou, Y., Dagenais, S., Glover, T. W., Suzuki, N., et al. (2007). A Gata2 intronic enhancer confers its pan-endothelia-specific regulation. *Development, 134*, 1703–1712.

Kieusseian, A., Brunet de la Grange, P., Burlen-Defranoux, O., Godin, I., & Cumano, A. (2012). Immature hematopoietic stem cells undergo maturation in the fetal liver. *Development, 139*, 3521–3530.

Kim, A. D., Melick, C. H., Clements, W. K., Stachura, D. L., Distel, M., Panakova, D., et al. (2014). Discrete Notch signaling requirements in the specification of hematopoietic stem cells. *The EMBO Journal, 33*, 2363–2373.

Kim, W.-Y., Sieweke, M., Ogawa, E., Wee, H.-J., Englmeier, U., Graf, T., et al. (1999). Mutual activation of Ets-1 and AML1 DNA binding by direct interaction of their autoinhibitory domains. *The EMBO Journal, 18*, 1609–1620.

Kim, A. D., Stachura, D. L., & Traver, D. (2014). Cell signaling pathways involved in hematopoietic stem cell specification. *Experimental Cell Research, 329*, 227–233.

Kim, I., Yilmaz, O. H., & Morrison, S. J. (2005). CD144 (VE-cadherin) is transiently expressed by fetal liver hematopoietic stem cells. *Blood, 106*, 903–905.

Kinder, S. J., Tsang, T. E., Quinlan, G. A., Hadjantonakis, A. K., Nagy, A., & Tam, P. P. (1999). The orderly allocation of mesodermal cells to the extraembryonic structures and the anteroposterior axis during gastrulation of the mouse embryo. *Development, 126*, 4691–4701.

Kingsley, P. D., Malik, J., Fantauzzo, K. A., & Palis, J. (2004). Yolk sac-derived primitive erythroblasts enucleate during mammalian embryogenesis. *Blood, 104*, 19–25.

Kissa, K., & Herbomel, P. (2010). Blood stem cells emerge from aortic endothelium by a novel type of cell transition. *Nature, 464*, 112–115.

Kissa, K., Murayama, E., Zapata, A., Cortes, A., Perret, E., Machu, C., et al. (2008). Live imaging of emerging hematopoietic stem cells and early thymus colonization. *Blood, 111*, 1147–1156.

Knezevic, K., Bee, T., Wilson, N. K., Janes, M. E., Kinston, S., Polderdijk, S., et al. (2011). A Runx1-Smad6 rheostat controls Runx1 activity during embryonic hematopoiesis. *Molecular and Cellular Biology, 31*, 2817–2826.

Kobayashi, M., Shelley, W. C., Seo, W., Vemula, S., Lin, Y., Liu, Y., et al. (2014). Functional B-1 progenitor cells are present in the hematopoietic stem cell-deficient embryo and depend on Cbfbeta for their development. *Proceedings of the National Academy of Sciences of the United States of America, 111*, 12151–12156.

Koh, C. P., Ng, C. E., Nah, G. S., Wang, C. Q., Tergaonkar, V., Matsumura, T., et al. (2015). Hematopoietic stem cell enhancer: A powerful tool in stem cell biology. *Histology and Histopathology, 30*, 661–672.

Kopan, R., & Ilagan, M. X. (2009). The canonical Notch signaling pathway: Unfolding the activation mechanism. *Cell, 137*, 216–233.

Krebs, L. T., Shutter, J. R., Tanigaki, K., Honjo, T., Stark, K. L., & Gridley, T. (2004). Haploinsufficient lethality and formation of arteriovenous malformations in Notch pathway mutants. *Genes & Development, 18*, 2469–2473.

Kumano, K., Chiba, S., Kunisato, A., Sata, M., Saito, T., Nakagami-Yamaguchi, E., et al. (2003). Notch1 but not Notch2 is essential for generating hematopoietic stem cells from endothelial cells. *Immunity, 18*, 699–711.

Kumaravelu, P., Hook, L., Morrison, A. M., Ure, J., Zhao, S., Zuyev, S., et al. (2002). Quantitative developmental anatomy of definitive haematopoietic stem cells/long-term repopulating units (HSC/RUs): Role of the aorta-gonad-mesonephros (AGM) region and the yolk sac in colonisation of the mouse embryonic liver. *Development, 129*, 4891–4899.

Lacaud, G., Gore, L., Kennedy, M., Kouskoff, V., Kingsley, P., Hogan, C., et al. (2002). Runx1 is essential for hematopoietic commitment at the hemangioblast stage of development in vitro. *Blood, 100*, 458–466.

Lacaud, G., Kouskoff, V., Trumble, A., Schwantz, S., & Keller, G. (2004). Haploinsufficiency of Runx1 results in the acceleration of mesodermal development and hemangioblast specification upon in vitro differentiation of ES cells. *Blood, 103*, 886–889.

Lai, E. C. (2004). Notch signaling: Control of cell communication and cell fate. *Development, 131*, 965–973.

Lam, E. Y., Chau, J. Y., Kalev-Zylinska, M. L., Fountaine, T. M., Mead, R. S., Hall, C. J., et al. (2009). Zebrafish runx1 promoter-EGFP transgenics mark discrete sites of definitive blood progenitors. *Blood, 113*, 1241–1249.

Lam, E. Y., Hall, C. J., Crosier, P. S., Crosier, K. E., & Flores, M. V. (2010). Live imaging of Runx1 expression in the dorsal aorta tracks the emergence of blood progenitors from endothelial cells. *Blood, 116*, 909–914.

Lam, K., & Zhang, D. E. (2012). RUNX1 and RUNX1-ETO: Roles in hematopoiesis and leukemogenesis. *Frontiers in Bioscience, 17*, 1120–1139.

Lancrin, C., Mazan, M., Stefanska, M., Patel, R., Lichtinger, M., Costa, G., et al. (2012). GFI1 and GFI1B control the loss of endothelial identity of hemogenic endothelium during hematopoietic commitment. *Blood, 120*, 314–322.

Lancrin, C., Sroczynska, P., Stephenson, C., Allen, T., Kouskoff, V., & Lacaud, G. (2009). The haemangioblast generates haematopoietic cells through a haemogenic endothelium stage. *Nature, 457*, 892–895.

Lawson, K. A., Meneses, J. J., & Pedersen, R. A. (1991). Clonal analysis of epiblast fate during germ layer formation in the mouse embryo. *Development, 113*, 891–911.

Lawson, N. D., Scheer, N., Pham, V. N., Kim, C. H., Chitnis, A. B., Campos-Ortega, J. A., et al. (2001). Notch signaling is required for arterial-venous differentiation during embryonic vascular development. *Development, 128*, 3675–3683.

Lawson, N. D., Vogel, A. M., & Weinstein, B. M. (2002). Sonic hedgehog and vascular endothelial growth factor act upstream of the Notch pathway during arterial endothelial differentiation. *Developmental Cell, 3*, 127–136.

Lebestky, T., Chang, T., Hartenstein, V., & Banerjee, U. (2000). Specification of *Drosophila* hematopoietic lineage by conserved transcription factors. *Science, 288*, 146–149.

Lebestky, T., Jung, S. H., & Banerjee, U. (2003). A Serrate-expressing signaling center controls Drosophila hematopoiesis. *Genes & Development, 17*, 348–353.

Levanon, D., Bernstein, Y., Negreanu, V., Ghozi, M. C., Bar-Am, I., Aloya, R., et al. (1996). A large variety of alternatively spliced and differentially expressed mRNAs are encoded by the human acute myeloid leukemia gene AML1. *DNA and Cell Biology, 15*, 175–185.

Levanon, D., & Groner, Y. (2004). Structure and regulated expression of mammalian RUNX genes. *Oncogene, 23*, 4211–4219.

Levantini, E., Lee, S., Radomska, H. S., Hetherington, C. J., Alberich-Jorda, M., Amabile, G., et al. (2011). RUNX1 regulates the CD34 gene in haematopoietic stem cells by mediating interactions with a distal regulatory element. *The EMBO Journal, 30*, 4059–4070.

Li, Z., Chen, M. J., Stacy, T., & Speck, N. A. (2006). Runx1 function in hematopoiesis is required in cells that express Tek. *Blood, 107*, 106–110.

Li, W., Ferkowicz, M. J., Johnson, S. A., Shelley, W. C., & Yoder, M. C. (2005). Endothelial cells in the early murine yolk sac give rise to CD41-expressing hematopoietic cells. *Stem Cells and Development, 14*, 44–54.

Li, Z., Lan, Y., He, W., Chen, D., Wang, J., Zhou, F., et al. (2012). Mouse embryonic head as a site for hematopoietic stem cell development. *Cell Stem Cell, 11*, 663–675.

Liakhovitskaia, A., Rybtsov, S., Smith, T., Batsivari, A., Rybtsova, N., Rode, C., et al. (2014). Runx1 is required for progression of CD41+ embryonic precursors into HSCs but not prior to this. *Development, 141*, 3319–3323.

Lichtinger, M., Ingram, R., Hannah, R., Muller, D., Clarke, D., Assi, S. A., et al. (2012). RUNX1 reshapes the epigenetic landscape at the onset of haematopoiesis. *EMBO Journal, 31*(22), 4318–4333.

Lie, A. L. M., Marinopoulou, E., Li, Y., Patel, R., Stefanska, M., Bonifer, C., et al. (2014). RUNX1 positively regulates a cell adhesion and migration program in murine hemogenic endothelium prior to blood emergence. *Blood, 124*, e11–e20.

Lim, K. C., Hosoya, T., Brandt, W., Ku, C. J., Hosoya-Ohmura, S., Camper, S. A., et al. (2012). Conditional Gata2 inactivation results in HSC loss and lymphatic mispatterning. *The Journal of Clinical Investigation, 122*, 3705–3717.

Lin, Y., Yoder, M. C., & Yoshimoto, M. (2014). Lymphoid progenitor emergence in the murine embryo and yolk sac precedes stem cell detection. *Stem Cells and Development, 23*, 1168–1177.

Liu, F., Walmsley, M., Rodaway, A., & Patient, R. (2008). Fli1 acts at the top of the transcriptional network driving blood and endothelial development. *Current Biology, 18*, 1234–1240.

Lorsbach, R. B., Moore, J., Ang, S. O., Sun, W., Lenny, N., & Downing, J. R. (2004). Role of Runx1 in adult hematopoiesis: Analysis of Runx1-IRES-GFP knock-in mice reveals differential lineage expression. *Blood, 103*, 2522–2529.

Mangan, J. K., & Speck, N. A. (2011). RUNX1 mutations in clonal myeloid disorders: From conventional cytogenetics to next generation sequencing, a story 40 years in the making. *Critical Reviews in Oncogenesis, 16*, 77–91.

Marks-Bluth, J., Khanna, A., Chandrakanthan, V., Thoms, J., Bee, T., Eich, C., et al. (2015). SMAD1 and SMAD5 expression is coordinately regulated by FLI1 and GATA2 during endothelial development. *Molecular and Cellular Biology, 35*, 2165–2172.

Marks-Bluth, J., & Pimanda, J. E. (2012). Cell signalling pathways that mediate haematopoietic stem cell specification. *The International Journal of Biochemistry & Cell Biology, 44*, 2175–2184.

Marshall, C. J., Moore, R. L., Thorogood, P., Brickell, P. M., Kinnon, C., & Thrasher, A. J. (1999). Detailed characterization of the human aorta-gonad-mesonephros region reveals morphological polarity resembling a hematopoietic stromal layer. *Developmental Dynamics, 215*, 139–147.

Matheny, C. J., Speck, M. E., Cushing, P. R., Zhou, Y., Corpora, T., Regan, M., et al. (2007). Disease mutations in RUNX1 and RUNX2 create nonfunctional, dominant-negative, or hypomorphic alleles. *Embo Journal, 26*, 1163–1175.

McGrath, K. E., Frame, J. M., Fegan, K. H., Bowen, J. R., Conway, S. J., Catherman, S. C., et al. (2015). Distinct sources of hematopoietic progenitors emerge before HSCs and provide functional blood cells in the mammalian embryo. *Cell Reports, 11*, 1892–1904.

Medvinsky, A., & Dzierzak, E. (1996). Definitive hematopoiesis is autonomously initiated by the AGM region. *Cell, 86*, 897–906.

Medvinsky, A. L., Samoylina, N. L., Müller, A. M., & Dzierzak, E. A. (1993). An early pre-liver intra-embryonic source of CFU-S in the developing mouse. *Nature, 364*, 64–66.

Mikkola, H. K., Fujiwara, Y., Schlaeger, T. M., Traver, D., & Orkin, S. H. (2003). Expression of CD41 marks the initiation of definitive hematopoiesis in the mouse embryo. *Blood, 101*, 508–516.

Minegishi, N., Ohta, J., Yamagiwa, H., Suzuki, N., Kawauchi, S., Zhou, Y., et al. (1999). The mouse GATA-2 gene is expressed in the para-aortic splanchnopleura and aorta-gonads and mesonephros region. *Blood, 93*, 4196–4207.

Mirshekar-Syahkal, B., Fitch, S. R., & Ottersbach, K. (2014). Concise review: From greenhouse to garden: The changing soil of the hematopoietic stem cell microenvironment during development. *Stem Cells*, *32*, 1691–1700.

Müller, A. M., Medvinsky, A., Strouboulis, J., Grosveld, F., & Dzierzak, E. (1994). Development of hematopoietic stem cell activity in the mouse embryo. *Immunity*, *1*, 291–301.

Murray, P. D. F. (1932). The development 'in vitro' of blood of the early chick embryo. *Proceedings. Royal Society of London*, *11*, 497–521.

Nadin, B. M., Goodell, M. A., & Hirschi, K. K. (2003). Phenotype and hematopoietic potential of side population cells throughout embryonic development. *Blood*, *102*, 2436–2443.

Nakagawa, M., Ichikawa, M., Kumano, K., Goyama, S., Kawazu, M., Asai, T., et al. (2006). AML1/Runx1 rescues Notch1-Null mutation-induced deficiency of para-aortic splanchnopleural hematopoiesis. *Blood*, *108*, 3329–3334.

Nakano, H., Liu, X., Arshi, A., Nakashima, Y., van Handel, B., Sasidharan, R., et al. (2013). Haemogenic endocardium contributes to transient definitive haematopoiesis. *Nature Communications*, *4*, 1564.

Ng, C. E., Yokomizo, T., Yamashita, N., Cirovic, B., Jin, H., Wen, Z., et al. (2010). A Runx1 intronic enhancer marks hemogenic endothelial cells and hematopoietic stem cells. *Stem Cells*, *28*, 1869–1881.

Niki, M., Okada, H., Takano, H., Kuno, J., Tani, K., Hibino, H., et al. (1997). Hematopoiesis in the fetal liver is impaired by targeted mutagenesis of a gene encoding a non-DNA binding subunit of the transcription factor, polyomavirus enhancer binding protein 2/core binding factor. *Proceedings of the National Academy of Sciences of the United States of America*, *94*, 5697–5702.

Nishikawa, S.-I., Nishikawa, S., Kawamoto, H., Yoshida, H., Kizumoto, M., Kataoka, H., et al. (1998). In vitro generation of lymphohematopoietic cells from endothelial cells purified from murine embryos. *Immunity*, *8*, 761–769.

North, T. E., de Bruijn, M. F., Stacy, T., Talebian, L., Lind, E., Robin, C., et al. (2002). Runx1 expression marks long-term repopulating hematopoietic stem cells in the midgestation mouse embryo. *Immunity*, *16*, 661–672.

North, T. E., Gu, T.-L., Stacy, T., Wang, Q., Howard, L., Binder, M., et al. (1999). *Cbfa2* is required for the formation of intra-aortic hematopoietic clusters. *Development*, *126*, 2563–2575.

North, T. E., Stacy, T., Matheny, C. J., Speck, N. A., & de Bruijn, M. F. (2004). Runx1 is expressed in adult mouse hematopoietic stem cells and differentiating myeloid and lymphoid cells, but not in maturing erythroid cells. *Stem Cells*, *22*, 158–168.

Nottingham, W. T., Jarratt, A., Burgess, M., Speck, C. L., Cheng, J. F., Prabhakar, S., et al. (2007). Runx1-mediated hematopoietic stem-cell emergence is controlled by a Gata/Ets/SCL-regulated enhancer. *Blood*, *110*, 4188–4197.

Ogawa, E., Inuzuka, M., Maruyama, M., Satake, M., Naito-Fujimoto, M., Ito, Y., et al. (1993). Molecular cloning and characterization of PEBP2b, the heterodimeric partner of a novel Drosophila runt-related DNA binding protein PEBP2a. *Virology*, *194*, 314–331.

Okuda, T., van Deursen, J., Hiebert, S. W., Grosveld, G., & Downing, J. R. (1996). AML1, the target of multiple chromosomal translocations in human leukemia, is essential for normal fetal liver hematopoiesis. *Cell*, *84*, 321–330.

Oren, T., Torregroza, I., & Evans, T. (2005). An Oct-1 binding site mediates activation of the gata2 promoter by BMP signaling. *Nucleic Acids Research*, *33*, 4357–4367.

Ottersbach, K., & Dzierzak, E. (2005). The murine placenta contains hematopoietic stem cells within the vascular labyrinth region. *Developmental Cell*, *8*, 377–387.

Padron-Barthe, L., Temino, S., Villa del Campo, C., Carramolino, L., Isern, J., & Torres, M. (2014). Clonal analysis identifies hemogenic endothelium as the source of the blood-endothelial common lineage in the mouse embryo. *Blood*, *124*, 2523–2532.

Palii, C. G., Perez-Iratxeta, C., Yao, Z., Cao, Y., Dai, F., Davison, J., et al. (2011). Differential genomic targeting of the transcription factor TAL1 in alternate haematopoietic lineages. *The EMBO Journal, 30*, 494–509.

Palis, J. (2014). Primitive and definitive erythropoiesis in mammals. *Frontiers in Physiology, 5*, 3.

Palis, J., Robertson, S., Kennedy, M., Wall, C., & Keller, G. (1999). Development of erythroid and myeloid progenitors in the yolk sac and embryo proper of the mouse. *Development, 126*, 5073–5084.

Parameswaran, M., & Tam, P. P. (1995). Regionalisation of cell fate and morphogenetic movement of the mesoderm during mouse gastrulation. *Developmental Genetics, 17*, 16–28.

Pimanda, J. E., Donaldson, I. J., de Bruijn, M. F., Kinston, S., Knezevic, K., Huckle, L., et al. (2007). The SCL transcriptional network and BMP signaling pathway interact to regulate RUNX1 activity. *Proceedings of the National Academy of Sciences of the United States of America, 104*, 840–845.

Pimanda, J. E., Ottersbach, K., Knezevic, K., Kinston, S., Chan, W. Y., Wilson, N. K., et al. (2007). Gata2, Fli1, and Scl form a recursively wired gene-regulatory circuit during early hematopoietic development. *Proceedings of the National Academy of Sciences of the United States of America, 104*, 17692–17697.

Porcher, C., Swat, W., Rockwell, K., Fujiwara, Y., Alt, F. W., & Orkin, S. H. (1996). The T-cell leukemia oncoprotein SCL/tal-1 is essential for development of all hematopoietic lineages. *Cell, 86*, 47–57.

Potts, K. S., Sargeant, T. J., Markham, J. F., Shi, W., Biben, C., Josefsson, E. C., et al. (2014). A lineage of diploid platelet-forming cells precedes polyploid megakaryocyte formation in the mouse embryo. *Blood, 124*, 2725–2729.

Pouget, C., Peterkin, T., Simoes, F. C., Lee, Y., Traver, D., & Patient, R. (2014). FGF signalling restricts haematopoietic stem cell specification via modulation of the BMP pathway. *Nature Communications, 5*, 5588.

Pozner, A., Goldenberg, D., Negreanu, V., Le, S. Y., Elroy-Stein, O., Levanon, D., et al. (2000). Transcription-coupled translation control of AML1/RUNX1 is mediated by cap- and internal ribosome entry site-dependent mechanisms. *Molecular and Cellular Biology, 20*, 2297–2307.

Pozner, A., Lotem, J., Xiao, C., Goldenberg, D., Brenner, O., Negreanu, V., et al. (2007). Developmentally regulated promoter-switch transcriptionally controls Runx1 function during embryonic hematopoiesis. *BMC Developmental Biology, 7*, 84.

Putz, G., Rosner, A., Nuesslein, I., Schmitz, N., & Buchholz, F. (2006). AML1 deletion in adult mice causes splenomegaly and lymphomas. *Oncogene, 25*, 929–939.

Quillien, A., Moore, J. C., Shin, M., Siekmann, A. F., Smith, T., Pan, L., et al. (2014). Distinct Notch signaling outputs pattern the developing arterial system. *Development, 141*, 1544–1552.

Rhodes, K. E., Gekas, C., Wang, Y., Lux, C. T., Francis, C. S., Chan, D. N., et al. (2008). The emergence of hematopoietic stem cells is initiated in the placental vasculature in the absence of circulation. *Cell Stem Cell, 2*, 252–263.

Richard, C., Drevon, C., Canto, P. Y., Villain, G., Bollerot, K., Lempereur, A., et al. (2013). Endothelio-mesenchymal interaction controls runx1 expression and modulates the notch pathway to initiate aortic hematopoiesis. *Developmental Cell, 24*, 600–611.

Robb, L., Lyons, I., Li, R., Hartley, L., Köntgen, F., Harvey, R. P., et al. (1995). Absence of yolk sac hematopoiesis from mice with a targeted disruption of the *scl* gene. *Proceedings of the National Academy of Sciences of the United States of America, 92*, 7075–7079.

Robert-Moreno, A., Espinosa, L., de la Pompa, J. L., & Bigas, A. (2005). RBPjkappa-dependent Notch function regulates Gata2 and is essential for the formation of intra-embryonic hematopoietic cells. *Development, 132*, 1117–1126.

Robert-Moreno, A., Guiu, J., Ruiz-Herguido, C., Lopez, M. E., Ingles-Esteve, J., Riera, L., et al. (2008). Impaired embryonic haematopoiesis yet normal arterial development in the absence of the Notch ligand Jagged1. *Embo Journal, 27*, 1886–1895.
Robertson, S. M., Kennedy, M., Shannon, J. M., & Keller, G. (2000). A transitional stage in the commitment of mesoderm to hematopoiesis requiring the transcription factor SCL/tal-1. *Development, 127*, 2447–2459.
Rybtsov, S., Batsivari, A., Bilotkach, K., Paruzina, D., Senserrich, J., Nerushev, O., et al. (2014). Tracing the origin of the HSC hierarchy reveals an SCF-dependent, IL-3-independent CD43(−) embryonic precursor. *Stem Cell Reports, 3*, 489–501.
Rybtsov, S., Sobiesiak, M., Taoudi, S., Souilhol, C., Senserrich, J., Liakhovitskaia, A., et al. (2011). Hierarchical organization and early hematopoietic specification of the developing HSC lineage in the AGM region. *The Journal of Experimental Medicine, 208*, 1305–1315.
Sabin, F. R. (1920). Studies on the origin of blood vessels and of red corpuscles as seen in the living blastoderm of the chick during the second day of incubation. *Contributions to Embryology, 9*, 213–262.
Sandler, V. M., Lis, R., Liu, Y., Kedem, A., James, D., Elemento, O., et al. (2014). Reprogramming human endothelial cells to haematopoietic cells requires vascular induction. *Nature, 511*, 312–318.
Sangle, N. A., & Perkins, S. L. (2011). Core-binding factor acute myeloid leukemia. *Archives of Pathology & Laboratory Medicine, 135*, 1504–1509.
Sasaki, K., Yagi, H., Bronson, R. T., Tominaga, K., Matsunashi, T., Deguchi, K., et al. (1996). Absence of fetal liver hematopoiesis in mice deficient in transcriptional coactivator core binding factor beta. *Proceedings of the National Academy of Sciences of the United States of America, 93*, 12359–12363.
Sato, Y. (2013). Dorsal aorta formation: Separate origins, lateral-to-medial migration, and remodeling. *Development, Growth & Differentiation, 55*, 113–129.
Schlaeger, T. M., Mikkola, H. K., Gekas, C., Helgadottir, H. B., & Orkin, S. H. (2005). Tie2Cre mediated gene ablation defines the Stem Cell Leukemia gene (SCL/tal1)-dependent window during hematopoietic stem cell development. *Blood, 105*, 3871–3874.
Schulz, C., Gomez Perdiguero, E., Chorro, L., Szabo-Rogers, H., Cagnard, N., Kierdorf, K., et al. (2012). A lineage of myeloid cells independent of Myb and hematopoietic stem cells. *Science, 336*, 86–90.
Song, W.-J., Sullivan, M. G., Legare, R. D., Hutchings, S., Tan, X., Kufrin, D., et al. (1999). Haploinsufficiency of *CBFA2* (*AML1*) causes familial thrombocytopenia with propensity to develop acute myelogenous leukamia (FPD/AML). *Nature Genetics, 23*, 166–175.
Sood, R., English, M. A., Belele, C. L., Jin, H., Bishop, K., Haskins, R., et al. (2010). Development of multilineage adult hematopoiesis in the zebrafish with a runx1 truncation mutation. *Blood, 115*, 2806–2809.
Spyropoulos, D. D., Pharr, P. N., Lavenburg, K. R., Jackers, P., Papas, T. S., Ogawa, M., et al. (2000). Hemorrhage, impaired hematopoiesis, and lethality in mouse embryos carrying a targeted disruption of the Fli1 transcription factor. *Molecular and Cellular Biology, 20*, 5643–5652.
Sroczynska, P., Lancrin, C., Pearson, S., Kouskoff, V., & Lacaud, G. (2009). In vitro differentiation of mouse embryonic stem cells as a model of early hematopoietic development. *Methods in Molecular Biology, 538*, 317–334.
Suonpaa, P., Kohonen, P., Koskela, K., Koskiniemi, H., Salminen-Mankonen, H., & Lassila, O. (2005). Development of early PCLP1-expressing haematopoietic cells within the avian dorsal aorta. *Scandinavian Journal of Immunology, 62*, 218–223.
Suzuki-Inoue, K., Inoue, O., Ding, G., Nishimura, S., Hokamura, K., Eto, K., et al. (2010). Essential in vivo roles of the C-type lectin receptor CLEC-2: Embryonic/neonatal

lethality of CLEC-2-deficient mice by blood/lymphatic misconnections and impaired thrombus formation of CLEC-2-deficient platelets. *The Journal of Biological Chemistry, 285*, 24494–24507.

Swiers, G., Baumann, C., O'Rourke, J., Giannoulatou, E., Taylor, S., Joshi, A., et al. (2013). Early dynamic fate changes in haemogenic endothelium characterized at the single-cell level. *Nature Communications, 4*, 2924.

Takakura, N., Watanabe, T., Suenobu, S., Yamada, Y., Noda, T., Ito, Y., et al. (2000). A role for hematopoietic stem cells in promoting angiogenesis. *Cell, 102*, 199–209.

Tam, P. P., & Beddington, R. S. (1987). The formation of mesodermal tissues in the mouse embryo during gastrulation and early organogenesis. *Development, 99*, 109–126.

Tamplin, O. J., Durand, E. M., Carr, L. A., Childs, S. J., Hagedorn, E. J., Li, P., et al. (2015). Hematopoietic stem cell arrival triggers dynamic remodeling of the perivascular niche. *Cell, 160*, 241–252.

Tanaka, Y., Joshi, A., Wilson, N. K., Kinston, S., Nishikawa, S., & Gottgens, B. (2012). The transcriptional programme controlled by Runx1 during early embryonic blood development. *Developmental Biology, 366*, 404–419.

Taoudi, S., Gonneau, C., Moore, K., Sheridan, J. M., Blackburn, C. C., Taylor, E., et al. (2008). Extensive hematopoietic stem cell generation in the AGM region via maturation of VE-cadherin+CD45+ pre-definitive HSCs. *Cell Stem Cell, 3*, 99–108.

Taoudi, S., & Medvinsky, A. (2007). Functional identification of the hematopoietic stem cell niche in the ventral domain of the embryonic dorsal aorta. *Proceedings of the National Academy of Sciences of the United States of America, 104*, 9399–9403.

Taoudi, S., Morrison, A. M., Inoue, H., Gribi, R., Ure, J., & Medvinsky, A. (2005). Progressive divergence of definitive haematopoietic stem cells from the endothelial compartment does not depend on contact with the foetal liver. *Development, 132*, 4179–4191.

Tavian, M., Coulombel, L., Luton, D., San Clemente, H., Dieterlen-Lièvre, F., & Peault, B. (1996). Aorta-associated $CD34^+$ hematopoietic cells in the early human embryo. *Blood, 87*, 67–72.

Telfer, J. C., Hedblom, E. E., Anderson, M. K., Laurent, M. N., & Rothenberg, E. V. (2004). Localization of the domains in Runx transcription factors required for the repression of CD4 in thymocytes. *The Journal of Immunology, 172*, 4359–4370.

Telfer, J. C., & Rothenberg, E. V. (2001). Expression and function of a stem cell promoter for the murine CBFa2 gene: Distinct roles and regulation in natural killer and T cell development. *Developmental Biology, 229*, 363–382.

Terriente-Felix, A., Li, J., Collins, S., Mulligan, A., Reekie, I., Bernard, F., et al. (2013). Notch cooperates with Lozenge/Runx to lock haemocytes into a differentiation programme. *Development, 140*, 926–937.

Tober, J., Koniski, A., McGrath, K. E., Vemishetti, R., Emerson, R., de Mesy-Bentley, K. K., et al. (2007). The megakaryocyte lineage originates from hemangioblast precursors and is an integral component both of primitive and of definitive hematopoiesis. *Blood, 109*, 1433–1441.

Tober, J., Yzaguirre, A. D., Piwarzyk, E., & Speck, N. A. (2013). Distinct temporal requirements for Runx1 in hematopoietic progenitors and stem cells. *Development, 140*, 3765–3776.

Tsai, J.-Y., Keller, T., Kuo, F. C., Weiss, M., Chen, J., Rosenblatt, M., et al. (1994). An early haematopoietic defect in mice lacking the transcription factor GATA-2. *Nature, 371*, 221–226.

Visvader, J. E., Fujiwara, Y., & Orkin, S. H. (1998). Unsuspected role for the T-cell leukemia protein SCL/tal-1 in vascular development. *Genes & Development, 12*, 473–479.

Vo, L. T., & Daley, G. Q. (2015). De novo generation of HSCs from somatic and pluripotent stem cell sources. *Blood, 125*, 2641–2648.

Vokes, S. A., & Krieg, P. A. (2002). Endoderm is required for vascular endothelial tube formation, but not for angioblast specification. *Development*, *129*, 775–785.

Vokes, S. A., Yatskievych, T. A., Heimark, R. L., McMahon, J., McMahon, A. P., Antin, P. B., et al. (2004). Hedgehog signaling is essential for endothelial tube formation during vasculogenesis. *Development*, *131*, 4371–4380.

Walmsley, M., Ciau-Uitz, A., & Patient, R. (2002). Adult and embryonic blood and endothelium derive from distinct precursor populations which are differentially programmed by BMP in Xenopus. *Development*, *129*, 5683–5695.

Wang, Q., Stacy, T., Binder, M., Marín-Padilla, M., Sharpe, A. H., & Speck, N. A. (1996a). Disruption of the *Cbfa2* gene causes necrosis and hemorrhaging in the central nervous system and blocks definitive hematopoiesis. *Proceedings of the National Academy of Sciences of the United States of America*, *93*, 3444–3449.

Wang, Q., Stacy, T., Miller, J. D., Lewis, A. F., Huang, X., Bories, J.-C., et al. (1996b). The CBFb subunit is essential for CBFa2 (AML1) function in vivo. *Cell*, *87*, 697–708.

Wang, S., Wang, Q., Crute, B. E., Melnikova, I. N., Keller, S. R., & Speck, N. A. (1993). Cloning and characterization of subunits of the T-cell receptor and murine leukemia virus enhancer core-binding factor. *Molecular and Cellular Biology*, *13*, 3324–3339.

Wang, H., Zang, C., Taing, L., Arnett, K. L., Wong, Y. J., Pear, W. S., et al. (2014). NOTCH1-RBPJ complexes drive target gene expression through dynamic interactions with superenhancers. *Proceedings of the National Academy of Sciences of the United States of America*, *111*, 705–710.

Wasteson, P., Johansson, B. R., Jukkola, T., Breuer, S., Akyurek, L. M., Partanen, J., et al. (2008). Developmental origin of smooth muscle cells in the descending aorta in mice. *Development*, *135*, 1823–1832.

Wiegreffe, C., Christ, B., Huang, R., & Scaal, M. (2009). Remodeling of aortic smooth muscle during avian embryonic development. *Developmental Dynamics*, *238*, 624–631.

Wiemels, J. L., Cazzaniga, G., Daniotti, M., Eden, O. B., Addison, G., Masera, G., et al. (1999). Prenatal origin of acute lymphoblastic leukaemia in children. *Lancet*, *354*, 1499–1503.

Wiemels, J. L., Ford, A. M., van Wering, E. R., Postma, A., & Greaves, M. (1999). Protracted and variable latency of acute lymphoblastic leukemia after TEL-AML1 gene fusion in utero. *Blood*, *94*, 1057–1062.

Wilkinson, R. N., Pouget, C., Gering, M., Russell, A. J., Davies, S. G., Kimelman, D., et al. (2009). Hedgehog and Bmp polarize hematopoietic stem cell emergence in the zebrafish dorsal aorta. *Developmental Cell*, *16*, 909–916.

Yoder, M. C. (2014). Inducing definitive hematopoiesis in a dish. *Nature Biotechnology*, *32*, 539–541.

Yoder, M. C., Hiatt, K., & Mukherjee, P. (1997). In vivo repopulating hematopoietic stem cells are present in the murine yolk sac at day 9.0 postcoitus. *Proceedings of the National Academy of Sciences of the United States of America*, *94*, 6776–6780.

Yokomizo, T., & Dzierzak, E. (2010). Three-dimensional cartography of hematopoietic clusters in the vasculature of whole mouse embryos. *Development*, *137*, 3651–3661.

Yokomizo, T., Hasegawa, K., Ishitobi, H., Osato, M., Ema, M., Ito, Y., et al. (2008). Runx1 is involved in primitive erythropoiesis in the mouse. *Blood*, *111*, 4075–4080.

Yokomizo, T., Ogawa, M., Osato, M., Kanno, T., Yoshida, H., Fujimoto, T., et al. (2001). Requirement of Runx1/AML1/PEBP2aB for the generation of haematopoietic cells from endothelial cells. *Genes to Cells*, *6*, 13–23.

Yokomizo, T., Osato, M., Takahashi, S., Kuroha, T., Yamamoto, M., & Ito, Y. (2000). *RUNX1/AML1/PEBP2aB* is involved in primitive hematopoiesis in addition to definitive hematopoiesis. *Blood*, *96*, 283a.

Yokomizo, T., Takahashi, S., Mochizuki, N., Kuroha, T., Ema, M., Wakamatsu, A., et al. (2007). Characterization of GATA-1(+) hemangioblastic cells in the mouse embryo. *Embo Journal*, *26*, 184–196.

Yokomizo, T., Yamada-Inagawa, T., Yzaguirre, A. D., Chen, M. J., Speck, N. A., & Dzierzak, E. (2012). Whole-mount three-dimensional imaging of internally localized immunostained cells within mouse embryos. *Nature Protocols*, *7*, 421–431.

Yokota, T., Huang, J., Tavian, M., Nagai, Y., Hirose, J., Zuniga-Pflucker, J. C., et al. (2006). Tracing the first waves of lymphopoiesis in mice. *Development*, *133*, 2041–2051.

Yokoyama, S., Ito, Y., Ueno-Kudoh, H., Shimizu, H., Uchibe, K., Albini, S., et al. (2009). A systems approach reveals that the myogenesis genome network is regulated by the transcriptional repressor RP58. *Developmental Cell*, *17*, 836–848.

Yoon, M. J., Koo, B. K., Song, R., Jeong, H. W., Shin, J., Kim, Y. W., et al. (2008). Mind bomb-1 is essential for intraembryonic hematopoiesis in the aortic endothelium and the subaortic patches. *Molecular and Cellular Biology*, *28*, 4794–4804.

Yoshimoto, M., Montecino-Rodriguez, E., Ferkowicz, M. J., Porayette, P., Shelley, W. C., Conway, S. J., et al. (2011). Embryonic day 9 yolk sac and intra-embryonic hemogenic endothelium independently generate a B-1 and marginal zone progenitor lacking B-2 potential. *Proceedings of the National Academy of Sciences of the United States of America*, *108*, 1468–1473.

Yoshimoto, M., Porayette, P., Glosson, N. L., Conway, S. J., Carlesso, N., Cardoso, A. A., et al. (2012). Autonomous murine T-cell progenitor production in the extra-embryonic yolk sac before HSC emergence. *Blood*, *119*, 5706–5714.

Zeigler, B. M., Sugiyama, D., Chen, M., Guo, Y., Downs, K. M., & Speck, N. A. (2006). The allantois and chorion, when isolated before circulation or chorio-allantoic fusion, have hematopoietic potential. *Development*, *133*, 4183–4192.

Zhang, C., Li, Q., Lim, C. H., Qiu, X., & Jiang, Y. J. (2007). The characterization of zebrafish antimorphic mib alleles reveals that Mib and Mind bomb-2 (Mib2) function redundantly. *Developmental Biology*, *305*, 14–27.

Zhang, C., Patient, R., & Liu, F. (2013). Hematopoietic stem cell development and regulatory signaling in zebrafish. *Biochimica et Biophysica Acta*, *1830*, 2370–2374.

Zovein, A. C., Hofmann, J. J., Lynch, M., French, W. J., Turlo, K. A., Yang, Y., et al. (2008). Fate tracing reveals the endothelial origin of hematopoietic stem cells. *Cell Stem Cell*, *3*, 625–636.

CHAPTER SIX

SCL/TAL1 in Hematopoiesis and Cellular Reprogramming

T. Hoang[1], J.A. Lambert, R. Martin
Laboratory of Hematopoiesis and Leukemia, Institute of Research in Immunology and Cancer (IRIC), University of Montreal, Montreal, QC, Canada
[1]Corresponding author: e-mail address: trang.hoang@umontreal.ca

Contents

Current Topics in Developmental Biology, Volume 118
ISSN 0070-2153
http://dx.doi.org/10.1016/bs.ctdb.2016.01.004

Abstract

SCL, a transcription factor of the basic helix–loop–helix family, is a master regulator of hematopoiesis. *Scl* specifies lateral plate mesoderm to a hematopoietic fate and establishes boundaries by inhibiting the cardiac lineage. A combinatorial interaction between *Scl* and *Vegfa/Flk1* sets in motion the first wave of primitive hematopoiesis. Subsequently, definitive hematopoietic stem cells (HSCs) emerge from the embryo proper via an endothelial-to-hematopoietic transition controlled by *Runx1*, acting with *Scl* and *Gata2*. Past this stage, *Scl* in steady state HSCs is redundant with *Lyl1*, a highly homologous factor. However, *Scl* is haploinsufficient in stress response, when a rare subpopulation of HSCs with very long term repopulating capacity is called into action. SCL activates transcription by recruiting a core complex on DNA that necessarily includes E2A/HEB, GATA1–3, LIM-only proteins LMO1/2, LDB1, and an extended complex comprising ETO2, RUNX1, ERG, or FLI1. These interactions confer multifunctionality to a complex that can control cell proliferation in erythroid progenitors or commitment to terminal differentiation through variations in single component. Ectopic *SCL* and *LMO1/2* expression in immature thymocytes activates of a stem cell gene network and reprogram cells with a finite lifespan into self-renewing preleukemic stem cells (pre-LSCs), an initiating event in T-cell acute lymphoblastic leukemias. Interestingly, fate conversion of fibroblasts to hematoendothelial cells requires not only *Scl* and *Lmo2* but also *Gata2*, *Runx1*, and *Erg*, indicating a necessary collaboration between these transcription factors for hematopoietic reprogramming. Nonetheless, full reprogramming into self-renewing multipotent HSCs may require additional factors and most likely, a permissive microenvironment.

1. INTRODUCTION

The stem cell leukemia gene (*SCL/TAL1*) is a master regulator of normal and aberrant hematopoiesis (Begley & Green, 1999; Lecuyer & Hoang, 2004). The gene, encoding a transcription factor of the basic helix–loop–helix (bHLH) family (Fig. 1), was independently cloned by several groups from a recurrent chromosomal translocation in T-cell acute lymphoblastic leukemia (T-ALL; Begley, Aplan, Denning, et al., 1989; Bernard et al., 1990; Chen, Cheng, et al., 1990; Finger et al., 1989). In the first case report, however, the disease initiated with an early T-cell phenotype and strikingly underwent phenotype conversion to a myeloid type leukemia following treatment (Begley, Aplan, Denning, et al., 1989). Because of this "stem cell leukemia," the *SCL* name was assigned to the gene. As it turned out, stem cell leukemia was an exception rather than the rule, although the

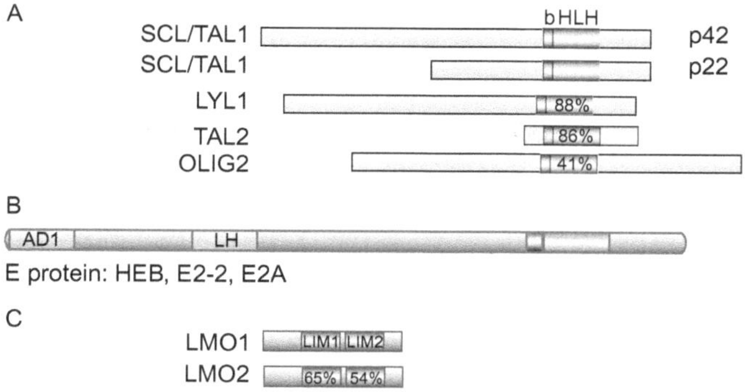

Fig. 1 *Schematic representation of hematopoietic transcription factors of the basic helix–loop–helix (bHLH) family and LIM-only (LMO) proteins.* (A) bHLH transcription factors encoded by genes that are activated by recurring chromosomal translocations in T-ALL. Two major protein isoforms are produced by translation regulation of *SCL*. Shown are the degrees of homology in the HLH domains of the various factors compared to SCL. (B) E proteins (HEB/TCF12, E2-2/TCF4 and E2A/TCF3) are relatively more ubiquitously expressed than hematopoietic bHLH factors. The *E2A* gene is involved in chromosomal translocations in B-ALL. (C) LMO1 and LMO2 proteins are encoded by genes that are activated in T-ALL, either by recurring chromosomal translocations or by retroviral integration. Shown are the homology within LIM1 or LIM2 between LMO1 and LMO2. (See the color plate.)

gene name was remarkably premonitory of SCL critical function at the onset of hematopoiesis and in stem cells, as shall be discussed later. The gene was also designated as *TAL1* because of its frequent occurrence in chromosomal translocations in T-ALL (Chen, Cheng, et al., 1990). Additional bHLH oncogenic transcription factors were later cloned from rare chromosomal translocations (Fig. 1).

2. SCL FUNCTION DURING DEVELOPMENT

2.1 Back to the Origin: Onset of Hematopoiesis

The development of the hematopoietic system occurs in at least three waves: a first wave of primitive erythropoiesis, followed by a second wave of transient multipotent progenitors, and a third wave of definitive or adult hematopoiesis. Exhaustive and critical analyses of the process can be found in recent reviews (Ciau-Uitz, Monteiro, Kirmizitas, & Patient, 2014; Palis, 2014). A brief overview is provided here as a framework for contextualizing SCL function. Primitive erythropoiesis consists in a rapid and transient

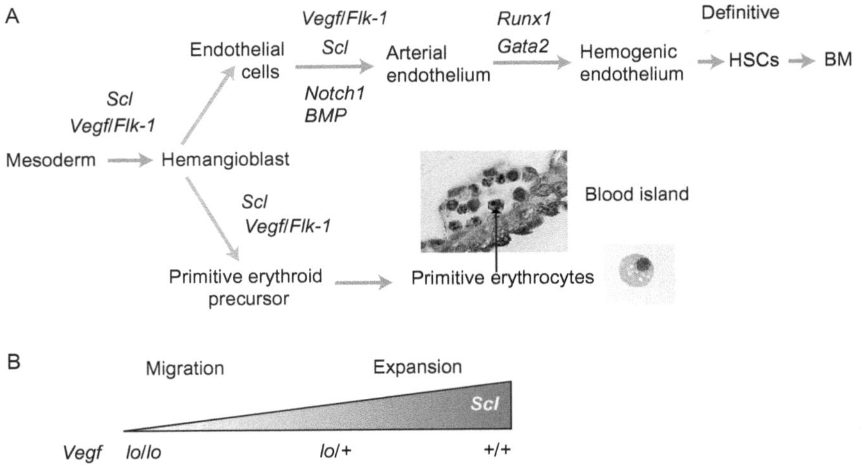

Fig. 2 *Role of Scl and Vegf during the ontogeny of the hematopoietic system.* (A) Spatial and temporal dynamics during hematopoietic development in the mouse. The lateral plate mesoderm gives rise FLK1$^+$ hemangioblasts from which originate both primitive and definitive lineages. The onset of hematopoiesis is characterized by the emergence of primitive erythrocytes within blood islands in the yolk sac (YS; primitive hematopoiesis). Hematopoietic stem cells (HSC) of the definitive myeloid and lymphoid lineages (definitive hematopoiesis) originate from the embryo proper via an endothelial to hematopoietic transition (EHT) to finally seed the bone marrow, the major site of hematopoiesis in the adult. *Vegf*: vascular endothelial growth factor. (B) Different *Vegf* thresholds control the migration of Flk1$^+$ hematopoietic and endothelial precursors to reach the yolk sac and their expansion during primitive erythropoiesis. *Scl* is also upregulated by VEGF signaling, in vitro and in vivo. *Vegf*$^{+/+}$: wild-type *Vegf* allele. *Vegf*^{lo}: hypomorphic allele. (See the color plate.)

generation of primitive erythrocytes, which are large, nucleated cells that express embryonic globins (Fig. 2; reviewed in Palis, 2014) and of primitive macrophages (Travnickova et al., 2015). Primitive erythrocytes emerge in the yolk sac after 7.5 days of gestation (E7.5) in the mouse, surrounded by endothelial cells, in structures known as blood islands (Fig. 2; Palis, 2014). At E9.5, erythro-myeloid progenitors (EMP) emerge in the yolk sac and migrate to the fetal liver to produce erythroid and myeloid lineages (McGrath et al., 2015). Definitive hematopoietic stem cells (HSCs) originate in the embryo proper, by budding from the aortic endothelium (Bertrand et al., 2010; Boisset et al., 2010) and begin to colonize the fetal liver at E11.5 (McGrath et al., 2015), which rapidly becomes the main site of hematopoiesis until birth, when self-renewing HSCs finally seed the bone marrow (Fig. 2; reviewed in Ciau-Uitz et al., 2014).

2.2 SCL Expression During Development

SCL, a bHLH transcription factor, is first expressed at E7.5, in groups of mesodermal cells of the yolk sac that are fated to generate blood islands (Elefanty, Begley, Hartley, Papaevangeliou, & Robb, 1999; Kallianpur, Jordan, & Brandt, 1994; Silver & Palis, 1997). As blood islands develop, SCL expression is maintained at high levels in primitive erythroid cells and at lower levels in surrounding endothelial cells (Drake & Fleming, 2000). In the developing embryo and in the adult, SCL is detected in all of the major hematopoietic organs, strongly in hematopoietic clusters emerging from the floor of the dorsal aorta (Labastie, Cortes, Romeo, Dulac, & Peault, 1998), fetal liver, spleen and bone marrow, and restricted to a few cells in the thymus (Elefanty et al., 1999; Green, Lints, Visvader, Harvey, & Begley, 1992; Herblot, Steff, Hugo, Aplan, & Hoang, 2000; Kallianpur et al., 1994; Mouthon et al., 1993; Pulford et al., 1995). In the adult, *Scl* levels are highest in HSCs (Lacombe et al., 2010; Moignard et al., 2013), although *Scl* is also present in CFU-S (Elefanty et al., 1998), in hematopoietic precursors of all lineages, including those of the T and B lymphoid pathways (Brady et al., 1995; Herblot, Aplan, & Hoang, 2002; Herblot et al., 2000; Moignard et al., 2013; Zhang, Mortazavi, Williams, Wold, & Rothenberg, 2012), and coexpressed with GATA1 during erythroid, megakaryocytic, and mast cell differentiation (Green et al., 1992; Hoang et al., 1996; Mouthon et al., 1993; Visvader, Begley, & Adams, 1991). Outside of the hematopoietic system, SCL is found within specific regions of the midbrain, hindbrain, and spinal cord (Sinclair et al., 1999; van Eekelen et al., 2003). Thus, the broad, yet specific, expression pattern of SCL suggests that it may be involved at the onset of hematopoiesis, in hematopoietic stem and progenitor cells (HSPCs) and during lineage differentiation.

2.3 Soloist of the First Wave: *Scl*

When *Scl* is inactivated, embryos die at E9.5 due to the complete absence of blood cells and to vascular defects (Robb et al., 1995; Shivdasani, Mayer, & Orkin, 1995; Visvader, Fujiwara, & Orkin, 1998). Conversely, *Scl* overexpression expands hematopoietic and endothelial cells at the expense of other mesoderm derived tissues (Ema et al., 2003; Gering, Rodaway, Gottgens, Patient, & Green, 1998; Martin et al., 2004; Mead, Deconinck, Huber, Orkin, & Zon, 2001; Mead, Kelley, Hahn, Piedad, & Zon, 1998). Together, these results clearly establish that *Scl* determines mesoderm to a hematopoietic fate and acts as master regulator of primitive erythropoiesis. With current understanding of gene ensemble and functional nodes within intricate gene networks, it is highly surprising that invalidation of a single gene should have such a dramatic effect at the onset of primitive hematopoiesis. As it turns out, this narrow window of non-redundancy at the onset of hematopoiesis is not the norm for *Scl* which has multiple functions in hematopoiesis, as shall be discussed later. But what acts upstream of *Scl* to activate the genetic program for hematopoiesis?

2.4 Combinatorial Interaction Between Cell Intrinsic and Noncell Intrinsic Processes at the Onset of Primitive Erythropoiesis: Role of *Scl* and VEGF/FLK1 Signaling

Hematopoiesis is activated in extraembryonic mesoderm by diffusable signals emanating from the visceral endoderm (VE) layer of the yolk sac. Ablation of VE through the genetic targeting of *Gata4* abrogates the development of hematopoietic cells during the in vitro differentiation of embryonic stem (ES) cells (Bielinska, Narita, Heikinheimo, Porter, & Wilson, 1996). Indian hedgehog (IHH) has emerged as a candidate VE hematopoietic-inducing signaling molecule, by reprogramming prospective neuroectoderm to hematopoietic and endothelial cell fates to induce blood formation in explant cultures (Dyer, Farrington, Mohn, Munday, & Baron, 2001). Since BMP4 is upregulated in extraembryonic mesoderm following treatment with IHH, it is possible that IHH mediates its effect on hematopoiesis in part through BMP signaling. BMP4 is required for the formation of blood islands (Winnier, Blessing, Labosky, & Hogan, 1995) and BMP signaling is known to enhance hematopoietic development in vitro and in vivo (Chadwick et al., 2003; Crisan et al., 2015; Johansson & Wiles, 1995; Kanatsu & Nishikawa, 1996; Mead et al., 1998; Nakayama, Lee, & Chiu, 2000; Park et al., 2004). Activin A (Kanatsu & Nishikawa, 1996), TGFβ (Dickson et al., 1995; Pardanaud & Dieterlen-Lievre, 1999), and bFGF (Faloon et al., 2000; Pardanaud & Dieterlen-Lievre, 1999) have also been implicated in blood island development, although their relationship to IHH signaling has yet to be defined. Interestingly, the hematopoietic defects caused by defective IHH signaling in culture of differentiating ES cells and in zebrafish can be rescued by Notch signaling or by *Scl* expression (Kim et al., 2013), thereby placing *Scl* downstream of *Hh* during embryonic EHT.

VEGF is also expressed by VE and extraembryonic mesoderm and can respecify somatopleural mesoderm for hematopoietic and endothelial development (Pardanaud & Dieterlen-Lievre, 1999). Through aggregation studies using tetraploid and wild-type embryos, Damert et al. demonstrated that the expression of VEGF by VE is prerequisite for the formation of blood islands and cannot be compensated for by expressing VEGF in the extra-embryonic mesoderm or in the embryo proper (Damert, Miquerol, Gertsenstein, Risau, & Nagy, 2002). Considering that VEGF and BMP4 synergize during the formation of hematopoietic cells in vitro (Nakayama et al., 2000; Park et al., 2004) and that hedgehog signaling activates the expression of VEGF in a mouse hindlimb ischemia model (Pola et al., 2001), it is conceivable that the effects of IHH at the onset of hematopoiesis are mediated through the combined action of BMP4 and VEGF.

Hematopoietic and endothelial cells develop from a common precursor, the hemangioblast (Keller, Lacaud, & Robertson, 1999; Kinder et al., 1999), expressing the VEGF receptor FLK1/KDR (Eichmann et al., 1997; Kabrun et al., 1997; Nishikawa, Nishikawa, Hirashima, Matsuyoshi, & Kodama, 1998; Ogawa et al., 1999; Pardanaud et al., 1996; Yamashita et al., 2000). At a clonal level, ES cell derived hemangioblasts (BL-CFC) originate exclusively from $FLK1^{+}$ cells and require VEGF for their proliferation (Choi, Kennedy, Kazarov, Papadimitriou, & Keller, 1998; Chung et al., 2002; Kennedy et al., 1997; Martin et al., 2004; Wang et al., 2004). Embryos lacking a single *Vegf* allele (Carmeliet et al., 1996; Ferrara et al., 1996) or *Flk1* (Shalaby et al., 1995) die between E8.5–9.5, due to the absence of blood islands (Fig. 2A). Furthermore, *Flk1*$^{-/-}$ ES cells fail to reach sites of hematopoiesis in chimeras (Shalaby et al., 1997), indicating a tight dose-dependent regulation of embryonic vessel and hematopoietic development by VEGF. Surprisingly, *Flk1*$^{-/-}$ ES cells generate BL-CFC colonies and differentiate into hematopoiesis in vitro, although BL-CFC are reduced in number and do not expand as efficiently in response to VEGF (Hidaka, Stanford, & Bernstein, 1999; Schuh, Faloon, Hu, Bhimani, & Choi, 1999). Therefore, VEGF/FLK1 signaling is not essential for hematopoietic and endothelial specification. Rather, VEGF is required for the migration of hemangioblasts from the mesoderm to sites that are permissive to hematopoiesis.

Once $FLK1^{+}$ hematopoietic and endothelial precursors have reached the yolk sac, they must expand and differentiate into blood islands. We showed that different thresholds of VEGF are required for the migration of hematopoietic and endothelial precursors to the yolk sac and for their subsequent expansion into primitive erythrocytes, through a dose-dependent anti-apoptotic effect of VEGF (Martin et al., 2004; Fig. 2B). In vitro, the addition of VEGF to differentiating ES cells expands the number of primitive erythroid precursors (Ery^{P}) in a graded fashion, whereas adult hematopoietic precursors remain largely unaffected, as reported for human ES cells (Cerdan, Rouleau, & Bhatia, 2004). Conversely, we observed a progressive reduction in the number of Ery^{P} in the yolk sacs of *Vegf* hypomorph heterozygous (*Vegf*$^{lo/wt}$) and homozygous (*Vegf*$^{lo/lo}$) embryos, as compared to wild-type littermates. Since *Flk1* expression is detected in the yolk sacs of homozygous (*Vegf*$^{lo/lo}$) embryos, these results suggest that low levels of VEGF is sufficient for the migration of FLK1^{+} hematopoietic precursors to the yolk sac, whereas higher levels of VEGF are required for their expansion. Remarkably, different *Vegfa* isoforms have different functions (Leung et al., 2013). Thus, the medium/long isoform controls *Notch1* expression in somites and specifies HSCs in dorsal aorta whereas the short isoform is sufficient for arteriogenesis. Even though both lineages originate from hemangioblasts via VEGF/FLK1 signaling, these observations uncouple *Vegfa* functions in the two lineages.

Several lines of evidence indicate that SCL functions downstream of VEGF/Flk-1 signaling in putative hemangioblasts. First, *Scl*, *Flk1*, and *Vegf* are required for hematopoietic and endothelial cell development in vivo and for hemangioblast development in vitro (BL-CFC; reviewed in Ciau-Uitz et al., 2014). Second, overexpression of *Scl* expands hematopoietic and endothelial cells at the expense of other mesodermal cell fates (Ema et al., 2003; Gering et al., 1998; Gering, Yamada, Rabbitts, & Patient, 2003; Mead et al., 2001; Mead et al., 1998). Third, during development, *Scl* expression follows that of *Flk1* (Drake & Fleming, 2000; Park et al., 2004; Robertson, Kennedy, Shannon, & Keller, 2000) and is absent in *Flk1*$^{-/-}$ embryos (Ema et al., 2003). Fourth, *Scl* can rescue the hematopoietic and vascular defects of the Zebrafish mutant cloche (Liao et al., 1998), which acts upstream of *Flk1* (Liao et al., 1997), and allows BL-CFC to grow in the absence of *Flk1* (Ema et al., 2003). Furthermore, *Scl* levels are increased by VEGF during the in vitro differentiation of ES cells (Cerdan et al., 2004; Martin et al., 2004) and severely compromised when VEGF activity is reduced, as seen in *Vegf* hypomorph embryos. Finally, a gain of function of *Scl* partially complements the hematopoietic defect caused by a *Vegf* hypomorph allele, to reestablish the survival of erythroid cells and the expression of primitive erythroid genes (Martin et al., 2004). Since increased levels of SCL do not enhance survival and the expression of hematopoietic genes in a wild-type background, the results therefore suggest that SCL functions downstream of VEGF signaling to ensure an expansion of the primitive hematopoietic compartment by sustaining the survival of primitive erythrocytes (Martin et al., 2004). Consistent with this hypothesis, *Scl* was subsequently placed downstream of FLK1 signaling and of *Etv2* (Liu et al., 2015), a transcription factor that marks the hemogenic endothelium (Wareing, Eliades, Lacaud, & Kouskoff, 2012). Indeed, deletion of *Etv2* abrogates *Flk1* expression and VEGF response as well as the hematopoietic and endothelial fates (Wareing, Mazan, et al., 2012). While *Etv2* controls the expression of *Scl*, *Gata2*, and *Fli1* (Liu et al., 2015), ectopic *Scl* expression was sufficient to rescue hematopoietic differentiation in $Etv2^{-/-}$ ES cells while *Gata2* and *Fli1* expression had a partial effect only (Wareing, Mazan, et al., 2012). Given the importance of VEGF in cell migration, it remains to be seen whether *Scl* rescue would be sufficient to reestablish hematopoiesis in vivo. Nonetheless, these rescue experiments confirm the primary and nonredundant role of *Scl* at the onset of hematopoiesis, as shown by gene invalidation (Porcher et al., 1996; Shivdasani et al., 1995; Visvader et al., 1998).

2.5 Controlling Boundaries: *Scl* Drives the Hematopoietic Fate at the Expense of the Cardiac Lineage

Blood cells, the vasculature and the cardiac lineage originate from the lateral plate mesoderm, most likely from a common FLK1^{+} multipotential progenitor (Kattman, Huber, & Keller, 2006). During embryogenesis, *Scl* acts as a master regulator that specifies the hematopoietic fate from mesoderm (Lancrin et al., 2009) and in the absence of *Scl*, mouse embryos die at E9.5 due to the complete absence of primitive erythrocytes and an abnormal vasculature (Porcher et al., 1996; Shivdasani et al., 1995; Visvader et al., 1998). These functional results confirm the common embryonic origin of endothelial cells and hematopoietic cells (Kattman et al., 2006) and a common role for *Scl* in both lineages (Deleuze et al., 2012; Ema et al., 2003; Lazrak et al., 2004; Porcher, Liao, Fujiwara, Zon, & Orkin, 1999). Contrary to the above, the cardiac and hematopoietic lineages proved to be antagonistic (Schoenebeck, Keegan, & Yelon, 2007) despite sharing a common ancestry (Kattman et al., 2006). Unexpectedly, this antagonism is controlled by *Scl* that acts to repress the cardiac fate (Org et al., 2015). Indeed, in the absence of *Scl*, cardiomyocytes develop ectopically in the yolk sac vasculature, normally a specialized hematopoietic site, and in the endocardium (Van Handel et al., 2012). These results are consistent with previous reports indicating that *Scl* has mesodermal patterning activity in murine ES cells in vitro (Ismailoglu, Yeamans, Daley, Perlingeiro, & Kyba, 2008) and that *Scl* and *Etv2/Er71* spatially restrict heart size in zebrafish in vivo (Schoenebeck et al., 2007). Remarkably, these results point to a recurring theme in lineage restriction in hematopoiesis, revealing an intrinsic plasticity in progenitors even after lineage diversification. Hence, *Scl* as a master regulator has multiple roles: in specifying the hematopoietic and endothelial fates from mesoderm; in lateral inhibition to prevent alternative fate(s); i.e., cardiomyogenesis; and in consolidating of the hematopoietic fate.

2.6 First Wave Duet: *Gata1* and *Gata2*

GATA1, a zinc finger transcription factor that drives erythroid gene expression, is essential for erythroid cell maturation (Pevny et al., 1991). Surprisingly, mouse embryos lacking *Gata1* die between E10.5 and E11.5 (Fujiwara, Browne, Cunniff, Goff, & Orkin, 1996), 1–2 days later than *Scli*$^{-/-}$ embryos (Shivdasani et al., 1995). In *Gata1*$^{-/-}$ embryos, primitive erythropoiesis was initiated but cells failed to progress beyond the proerythroblast stage (Fujiwara et al., 1996) while erythropoiesis was completely blocked in *Scl*$^{-/-}$ embryos (Shivdasani et al., 1995). This is because of a functional overlap between *Gata2* and *Gata1* in the yolk sac as the combined absence of both *Gata* genes completely ablated primitive erythroid cells (Fujiwara, Chang, Williams, & Orkin, 2004). This raises the question whether *Scl* is upstream of *Gata1* and *Gata2* at the onset of primitive erythropoiesis. Genome-wide occupancy determined by ChIP-seq analysis later revealed that SCL binds to the promoter region close to the transcription start site (TSS) of *Gata1* in Lin^{-} fetal liver hematopoietic progenitors (Kassouf et al., 2010) and both *Gata1* and *Gata2* in multipotent hematopoietic progenitor cells (HPC; Wilson et al., 2010). While genome occupancy does not warrant functional consequence, Org et al. showed that *Gata1/2* expression is nearly abolished in FLK1^{+} cells in the absence of *Scl*, while *Tal1/Scl* expression is unaffected by *Gata1/2* inactivation (Org et al., 2015). These observations suggest that at this time point in development, *Scl* acts upstream of both *Gata1* and *Gata2* to initiate primitive erythropoiesis.

2.7 Third Wave Trio: The *Runx1–Gata2–Scl* Connection Unfolded

The evidence that HSCs emerge from the AGM region to seed the fetal liver was paradigm shifting (Bertrand et al., 2010; Medvinsky & Dzierzak, 1996), and this endothelial-to-hematopoietic transition (EHT) was attributed to *Runx1* at the onset of definitive hematopoiesis (Chen,

Yokomizo, Zeigler, Dzierzak, & Speck, 2009). It is now recognized that GATA2 and SCL share this privilege with RUNX1. Indeed, *Runx1* is not required during primitive erythropoiesis while the deletion of *Runx1* in vascular endothelial cells prevented the emergence of HSCs from endothelium. These observations established the functional importance of *Runx1* during EHT (Chen et al., 2009). In contrast, the early lethality of *Scl*$^{-/-}$ or *Gata2*$^{-/-}$ embryos hindered a precise assessment of their function at later developmental stages. Interestingly, *Gata2* expression during development was assigned in part to a cis-regulatory enhancer at +9.5, which then controls *Scl* et *Gfi1* expression in the AGM (Gao et al., 2013). Moreover, the generation of HSCs from the AGM depends on the integrity of the +9.5 element in vivo. Concurring with the view that RUNX1, GATA2, and SCL are often found in association in ChIP-seq experiments (Wilson et al., 2010), the above functional studies point to the importance of all three transcription factors for the emergence of definitive HSCs from endothelium (Fig. 2A). Additionally *Scl* expression and the onset of definitive HSCs is also controlled by transcription factor genes of the Ets family, *Etv2* (Liu et al., 2015), *Etv6*, and Fli1 (Ciau-Uitz, Pinheiro, Gupta, Enver, & Patient, 2010; Ciau-Uitz, Pinheiro, Kirmizitas, Zuo, & Patient, 2013).

These results also suggest that *Gata2* is upstream of *Scl* during EHT (Gao et al., 2013). Since master regulators are often autoregulatory, or involved in positive feed-forward loops, it remains to be seen whether *Scl* may in turn upregulate *Gata2* during EHT. Indeed, gene regulatory networks inferred enhancer sequences driving *Scl*, *Gata2*, and *Fli1* in embryonic day-11.5 dorsal aorta in the mouse during EHT that implicate all three genes in a recursive wiring (Pimanda et al., 2007; Sive & Gottgens, 2014). Moreover, the +9.5 sequence controlling *Gata2* expression during EHT was shown to be occupied by SCL in fetal liver hematopoietic progenitors, in a binding site-dependent manner (Sanalkumar et al., 2014). These observations suggest that *Scl* and partners may indeed form a coherent feed-forward loop with *Gata2* during EHT. Gene knockout and conditional invalidation have revealed major insight into gene function. In addition to these gene targeted approaches, the report by Gao et al. suggests that modulation of cis-regulatory sequences based on the specificity of protein DNA interactions may bring more fine-tuned instruments into the study of gene function, by integrating the timing of expression as well as spatial regulation (Gao et al., 2013; Hewitt et al., 2015).

3. THE LONG, THE INTERMEDIATE, AND THE SHORT (TERM): HSC FUNCTION AND *SCL*

3.1 A Hierarchy Within HSCs

The life-long production of blood cells depends on the dual capacity of HSCs to commit to multilineage differentiation and to self-renewal. Both properties are assessed in transplantation assays which also unraveled intrinsic heterogeneity within HSCs based on their differing capacities to sustain hematopoiesis in host mice (reviewed in Rojas-Sutterlin & Hoang, 2013). Initially, HSCs were classified as short term (ST-HSCs) and long term (LT-HSCs), based on the durability of engraftment (Fig. 3). ST-HSCs

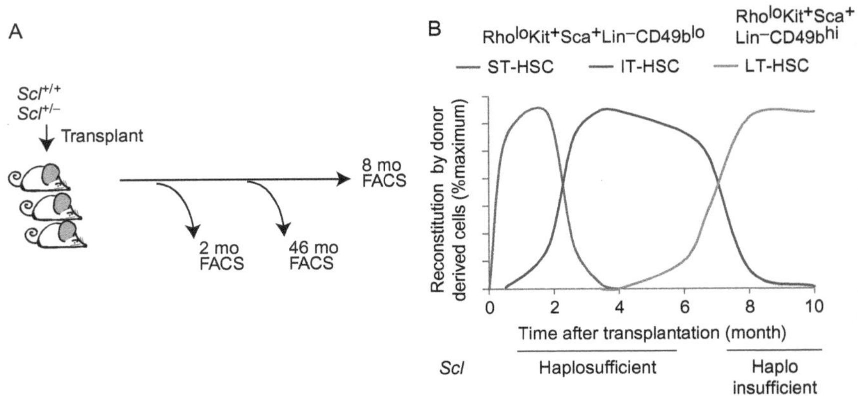

Fig. 3 *Time-line of hematopoietic reconstitution by transplanted hematopoietic stem cells.* (A) Mice are transplanted with adult bone marrow cells of different genotypes. Illustrated is a competitive assay in which HSCs of two different genotypes are transplanted in competition, typically at 1:1 ratio. Mice are sacrificed at the indicated times and analyzed for reconstitution in all hematopoietic organs, blood, bone marrow, thymus, and spleen. Test cells and competitor cells can be distinguished on the basis of CD45 alleles. (B) HSCs can be resolved into different populations based on the duration of reconstitution posttransplant. Short term HSCs (ST-HSCs) provide a transient wave of hematopoietic reconstitution (1–2 month) while long term HSCs (LT-HSCs, >4 month) can confer more durable reconstitution. These two cell types can be separated based on surface markers. Phenotypically, HSCs are enriched in the Kit$^+$Sca$^+$Lin$^-$ (KSL). ST-HSCs and LT-HSCs are found within the CD48$^+$CD150$^-$ and CD48$^-$CD150$^+$ fractions, respectively (Kiel et al., 2005). More recent evidence indicates that the population of LT-HSCs is heterogenous, containing CD49b^{lo} cells that indeed provide sustained reconstitution (>8 month), but mostly CD49b^{hi} cells providing intermediate reconstitution (IT-HSCs, <6 month) (Benveniste et al., 2010). *Scl* is haploinsufficient in these very long term HSCs (>8 month) while haplosufficient in IT-HSCs (Lacombe et al., 2010). (See the color plate.)

provided lymphomyeloid reconstitution that peaked at 3–4 weeks while LT-HSCs were defined as stem cells that generated circulating blood cells beyond 16 weeks. These cells are separable based on cell surface markers, in particular SLAM family members (Kiel et al., 2005). More recent evidence indicates that the population previously considered as LT-HSCs indeed contains cells that provide permanent and sustained life-time reconstitution but these rare cells are outnumbered by intermediate clones that self-maintained for 6–8 months and disappeared afterward (Benveniste et al., 2010).

3.2 Hierarchy Within a Hierarchy: Very Long Term HSCs Unraveled by *Scl* Haploinsufficiency—Role in Stress Response

The enhanced reconstitution ability of *SCL* overexpressing-human HSCs (Kunisato et al., 2004; Reynaud et al., 2005) suggested a role for *SCL* in HSCs, an observation that was not mirrored by loss of function studies. Although *Scl* is required for the generation of all hematopoietic cells during ontogeny (Porcher et al., 1996; Robb et al., 1996), analysis of conditional knockouts indicated that the loss of *Scl*, specifically within the adult hematopoietic compartment, did not compromise the ability of HSCs to self-renew and to reconstitute lethally irradiated recipients (Curtis et al., 2004; Mikkola et al., 2003). This unexpected observation suggests several possibilities that need not be mutually exclusive. First, the genetic program for the specification and function of a tissue may differ, implying the existence of distinct transcription factors which stabilize the hematopoietic fate, similar to the role of *Pax-5* during B cell commitment (Nutt, Heavey, Rolink, & Busslinger, 1999; Schaniel, Bruno, Melchers, & Rolink, 2002; Schaniel, Gottar, Roosnek, Melchers, & Rolink, 2002), and others that regulate self-renewal, such as *HoxB4* (Sauvageau et al., 1995) and *Bmi-1* (Lessard & Sauvageau, 2003; Park et al., 2003). Along these lines, *Runx1* and *Gata2* are shown to be essential for adult HSCs. Second, *Scl* may be redundant with other factors. Indeed, HSC survival requires both *Scl* and *Lyl1* in a gene dosage-dependent manner (Souroullas, Salmon, Sablitzky, Curtis, & Goodell, 2009). This functional redundancy between two highly homologous proteins is not unprecedented, and was previously reported for *E2a* and *Heb* during thymocyte development (Jones & Zhuang, 2007). Interestingly, the *Scl*–*Lyl1* redundancy occurs in adult HSCs but not in hemangioblasts in which reexpression of *Lyl1* cannot rescue the lethality of $Scl^{-/-}$ embryos (Chan et al., 2007). Finally, a third alternative is the possibility that different classes of HSCs may have different requirements for *Scl* gene dosage (Fig. 3). With the recognition of heterogeneity within HSCs and the existence of a highly dormant subclass that may take up to 90 days per cell division (Trumpp, Essers, & Wilson, 2010), we reasoned that the activity of these highly dormant HSCs may be revealed only after very long term reconstitution or after a prolonged proliferative stress.

We compared host reconstitution by donor cells at 4–8 months and found that stem cell functions at 4 months posttransplantation are independent of *Scl* gene dosage. However, *Scl* is haploinsufficient when HSC functions were monitored at 8 months posttransplant (Lacombe et al., 2010). Thus, our observations provide genetic evidence for a difference between intermediate term- and LT-HSCs. Furthermore, the transcription of the *Scl gene itself* was highest in LT-HSCs that are in G0 and decreased in ST-HSCs that are committed into the cell cycle. This conclusion was independently confirmed through the study of gene expression in LTR-HSCs that were purified through another protocol (Benveniste et al., 2010). Interestingly, removing one allele of *Scl* was sufficient to allow dormant LT-HSCs to transit through G0–G1, and the cells undergo exhaustion upon serial transplantation under conditions where wild-type HSCs do not. These observations suggest that quiescence control is an important determinant of the maintenance of the stem cell pool in vivo (reviewed in Nakamura-Ishizu, Takizawa, & Suda, 2014; Rojas-Sutterlin & Hoang, 2013; Sorrentino, 2010).

3.3 Role of SCL During Lineage Differentiation

The role of SCL during lineage commitment and differentiation remains unclear, mainly as a consequence of the embryonic lethality of $Scl^{-/-}$ embryos. The generation of conditional knockouts has clearly established an essential role for SCL during erythroid and megakaryocytic differentiation, as the loss of SCL results in a complete lack of differentiated cells in both lineages (Hall et al., 2003; Mikkola et al., 2003). In agreement, overexpression of SCL in immature primary hematopoietic cells enhances the output in erythroid and megakaryocytic cells (Elwood, Zogos, Pereira, Dick, & Begley, 1998; Ravet et al., 2004; Valtieri et al., 1998). These findings also concur with earlier studies describing an upregulation of *Scl* during erythroid differentiation and enhanced spontaneous erythroid differentiation upon overexpression of SCL in cell lines (Aplan, Nakahara, Orkin, & Kirsch, 1992; Green, Salvaris, & Begley, 1991; Hoang et al., 1996). The question arising is whether SCL operates at lineage branchpoints. Initially, overexpression of SCL in cell lines enhances erythroid differentiation, as seen in primary cells, but blocks myeloid differentiation

(Condorelli et al., 1997; Hoang et al., 1996; Tanigawa et al., 1993; Tanigawa, Nicola, McArthur, Strasser, & Begley, 1995). However, observations in cell lines should be interpreted with caution, as cell lines often have an ambiguous lineage status and a poor differentiation response (Socolovsky, Lodish, & Daley, 1998; Wu, Liu, Jaenisch, & Lodish, 1995). Indeed, increasing levels of SCL in primary hematopoietic cells expand the erythroid and megakaryocytic compartments, without affecting myeloid development (Elwood et al., 1998; Hall et al., 2003; Ravet et al., 2004; Valtieri et al., 1998). *Scl* was later shown to control the expansion of primary monocyte progenitors (Dey, Curtis, Jane, & Brandt, 2010). Therefore, unlike the "lateral" inhibition of the cardiac fate during development (Van Handel et al., 2012), *Scl* is permissive for both myeloid and lymphoid lineage commitment.

Within the lymphoid compartment, SCL is expressed in the earliest committed precursors of the B and T cells lineages, but is rapidly downregulated as cells mature (Herblot et al., 2002; Herblot et al., 2000), suggesting that SCL downregulation may be prerequisite to lymphoid differentiation. Indeed, the analysis of *SIL–SCL* transgenic mice (SCL^{tg}), which express SCL ubiquitously, or in *Ly6–Scl* transgenic mice, indicated that sustained expression of SCL severely compromises the earliest committed precursor of the B (Herblot et al., 2002), but not the T-cell lineages (Herblot et al., 2000). Since the number of common lymphoid progenitors (CLPs) in Scl^{tg} mice are unaffected (R. Martin & T. Hoang, unpublished data), the results suggest that SCL specifically hinders B cell differentiation after the commitment stage to CLPs.

With regards to the initial decision of HSCs to commit to either myeloid or lymphoid compartments, Kunisato et al. proposed that SCL levels skews differentiation of HSCs toward common myeloid progenitors (CMPs), at the expense of CLPs (Kunisato et al., 2004). Nonetheless, unlike its decisive role in favoring the hematopoietic fate at the expense of the cardiac fate during embryonic development, current evidence does not favor such a decisive role for SCL in establishing lineages in HSCs. Rather, as exemplified by the ability of SCL to specifically block B cell differentiation at the pro-B stage (Herblot et al., 2002), SCL may be acting not at the level of lineage choice in HSCs but later during differentiation.

3.4 Erythroid and Megakaryocyte Lineages: A Matter of SCL Isoforms?

The *Scl* gene encodes different protein isoforms (Aplan et al., 1990; Calkhoven et al., 2003). The major product is a 42-kD full length SCL protein, which contains an N-terminal transactivation domain (Hsu, Wadman, Tsan, & Baer, 1994; Sanchez-Garcia & Rabbitts, 1994; Wadman, Hsu, Cobb, & Baer, 1994), a basic domain which binds to specific DNA sequences known as E-boxes (CANNTG; Hsu, Huang, Tsan, et al., 1994) and an HLH domain (Fig. 1A) required for heterodimerization with widely expressed class-I bHLH proteins, which include E2A, HEB, and E2-2 (Goldfarb, Lewandowska, & Shoham, 1996; Hsu, Cheng, Chen, & Baer, 1991; Hsu, Wadman, & Baer, 1994; Voronova & Lee, 1994; Fig. 1B). Shorter SCL isoforms (40, 34, and 22 kD) retain the bHLH domain, but present N-terminal deletions of the transactivation domain (Aplan et al., 1990; Calkhoven et al., 2003; Cheng, Hsu, Hwang, & Baer, 1993; Pulford et al., 1995). The expression of SCL isoforms are regulated primarily through translational control (Calkhoven et al., 2003).

Structure function studies suggest that different domains of SCL are required at distinct stages of hematopoiesis. While the N-terminal and basic domains of SCL are dispensable for hematopoietic specification (Porcher et al., 1999) and transactivation of the c-kit gene in immature cells (Lecuyer et al., 2002), binding of SCL to DNA is required for the survival of HSPCs (Krosl et al., 1998; Lacombe et al., 2013; Ravet et al., 2004), for the activation of the erythroid gene *glycophorin A* (Lahlil, Lecuyer, Herblot, & Hoang, 2004; Ravet et al., 2004) and for erythroid and megakaryocytic differentiation (Porcher et al., 1999). Our observations suggest that SCL isoforms differentially regulate hematopoiesis. We have found that short isoforms are preferentially upregulated during erythroid differentiation, while remaining at low levels during megakaryocytic differentiation (Calkhoven et al., 2003). When expressed in bone marrow cells, short isoforms favor erythroid differentiation, whereas longer isoforms are favor megakaryocytic differentiation (Calkhoven et al., 2003). It would be interesting to examine the expression pattern of SCL isoforms throughout the hematopoietic hierarchy in order to determine whether specific isoforms may influence lineage output at different branchpoints.

4. THE SCL COMPLEX

4.1 Networking via LMO2

SCL does not bind DNA on its own but requires heterodimerization with E proteins, preferentially E2A/TCF3 and HEB/TCF12, for DNA binding (Hsu et al., 1991; Lecuyer & Hoang, 2004). Surprisingly, despite binding with high affinity to a consensus E box (Hsu, Huang, Tsan, et al., 1994), the SCL–E2A heterodimer does not activate transcription but instead, tempers transcription activation by E2A homodimers or E2A–HEB heterodimers (Hsu, Wadman, Tsan, et al., 1994). It was later found that the missing partner in these experiments is LMO2 (Fig. 1C) that directly interacts with SCL, GATA1, and LDB1 (El Omari et al., 2013; Wadman et al., 1997; Xu, Huang, Chang, Agulnick, & Brandt, 2003) to activate transcription. In the erythroid lineage, we showed that transcription activation of erythroid genes occurs if and only if all components of the complex are properly assembled on the promoter (Lahlil et al., 2004), resulting in a tight control of gene expression(Lecuyer & Hoang, 2004). Interestingly, the SCL–LMO2 interaction is essential for nucleation of this pentameric SCL complex (Fig. 4), a specific property that clearly distinguishes SCL from the neuronal homologue NSCL/NHLH1 (Lecuyer et al., 2007; Schlaeger et al., 2004). Furthermore, interaction with SCL protects LMO2 from proteasomal degradation (Lecuyer et al., 2007), with the resulting consequence that LMO2 is always stoichiometric with SCL and involved in complex formation.

Analysis of preferred binding sites of the SCL complex ChIP-seq indicated preferred binding to a composite E box-GATA motif (Kassouf et al., 2010; Wilson et al., 2010), a modest variation from the initial motif identified by in vitro selection (Wadman et al., 1997). The motif was found

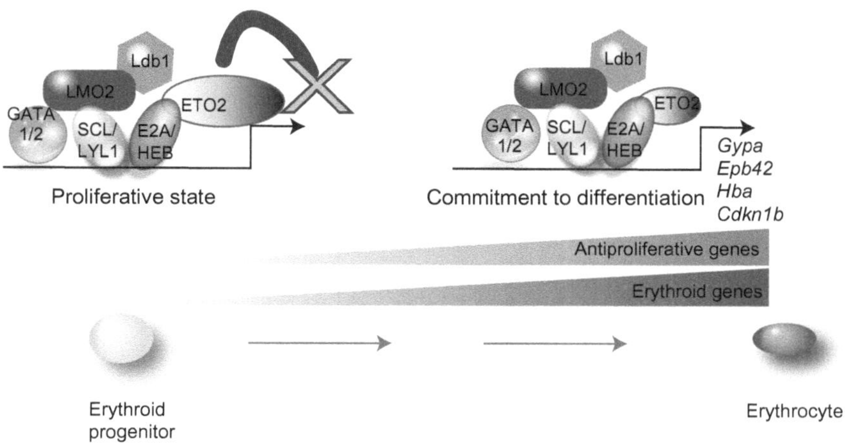

Fig. 4 *Dynamic variations in the stoichiometry of the ETO2/MTGR1 corepressor and SCL/E2A* within the SCL complex determine whether erythroid progenitors proliferate or commit to terminal differentiation. *Cdkn1b*: p27Kip; *Gypa*: glycophorin A; *Epb42*: p4.2; *Hba*: hemoglobin α (Goardon et al., 2006). (See the color plate.)

in individual SCL targets (Cohen-Kaminsky et al., 1998; Vyas et al., 1999) or more recently, via genome-wide cistrome analysis (Hewitt et al., 2015). The complex can also assemble on GATA motif in the absence of an Ebox (Kassouf et al., 2010; Lahlil et al., 2004; Wilson et al., 2010) or CACC box (Kassouf et al., 2010). Using a mutant devoid of the basic domain, Kassouf et al. showed that one fifth of SCL targets does not require direct SCL DNA binding (Kassouf et al., 2010), indicating that SCL can be recruited to DNA via protein–protein interactions (Kassouf, Chagraoui, Vyas, & Porcher, 2008; Lecuyer et al., 2002; reviewed by Lecuyer & Hoang, 2004). Resolution of the structure of the SCL complex at the atomic level revealed an unexpected function for LMO2 in strengthening the SCL–E2A heterodimer that reduces DNA binding and favors additional protein–protein interactions (El Omari et al., 2013). Among SCL direct targets, *Kit* (Kassouf et al., 2010; Lecuyer et al., 2002; Wilson et al., 2010) establishes with *Scl* a coherent feed-forward loop in multipotent and megakaryocyte-erythroid progenitors (Krosl et al., 1998; Lacombe et al., 2013; reviewed in Rojas-Sutterlin, Lecuyer, & Hoang, 2014).

4.2 Building Multifunctionality

The concept of "building multifunctionality" (Fujiwara, Lee, Sanalkumar, & Bresnick, 2010) is illustrated by the proteomics of the SCL complex that confirmed members of the core complex and revealed the recruitment of novel corepressors, in particular ETO2 (Cai et al., 2009; Goardon et al., 2006; Kolodziej et al., 2014; Schuh et al., 2005). Since SCL can both activate and inhibit gene expression, dynamic change in the stoichiometry of SCL–ETO2 could govern the switch between an activating to a repressive complex in transcription regulation. In the erythroid lineage, this switch was shown to govern the transition between a proliferative state in progenitors and growth cessation required for commitment to terminal differentiation (Goardon et al., 2006). That ETO2 should be a corepressor was not a surprise and *Eto2* knockdown resulted mostly in upregulation of gene expression in G1E erythroid cells (Fujiwara et al., 2010). The repressive function of ETO2 is well documented and occurs via HDAC1 and 2, causing E protein silencing (Zhang, Kalkum, Yamamura, Chait, & Roeder, 2004). The identification of ETO2 within the SCL complex raised much interest, since SCL also inhibits E protein activity (Chervinsky et al., 1999; Goardon et al., 2002; Herblot et al., 2000; O'Neil, Shank, Cusson, Murre, & Kelliher, 2004; Sanda et al., 2012). Contrary to ETO2, LMO2 is expected to convert this inhibitory state into an activator of transcription via its capacity to recruit transcription factors such as GATA1/2 and coactivators into the SCL pentameric complex. Unexpectedly, knocking down *Lmo2* in the same cells also derepressed gene expression (Fujiwara et al., 2010). It is possible that knocking down *Lmo2* prevents the assembly of the pentameric complex, leaving SCL in heterodimer with E2A and thereby, inhibiting E2A target genes. Nonetheless, the gene ensembles controlled by *Lmo2* and *Eto2* are significantly divergent (Fujiwara et al., 2010). These results bring into light the multifunctionality of protein complexes formed on DNA. It is possible that specific genes may have differing tolerance to a twofold decrease dosage of ETO2 and LMO2, depending on the chromatin context or on the local composition of the complex formed on DNA. Notwithstanding the exact molecular mechanism, these observations call for a reassessment of the activity of the SCL complex that may have more functions beyond the inhibition of E proteins.

SCL can also repress or activate gene expression (Lecuyer & Hoang, 2004; Palomero et al., 2006) depending on interaction with the coactivator proteins CBP/P300, P/CAF (Huang, Qiu, Shi, Xu, & Brandt, 2000; Huang, Qiu, Stein, & Brandt, 1999) or the corepressor LSD1 (Hu et al., 2009). More recent affinity purification of SCL/TAL1 coupled with quantitative mass spectrometry revealed the distinctive association of SCL with the peptidylarginine deiminase IV (PADI4) that counteracts the repressive histone mark induced by PRMT6 to enhance the active H3K4me3 mark (Kolodziej et al., 2014).

4.3 A Pentameric Complex Evolves into a Heptad

Novel technological advance brings novel insights into gene regulation. The capacity to systematically map genome-wide binding profiles of 10 hematopoietic transcription regulators via ChIP-sequencing in a multipotent model cell line (HPC7) confirmed the proximity of SCL and LMO2 on DNA (highest Z' score). Interestingly, LMO2 was found to be the major nucleation factor with pair-wise proximity to GATA2 and LYL1 as expected, and additionally, with two Ets transcription factors, ERG and FLI1, as well as RUNX1 (Wilson et al., 2010). SCL also mapped close to the same factors, although Z' scores were slightly lower. These results led to the identification of a "heptad" as illustrated in Fig. 5. Since the genome binding profile of E proteins and of LDB1 remained to be assessed in the same cells, it is predicted that the heptad may turn out to be a Decad, a composition of 10 transcription factors that control hematopoietic cell fate.

Indeed, nine of these ten factors have been shown individually to promote HSC maintenance(Knudsen et al., 2015; Kruse et al., 2009; Lacombe

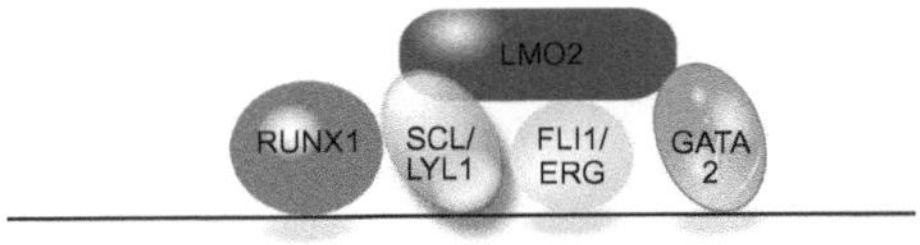

Fig. 5 *Transcription factor heptad,* as revealed by the proximity of DNA-bound peaks in multipotent HPC7 cells (Wilson et al., 2010). (See the color plate.)

et al., 2010; Ravet et al., 2004; Rodrigues et al., 2005; Semerad, Mercer, Inlay, Weissman, & Murre, 2009; Souroullas et al., 2009; Taoudi et al., 2011). These observations are consistent with the view that an extensive transcription factor network controls HSC function, thereby explaining HSC robustness and tolerance to fluctuations in *Scl* alleles (Mikkola et al., 2003), except under extensive proliferative stress when *Scl* turned out to be haploinsufficient (Lacombe et al., 2010). Similarly, *Runx1* is not essential in adult HSCs but required in stress response (Cai et al., 2015). Finally, SCL complexes can form higher order structures via the self-association capacity of LDB1, which could lead to DNA looping and thus facilitating communication between enhancers and promoters (Love, Warzecha, & Li, 2014; Yun et al., 2014).

4.4 The Importance of Being Noncell Autonomous

In another series of experiments in zebrafish embryos, Gering et al. showed that coinjection of *Scl* and *Lmo2* is sufficient to induce blood and endothelial genes along the anteroposterior axis (Gering et al., 2003). Nonetheless, erythroid cells were not produced in areas such as the head and heart, and this is due to the absence of *Gata1*. Together, these experiments indicate that transcription activation by SCL minimally requires a molecular context whereby LMO2 and GATA factors are present, possibly requiring a permissive microenvironment as well, as discussed later.

5. SCL AND CELLULAR REPROGRAMMING

5.1 The SCL Complex in Thymocyte Reprogramming

A number of functional studies indicate a role for SCL/TAL1 in suppressing apoptosis in erythroid cells during development (Martin et al., 2004) or in the adult (Palii et al., 2011), in normal HSPC (Krosl et al., 1998; Lacombe et al., 2013) and in leukemic T cells (Leroy-Viard et al., 1995; Sanda et al., 2012). In addition, SCL also exerts cell cycle control (Chagraoui et al., 2011; Lacombe et al., 2010; Palii et al., 2011), indicating a multifunctional control of hematopoietic cell fate. With the possibility of somatic cell reprogramming and given the importance of SCL in HSCs, the question arises whether SCL or its protein partners can confer de novo stem cell properties to nonstem cells? We will examine this question in the T-lineage in which SCL and LMO1 or LMO2 act as an oncogene, and extend this notion to somatic cell reprogramming. The first area addresses mechanism(s) of cell transformation, the second one would have a major impact on cell-based therapy. Indeed, bone marrow transplantation has emerged as the most promising therapeutic approach in adult acute leukemias. Nonetheless, this is limited by the availability a compatible HLA-matched donor as well as the number of cells that can be obtained for transplantation. The reader is referred to several exhaustive reviews on the molecular basis of T-ALL (Aifantis, Raetz, & Buonamici, 2008; Look, 1997; Sive & Gottgens, 2014; Van Vlierberghe & Ferrando, 2012; Van Vlierberghe, Pieters, Beverloo, & Meijerink, 2008), on preleukemia (Greaves & Maley, 2012; Nowell, 1977; Shlush & Minden, 2015) and hemogenic reprogramming (Ebina & Rossi, 2015; Vo & Daley, 2015). The current section is focused on SCL.

5.1.1 Thymocyte Differentiation

Thymocytes lack intrinsic self-renewal capacities (Bhandoola & Sambandam, 2006). Thymic output depends on the continuous influx of multipotent progenitors that are released from the bone marrow and settle in the thymus where they are directed toward the T-lineage by the thymic microenvironment, in particular Notch1 signaling (Allman et al., 2001; Petrie & Zuniga-Pflucker, 2007; Fig. 6). Rearrangement of *Tcr* genes represent the first commitment step in the thymus, which is initiated at the $CD4^-CD8^-$ double

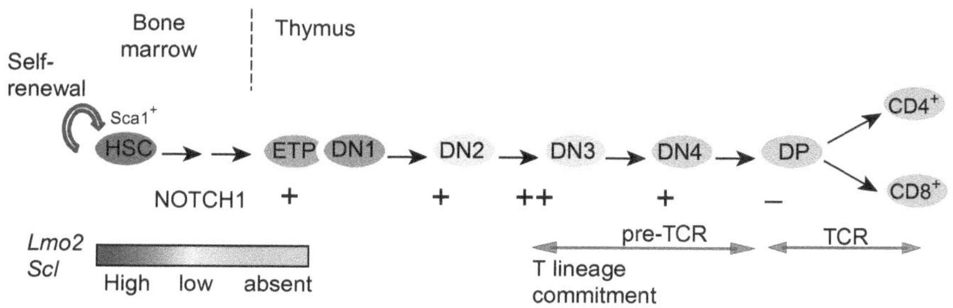

Fig. 6 *Schematic diagram of thymocyte differentiation.* Upon entering the thymus, multipotent hematopoietic progenitors that are produced in the bone marrow are guided toward a T-lineage commitment point controlled by the pre-TCR in concert with elevated Notch1 signaling at the DN3 stage ($Thy1^+CD4^-CD8^-CD44^-CD25^+$). (See the color plate.)

negative 2 (DN2) stage. *Trb/Tcrβ* rearrangement is completed at the DN3 stage. Subsequently, signaling through a functional pre-TCR and Notch1 represents an irreversible T-lineage commitment into the α/β lineage, associated with cell survival and proliferation. Cells having passed this beta selection checkpoint progress through the DN4 stage to the double positive (DP) stage to acquire CD4 and CD8 expression while rearranging their *TRA/TCRA* genes (reviewed in Ciofani & Zuniga-Pflucker, 2010; Rothenberg, 2014; Yang, Jeremiah Bell, & Bhandoola, 2010).

Scl and *Lmo2* mRNA levels are highest in DN1 thymocytes and decrease to levels that are below the limit of detection in DN3 thymocytes (Herblot et al., 2000; Zhang et al., 2012; Fig. 6). Chromosomal translocations driven by illegitimate RAG activity associated with t(1;14), t(1;7) juxtapose *TRA/TCRA or TRB/TCRB* regulatory sequences to the *SCL* gene respectively, thereby driving *SCL* transcription in all *TRA or TRB* expressing thymocyte populations (Begley, Aplan, Davey, et al., 1989; Bernard et al., 1990; Chen, Yang, et al., 1990). Another mechanism in t(1;3) translocation (Aplan et al., 1992) possibly involves Z DNA structure associated with alternating purine and pyrimidine residues on chromosome 3 together with illegitimate recombination events. These recurring chromosomal translocations never altered *SCL* coding sequences but caused ectopic *SCL* expression in thymocytes.

5.1.2 SCL *Collaborates with* LMO1 *to Reprogram Thymocytes into Preleukemic Stem Cells*

Transgenic mouse model expressing the human *SCL* and *LMO1* or *LMO2* oncogenes in the thymus closely reproduces the natural history of the human disease, with a preleukemic state evolving into aggressive T-ALL that recapitulates the same mutations found in patients' blasts (Lin, Nichols, Letterio, & Aplan, 2006; Tremblay et al., 2010). Using this mouse model, McCormack et al. and our group showed that *LMO2* or *SCL–LMO1* establish a preleukemic state in thymocytes by conferring de novo self-renewal activity to DN3 thymocytes and converting them into preleukemic stem cells (pre-LSCs; Gerby et al., 2014; Hoang, 2010; McCormack et al., 2010; Fig. 7). SCL has a dual activity in transcription regulation, either to inhibit the activity of the HEB–E2A transcription factors (Hsu, Wadman, Tsan, et al., 1994; Murre, 2000) that control normal thymocyte differentiation (Jones & Zhuang, 2007) or to activate transcription via direct association with LMO2 to nucleate the assembly of the SCL core complex (Lecuyer & Hoang, 2004). To distinguish between these two properties

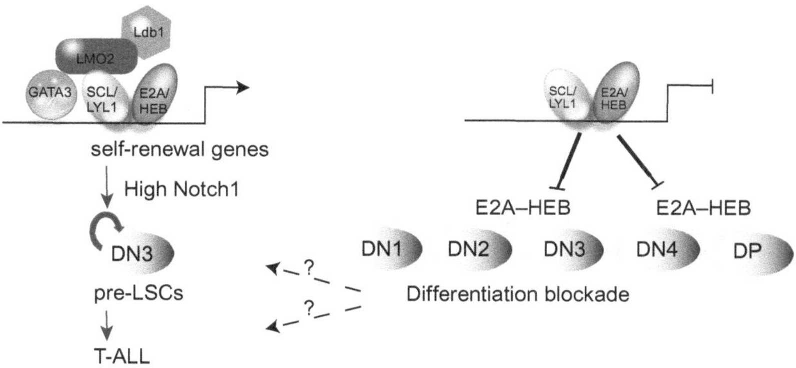

Fig. 7 *SCL acts in two parallel pathways in DN thymocytes.* SCL activates the transcription of self-renewal genes if and only if all members of the transcription complex are present. The SCL complex collaborates with high levels of Notch1 in DN3 thymocytes to reprogram these cells into self-renewing preleukemic stem cells (pre-LSCs). DN3 cells are the only targets of transformation by the SCL complex because Notch1 levels in other DN subpopulations are insufficient for collaboration. These long lasting cells acquire collaborating mutations leading to T-ALL. When bound to E2A or HEB in the absence of other transcription partners, SCL inhibits the activity of E2A/HEB heterodimers or homodimers, thereby causing differentiation blockade via decreased expression levels of E protein target genes that are required for thymocyte differentiation. (See the color plate.)

in thymocyte reprogramming, we designed an LMO2-binding defective SCL mutant that no longer activates transcription but retains its capacity to inhibit E2A, based on the interface of the SCL–LMO2 interaction (El Omari et al., 2013; Lecuyer et al., 2007). When expressed in transgenic mice, the SCL mutant fails to reprogram thymocytes and exhibits much impaired capacity to induce T-ALL, as assessed by decreased penetrance and a threefold delay in time of disease onset. Therefore, acquisition of self-renewal activity is a major determinant of T-ALL. Our genetic and genomic approaches indicate that self-renewal activity depends not on inhibition of HEB–E2A but on transcription activation by the SCL complex to drive expression of a self-renewal gene network in thymocytes. We do not rule out the possibility that SCL inhibits HEB–E2A, but this leads to differentiation blockade (Chervinsky et al., 1999; Goardon et al., 2002; Herblot et al., 2000) and not to self-renewal activity (Gerby et al., 2014; Fig. 7), nor the hypothesis that inhibition of E2A could facilitate T-ALL since removal of one *E2a* or one *Heb* allele accelerates *SCL/TAL1*-induced T-ALL (O'Neil et al., 2004). Interestingly, this genetic complementation suggests that both *E2a* and *Heb* function in a parallel pathway to *SCL*, not downstream as initially proposed by Hsu, Wadman, and Baer (1994). This is further supported by the fact that ectopic E47 expression in four LMO2-induced leukemic T-cell lines caused proliferation arrest in two of lines only, indicating that inhibition of E proteins is not a consistent feature (Goodings et al., 2015), as would be the case if *Lmo2* was to enforce E2A deficiency. Rather, these observations suggest that E2A deficiency is a collaborating event, frequently occurring in *Lmo2*-induced leukemias.

5.1.3 A Notch Up from the Microenvironment

The cell of origin of acute leukemia, which is distinct from leukemic stem cells, has been a matter of debate as it could be a self-renewing HSC or a progenitor. Our results show that the target cells of transformation by these two oncogenes are immature DN3 thymocytes, concurring with the stunning report by McCormack et al. for LMO2 (McCormack et al., 2010) which was independently confirmed in another mouse model (Cleveland et al., 2013). Consequently, the question becomes: what determines such specific targeting of DN3 cells, given that both transgenes are well expressed in all thymocyte populations (Tremblay et al., 2010)? By exploiting the ImmGen database (Heng, Painter, & Immunological Genome Project Consortium, 2008) and the Haemcode/Codex ChIP-seq resource (Wilson et al., 2010), we analyzed gene sets that are significantly increased at each transition, e.g., DN1–DN2, DN2a–DN2b, for enrichment in particular transcription factor targets. We identified *Notch1*-regulated genes as the most highly induced at the DN2b–DN3a transition. Accordingly, overexpressing an activated *Notch1* gene in transgenic mice converts all *SCL–LMO1*-expressing DN1–ISP8 thymocyte subsets into self-renewing cells. Therefore, NOTCH1 activation levels determine the susceptibility of thymocytes to reprogramming by SCL–LMO1.

These self-renewing thymocytes are considered preleukemic because they engraft the thymus and give rise to functional T cells. Furthermore, pre-LSCs can be serially transplanted 3 × without giving rise to T-ALL in most cases. Finally, pre-LSCs remain T-lineage restricted and exclusively engraft in the thymus of recipient mice. Therefore, SCL–LMO1 reprogram DN3 thymocytes into T-lineage committed pre-LSCs, not into multipotent HSCs (Gerby et al., 2014). These long lasting pre-LSCs subsequently acquire *Notch1* mutations and a competitive advantage to become leukemia propagating cells (Tremblay et al., 2010). Interestingly, in leukemic T cells, SCL forms a regulatory circuit involving HEB, E2A, LMO1/2, and RUNX1, much like the complex found in normal HSPC (Sanda et al.,

2012). In the T-lineage, GATA3 is associated with the core complex instead of GATA1 or GATA2 (Sanda et al., 2012), suggesting that GATA3 can compensate for the lack of GATA1/2 in these cells.

In summary, SCL and LMO1 operate cell autonomously to activate a self-renewal program in differentiating thymocytes, and this reprogramming activity requires high levels of Notch1 activation in DN3 cells via noncell autonomous pathways.

5.2 Hemogenic Reprogramming

Can SCL confer de novo hemogenic potential to nonhematopoietic cells? Recent evidence (Doulatov et al., 2013; Riddell et al., 2014) indicates that there are at least "Two routes to make blood" (Singbrant et al., 2015).

SCL overexpression in human ES cells accelerates hemangioblast formation while *SCL* knockdown prevented this process, as described in murine ES cells (Real et al., 2012). Furthermore, transduction of human pluripotent stem cells in culture with either *Etv2/Gata2* or *Gata2/Scl* induced robust hematopoiesis which was predominantly myeloid or erythroid/megakaryocytic, respectively (Elcheva et al., 2014; Fig. 8). Nonetheless this was not sufficient for engraftment, which requires a tailored set of transcription factors to produce mature hematopoietic cells from induced pluripotent cells

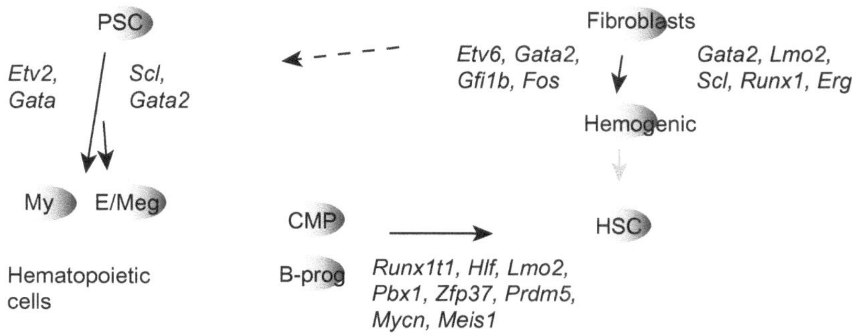

Fig. 8 *Hemogenic reprogramming*. Different starting cell types can be converted to a hematopoietic fate via the expression of a limited number of transcription factors. Human pluripotent stem cells (embryonic or induced) can be converted to endothelium to produce myeloid cells (My) by ETV2 and GATA2, or erythro-megakaryocytic (E/Meg) cells by GATA2 and SCL (Elcheva et al., 2014). In parallel, fibroblasts are directed toward hematoendothelial cells by *Etv6*, *Gata2*, *Gfi1b*, and *Fos* (Pereira et al., 2013), or by *Gata2*, *Lmo2*, *Scl*, *Runx1*, and *Erg* (Batta, Florkowska, Kouskoff, & Lacaud, 2014). Nonetheless, these cells do not engraft. Committed hematopoietic cells, i.e., common myeloid progenitors (CMPs) or B-progenitor cells (B-prog) can be reprogrammed toward HSCs that engraft in vivo by allowing for HSC selection in their niches (Riddell et al., 2014).

(Doulatov et al., 2013). In parallel, a screen for transcription factor combinations that can induce hemogenic reprogramming in murine fibroblasts revealed a combination of four factors that did not include *Scl* : *Gata2*, *Gfi1b*, *cFos*, and *Etv6* (Pereira et al., 2013; Fig. 8). Given the critical importance of *Scl* at the onset of hematopoiesis, the fact that *Scl* did not score as a hit in the screen was unexpected. It is conceivable that loss of function studies and gain of function studies do not assess the same properties. Hence, *Scl* gene knockout revealed the nonredundant function of *Scl* at a specific time point, whereas gain of function studies indicate that *Scl* might be necessary but not sufficient. In addition, the first wave hematopoiesis is geared mostly toward the production of primitive erythroid cells while the establishment of a complex hemogenic program of the third wave is more complex and requires an intricate gene network. Given that transcription factors and SCL in particular, do not function in solo but within ensembles and that networking (Kerenyi & Orkin, 2010) may be a recurring theme in cell fate decision that extends well beyond erythropoiesis, perhaps the missing factor in experiments relying on ectopic expression is the timing of SCL expression and/or the cellular/molecular context for SCL activity.

Since *Etv6* is upstream of *Scl* during hemangioblast formation (Ciau-Uitz et al., 2013) and that *Gata2* is upstream of *Scl* and *Gfi1b* (Gao et al., 2013) during EHT, these transcription factors may induce *Scl* as part of their hemogenic specification, a possibility that remains to be addressed. However, *Etv6* or *Gata2* have additional targets that may complement *Scl* function. Indeed, the study of thymocyte reprogramming indicates that SCL–LMO activity strictly depends on the microenvironment of target cells, suggesting that hemogenic reprogramming of fibroblasts may also require a permissive microenvironment. Consistent with this hypothesis, *Etv6* activates *Vegfa* expression in lateral plate mesoderm and in somites, which then upregulates *Scl* expression via FLK1 signaling, indicating a noncell autonomous role for *Etv6* in HSC development (Ciau-Uitz et al., 2010). The upregulation of *Scl* by *Vegf* in HSC development is reminiscent of the genetic interaction between *Vegf* and *Scl* at the onset of the primitive erythroid lineage (Ema et al., 2003; Martin et al., 2004). Together, these results suggest a complementation of cell autonomous and noncell autonomous pathways converging on *Scl* induction during hemogenic reprogramming. Furthermore, these results clearly indicate that fate conversion can be dictated by a limited number of transcription factors (Pereira et al., 2013), possibly through modification of the gene expression landscape and the chromatin architecture. Reprogrammed cells nonetheless did not engraft in vivo.

Another approach using a combination of 19 transcription factors delivered into E14.5 murine embryonic fibroblasts identified a minimal combination of five factors, *Erg*, *Gata2*, *Lmo2*, *Runx1c*, and *Scl* to generate multipotent hematopoietic progenitors (Batta et al., 2014; Fig. 8). In this experiment, three hematopoietic-specific components of the SCL complex scored as hits. This approach did not assess more ubiquitous protein partners encoded by *Tcf3* or *Tcf12* and *Ldb1* that are required for the activity of the SCL complex. It may be postulated that these factors are expressed at sufficiently high levels in fibroblasts to be recruited within the SCL complex. In addition, complementation by *Erg* and *Runx1c* suggests that these factors are not redundant with the SCL complex, a possibility that remains to be addressed. Finally, these reprogrammed fibroblasts showed limited self-renewal activity compared to committed common myeloid progenitors (CMP) or B-progenitor that are reprogrammed by a set of six factors (Riddell et al., 2014; Fig. 8). The latter cocktail includes *Lmo2* but not *Scl*. Since these progenitors express *Scl* (Herblot et al., 2002), it is possible that *Lmo2* is the limiting factor that needs to be supplied in trans. Interestingly, Riddell et al. took into account the importance of the microenvironment for the production of HSCs and, therefore, transplanted cells immediately after gene transfer for selection of HSCs in their physiological niches. As a consequence, the authors were able to select for HSCs that are capable of engraftment in vivo, although the efficiency of gene transfer was low (Riddell et al., 2014).

Akin to the inhibition of the cardiac fate at the onset of hematopoiesis, SCL together with LMO2 may direct cell fate conversion by driving a hematopoietic gene expression program and by inhibiting the fate of the starting cell type, be it a fibroblast or a B-progenitor. Furthermore, the combinatorial interaction of *Scl* and *Vegf* or other signaling pathways during embryonic development as well as the reliance of *SCL–LMO1* reprogrammed pre-LSCs on high Notch1 signaling indicate that the microenvironment is an important determinant in the reprogramming equation, as suggested by the successful reprogramming of B-progenitors into self-renewing HSCs via in vivo selection (Riddell et al., 2014). In addition, the requirement for coincidence detection (if and only if) of the necessary factors on DNA and the need for stoichiometry may be met at low frequency and therefore impact on the efficiency of conversion. These factors may also explain the difference in fate conversion between fibroblasts and B-progenitors as starting cell types. Finally, if cellular reprogramming is to be considered as a source of stem cells for transplantation, a strategy for regulated self-renewal activity may be needed, as the precedence in the T-lineage clearly point to an increased probability of cell transformation when these transcription factors endowed with reprogramming activity are constitutively expressed.

6. CONCLUDING REMARKS

Multidisciplinary approaches targeting *Scl* have revealed major molecular and biological insights. Molecular studies revealed that SCL interacts with an extensive network of transcription regulators, coactivators, and corepressors via its association with LMO2. These multiprotein complexes confer surprising flexibility in DNA binding, which is not limited to the canonical E box and may explain the multifaceted properties of SCL at the molecular level and in multiple hematopoietic cell types. One of the distinctive properties of this complex is the requirement for coincidence detection of at least five protein partners for transcription to occur. Since several members of the initial pentameric complex are well-established transcription factors, GATA1/2 and E2A/HEB, the requirement for all five binding to promoters for activation was initially surprising, but later explained by the structure of the complex. These extensive protein–protein interactions could explain the large numbers of SCL- or LMO2-bound sequences in the genome. The significance of these large numbers of peaks remain to be assessed, as only a limited subset of genes that are in the vicinity of these peaks are regulated by SCL levels. With the advent of quantitative proteomics, it will also be interesting to determine dynamic variations in the composition of the complex formed on DNA, not only in terms of which proteins are present or absent, but with regards to the stoichiometry of protein partners during processes such as erythroid differentiation and in T-cell transformation. Another open question is the different protein complexes in which SCL and its protein partner LMO2 are involved. For example, RUNX1, ERG, and FLI1 are often found in the vicinity of LMO2 on DNA. It is not known whether these factors associate with SCL–LMO2 in the same complex or in distinct complexes. Is there a role beyond transcription regulation for those genes that are bound by SCL or LMO2 but are not affected by *Scl* or *Lmo2* knockdown? Another aspect that remains to be documented is the importance of SCL isoforms.

Functional studies with *Scl* revealed the surprising robustness of HSCs during steady state. These genetic studies also unraveled an unexpected hierarchy within LT-HSCs and unearthed a rare subpopulation of very long term and dormant HSCs that are recruited into action under stress conditions. These cells represent the only subpopulation of LT-HSCs that depends on *Scl* gene dosage. Finally, the study of transgenic mice expressing the human *SCL* and *LMO1* oncogenes provides evidence for a preleukemic state which is initiated by the reprogramming of cells with a finite life span into self-renewing preleukemic stem cells. These cells eventually evolve into leukemia propagating cells. *Scl* and *Lmo2* are identified in strategies aimed at hematopoietic reprogramming of diverse cell types into blood or blood stem cells, consistent with the role of *Scl* as a master regulator that specifies the hematopoietic fate and prevents alternate fates. These successful approaches indicate that fate conversion can be achieved and open highly promising perspectives in cell therapy. The next question is how to achieve regulated expression, in particular for transcription factors that can confer de novo self-renewal properties.

ACKNOWLEDGMENTS

This work was funded by grants from the Canadian Institutes for Health Research (CIHR), the Canadian Cancer Society Research Institute (CCSRI), the Leukemia Lymphoma Society of Canada, the Cancer Research Society Inc., National Science and Engineering Council of Canada (NSERC), the MERST Québec and Fonds de Recherche du Quebec (FRQ)-Nature et Technologie (T.H.), and by studentships from the FRQ-Santé (FRQS) and NSERC (R.M., J.L.). The infrastructure of IRIC is supported in part by a group grant from FRQS. The funders had no role in study design, data collection and analysis, decision to publish, or preparation of the manuscript.

REFERENCES

Aifantis, I., Raetz, E., & Buonamici, S. (2008). Molecular pathogenesis of T-cell leukaemia and lymphoma. *Nature Reviews. Immunology*, *8*(5), 380–390.

Allman, D., Karnell, F. G., Punt, J. A., Bakkour, S., Xu, L., Myung, P., … Pear, W. S. (2001). Separation of Notch1 promoted lineage commitment and expansion/transformation in developing T cells. *The Journal of Experimental Medicine*, *194*(1), 99–106.

Aplan, P. D., Begley, C. G., Bertness, V., Nussmeier, M., Ezquerra, A., Coligan, J., & Kirsch, I. R. (1990). The SCL gene is formed from a transcriptionally complex locus. *Molecular and Cellular Biology*, *10*(12), 6426–6435.

Aplan, P. D., Nakahara, K., Orkin, S. H., & Kirsch, I. R. (1992). The SCL gene product: A positive regulator of erythroid differentiation. *The EMBO Journal*, *11*(11), 4073–4081.

Batta, K., Florkowska, M., Kouskoff, V., & Lacaud, G. (2014). Direct reprogramming of murine fibroblasts to hematopoietic progenitor cells. *Cell Reports*, *9*(5), 1871–1884. http://dx.doi.org/10.1016/j.celrep.2014.11.002.

Begley, C. G., Aplan, P. D., Davey, M. P., Nakahara, K., Tchorz, K., Kurtzberg, J., … Waldmann, T. A. (1989a). Chromosomal translocation in a human leukemic stem-cell line disrupts the T-cell antigen receptor delta-chain diversity region and results in a previously unreported fusion transcript. *Proceedings of the National Academy of Sciences of the United States of America*, *86*(6), 2031–2035.

Begley, C. G., Aplan, P. D., Denning, S. M., Haynes, B. F., Waldmann, T. A., & Kirsch, I. R. (1989b). The gene SCL is expressed during early hematopoiesis and encodes a differentiation-related DNA-binding motif. *Proceedings of the National Academy of Sciences of the United States of America*, *86*(24), 10128–10132.

Begley, C. G., & Green, A. R. (1999). The SCL gene: From case report to critical hematopoietic regulator. *Blood*, *93*(9), 2760–2770.

Benveniste, P., Frelin, C., Janmohamed, S., Barbara, M., Herrington, R., Hyam, D., & Iscove, N. N. (2010). Intermediate-term hematopoietic stem cells with extended but time-limited reconstitution potential. *Cell Stem Cell*, *6*(1), 48–58. http://dx.doi.org/10.1016/j.stem.2009.11.014.

Bernard, O., Guglielmi, P., Jonveaux, P., Cherif, D., Gisselbrecht, S., Mauchauffe, M., … Mathieu-Mahul, D. (1990). Two distinct mechanisms for the SCL gene activation in the t(1;14) translocation of T-cell leukemias. *Genes, Chromosomes & Cancer*, *1*(3), 194–208.

Bertrand, J. Y., Chi, N. C., Santoso, B., Teng, S., Stainier, D. Y., & Traver, D. (2010). Haematopoietic stem cells derive directly from aortic endothelium during development. *Nature*, *464*(7285), 108–111. http://dx.doi.org/10.1038/nature08738.

Bhandoola, A., & Sambandam, A. (2006). From stem cell to T cell: One route or many? *Nature Reviews. Immunology*, *6*(2), 117–126.

Bielinska, M., Narita, N., Heikinheimo, M., Porter, S. B., & Wilson, D. B. (1996). Erythropoiesis and vasculogenesis in embryoid bodies lacking visceral yolk sac endoderm. *Blood*, *88*(10), 3720–3730.

Boisset, J. C., van Cappellen, W., Andrieu-Soler, C., Galjart, N., Dzierzak, E., & Robin, C. (2010). In vivo imaging of haematopoietic cells emerging from the mouse aortic endothelium. *Nature*, *464*(7285), 116–120.

Brady, G., Billia, F., Knox, J., Hoang, T., Kirsch, I. R., Voura, E. B., … Siminovitch, K. (1995). Analysis of gene expression in a complex differentiation hierarchy by global amplification of cDNA from single cells. *Current Biology*, *5*(8), 909–922.

Cai, X., Gao, L., Teng, L., Ge, J., Oo, Z. M., Kumar, A. R., … Speck, N. A. (2015). Runx1 deficiency decreases ribosome biogenesis and confers stress Resistance to hematopoietic stem and progenitor cells. *Cell Stem Cell*, *17*(2), 165–177.

Cai, Y., Xu, Z., Xie, J., Ham, A. J., Koury, M. J., Hiebert, S. W., & Brandt, S. J. (2009). Eto2/MTG16 and MTGR1 are heteromeric corepressors of the TAL1/SCL transcription factor in murine erythroid progenitors. *Biochemical and Biophysical Research Communications*, *390*(2), 295–301.

Calkhoven, C. F., Muller, C., Martin, R., Krosl, G., Pietsch, H., Hoang, T., & Leutz, A. (2003). Translational control of SCL-isoform expression in hematopoietic lineage choice. *Genes & Development*, *17*(8), 959–964.

Carmeliet, P., Ferreira, V., Breier, G., Pollefeyt, S., Kieckens, L., Gertsenstein, M., … Nagy, A. (1996). Abnormal blood vessel development and lethality in embryos lacking a single VEGF allele. *Nature*, *380*(6573), 435–439. http://dx.doi.org/10.1038/380435a0.

Cerdan, C., Rouleau, A., & Bhatia, M. (2004). VEGF-A165 augments erythropoietic development from human embryonic stem cells. *Blood*, *103*(7), 2504–2512. http://dx.doi.org/10.1182/blood-2003-07-2563.

Chadwick, K., Wang, L., Li, L., Menendez, P., Murdoch, B., Rouleau, A., & Bhatia, M. (2003). Cytokines and BMP-4 promote hematopoietic differentiation of human embryonic stem cells. *Blood*, *102*(3), 906–915. http://dx.doi.org/10.1182/blood-2003-03-0832.

Chagraoui, H., Kassouf, M., Banerjee, S., Goardon, N., Clark, K., Atzberger, A., … Porcher, C. (2011). SCL-mediated regulation of the cell-cycle regulator p21 is critical for murine megakaryopoiesis. *Blood*, *118*(3), 723–735.

Chan, W. Y., Follows, G. A., Lacaud, G., Pimanda, J. E., Landry, J. R., Kinston, S., ... Gottgens, B. (2007). The paralogous hematopoietic regulators Lyl1 and Scl are coregulated by Ets and GATA factors, but Lyl1 cannot rescue the early Scl −/− phenotype. *Blood*, *109*(5), 1908–1916.

Chen, Q., Cheng, J. T., Tasi, L. H., Schneider, N., Buchanan, G., Carroll, A., ... Baer, R. (1990). The tal gene undergoes chromosome translocation in T cell leukemia and potentially encodes a helix-loop-helix protein. *The EMBO Journal*, *9*(2), 415–424.

Chen, Q., Yang, C. Y., Tsan, J. T., Xia, Y., Ragab, A. H., Peiper, S. C., ... Baer, R. (1990). Coding sequences of the tal-1 gene are disrupted by chromosome translocation in human T cell leukemia. *The Journal of Experimental Medicine*, *172*(5), 1403–1408.

Chen, M. J., Yokomizo, T., Zeigler, B. M., Dzierzak, E., & Speck, N. A. (2009). Runx1 is required for the endothelial to haematopoietic cell transition but not thereafter. *Nature*, *457*(7231), 887–891.

Cheng, J. T., Hsu, H. L., Hwang, L. Y., & Baer, R. (1993). Products of the TAL1 oncogene: Basic helix-loop-helix proteins phosphorylated at serine residues. *Oncogene*, *8*(3), 677–683.

Chervinsky, D. S., Zhao, X. F., Lam, D. H., Ellsworth, M., Gross, K. W., & Aplan, P. D. (1999). Disordered T-cell development and T-cell malignancies in SCL LMO1 double-transgenic mice: Parallels with E2A-deficient mice. *Molecular and Cellular Biology*, *19*(7), 5025–5035.

Choi, K., Kennedy, M., Kazarov, A., Papadimitriou, J. C., & Keller, G. (1998). A common precursor for hematopoietic and endothelial cells. *Development*, *125*(4), 725–732.

Chung, Y. S., Zhang, W. J., Arentson, E., Kingsley, P. D., Palis, J., & Choi, K. (2002). Lineage analysis of the hemangioblast as defined by FLK1 and SCL expression. *Development*, *129*(23), 5511–5520.

Ciau-Uitz, A., Monteiro, R., Kirmizitas, A., & Patient, R. (2014). Developmental hematopoiesis: Ontogeny, genetic programming and conservation. *Experimental Hematology*, *42*(8), 669–683.

Ciau-Uitz, A., Pinheiro, P., Gupta, R., Enver, T., & Patient, R. (2010). Tel1/ETV6 specifies blood stem cells through the agency of VEGF signaling. *Developmental Cell*, *18*(4), 569–578.

Ciau-Uitz, A., Pinheiro, P., Kirmizitas, A., Zuo, J., & Patient, R. (2013). VEGFA-dependent and -independent pathways synergise to drive Scl expression and initiate programming of the blood stem cell lineage in Xenopus. *Development*, *140*(12), 2632–2642.

Ciofani, M., & Zuniga-Pflucker, J. C. (2010). Determining gammadelta versus alphass T cell development. *Nature Reviews. Immunology*, *10*(9), 657–663.

Cleveland, S. M., Smith, S., Tripathi, R., Mathias, E. M., Goodings, C., Elliott, N., ... Dave, U. P. (2013). Lmo2 induces hematopoietic stem cell-like features in T-cell progenitor cells prior to leukemia. *Stem Cells*, *31*(5), 882–894. http://dx.doi.org/10.1002/stem.1345.

Cohen-Kaminsky, S., Maouche-Chretien, L., Vitelli, L., Vinit, M. A., Blanchard, I., Yamamoto, M., ... Romeo, P. H. (1998). Chromatin immunoselection defines a TAL-1 target gene. *The EMBO Journal*, *17*(17), 5151–5160.

Condorelli, G. L., Tocci, A., Botta, R., Facchiano, F., Testa, U., Vitelli, L., ... Peschle, C. (1997). Ectopic TAL-1/SCL expression in phenotypically normal or leukemic myeloid precursors: Proliferative and antiapoptotic effects coupled with a differentiation blockade. *Molecular and Cellular Biology*, *17*(5), 2954–2969.

Crisan, M., Kartalaei, P. S., Vink, C., Yamada-Inagawa, T., Bollerot, K., van, I. W., & Dzierzak, E. (2015). BMP signalling differentially regulates distinct haematopoietic stem cell types. *Nature Communications*, *6*, 8040.

Curtis, D. J., Hall, M. A., Van Stekelenburg, L. J., Robb, L., Jane, S. M., & Begley, C. G. (2004). SCL is required for normal function of short-term repopulating hematopoietic stem cells. *Blood, 103*(9), 3342–3348.

Damert, A., Miquerol, L., Gertsenstein, M., Risau, W., & Nagy, A. (2002). Insufficient VEGFA activity in yolk sac endoderm compromises haematopoietic and endothelial differentiation. *Development, 129*(8), 1881–1892.

Deleuze, V., El-Hajj, R., Chalhoub, E., Dohet, C., Pinet, V., Couttet, P., & Mathieu, D. (2012). Angiopoietin-2 is a direct transcriptional target of TAL1, LYL1 and LMO2 in endothelial cells. *PLoS One,* 7(7), e40484.

Dey, S., Curtis, D. J., Jane, S. M., & Brandt, S. J. (2010). The TAL1/SCL transcription factor regulates cell cycle progression and proliferation in differentiating murine bone marrow monocyte precursors. *Molecular and Cellular Biology, 30*(9), 2181–2192.

Dickson, M. C., Martin, J. S., Cousins, F. M., Kulkarni, A. B., Karlsson, S., & Akhurst, R. J. (1995). Defective haematopoiesis and vasculogenesis in transforming growth factor-beta 1 knock out mice. *Development, 121*(6), 1845–1854.

Doulatov, S., Vo, L. T., Chou, S. S., Kim, P. G., Arora, N., Li, H., … Daley, G. Q. (2013). Induction of multipotential hematopoietic progenitors from human pluripotent stem cells via respecification of lineage-restricted precursors. *Cell Stem Cell, 13*(4), 459–470.

Drake, C. J., & Fleming, P. A. (2000). Vasculogenesis in the day 6.5 to 9.5 mouse embryo. *Blood, 95*(5), 1671–1679.

Dyer, M. A., Farrington, S. M., Mohn, D., Munday, J. R., & Baron, M. H. (2001). Indian hedgehog activates hematopoiesis and vasculogenesis and can respecify prospective neurectodermal cell fate in the mouse embryo. *Development, 128*(10), 1717–1730.

Ebina, W., & Rossi, D. J. (2015). Transcription factor-mediated reprogramming toward hematopoietic stem cells. *The EMBO Journal, 34*(6), 694–709.

Eichmann, A., Corbel, C., Nataf, V., Vaigot, P., Breant, C., & Le Douarin, N. M. (1997). Ligand-dependent development of the endothelial and hemopoietic lineages from embryonic mesodermal cells expressing vascular endothelial growth factor receptor 2. *Proceedings of the National Academy of Sciences of the United States of America, 94*(10), 5141–5146.

Elcheva, I., Brok-Volchanskaya, V., Kumar, A., Liu, P., Lee, J. H., Tong, L., … Slukvin, I. (2014). Direct induction of haematoendothelial programs in human pluripotent stem cells by transcriptional regulators. *Nature Communications, 5*, 4372.

Elefanty, A. G., Begley, C. G., Hartley, L., Papaevangeliou, B., & Robb, L. (1999). SCL expression in the mouse embryo detected with a targeted lacZ reporter gene demonstrates its localization to hematopoietic, vascular, and neural tissues. *Blood, 94*(11), 3754–3763.

Elefanty, A. G., Begley, C. G., Metcalf, D., Barnett, L., Kontgen, F., & Robb, L. (1998). Characterization of hematopoietic progenitor cells that express the transcription factor SCL, using a lacZ "knock-in" strategy. *Proceedings of the National Academy of Sciences of the United States of America, 95*(20), 11897–11902.

El Omari, K., Hoosdally, S. J., Tuladhar, K., Karia, D., Hall-Ponsele, E., Platonova, O., … Mancini, E. J. (2013). Structural basis for LMO2-driven recruitment of the SCL:E47bHLH heterodimer to hematopoietic-specific transcriptional targets. *Cell Reports, 4*(1), 135–147.

Elwood, N. J., Zogos, H., Pereira, D. S., Dick, J. E., & Begley, C. G. (1998). Enhanced megakaryocyte and erythroid development from normal human CD34(+) cells: Consequence of enforced expression of SCL. *Blood, 91*(10), 3756–3765.

Ema, M., Faloon, P., Zhang, W. J., Hirashima, M., Reid, T., Stanford, W. L., … Rossant, J. (2003). Combinatorial effects of Flk1 and Tal1 on vascular and hematopoietic development in the mouse. *Genes & Development, 17*(3), 380–393.

Faloon, P., Arentson, E., Kazarov, A., Deng, C. X., Porcher, C., Orkin, S., & Choi, K. (2000). Basic fibroblast growth factor positively regulates hematopoietic development. *Development, 127*(9), 1931–1941.
Ferrara, N., Carver-Moore, K., Chen, H., Dowd, M., Lu, L., O'Shea, K. S., ... Moore, M. W. (1996). Heterozygous embryonic lethality induced by targeted inactivation of the VEGF gene. *Nature, 380*(6573), 439–442.
Finger, L. R., Kagan, J., Christopher, G., Kurtzberg, J., Hershfield, M. S., Nowell, P. C., & Croce, C. M. (1989). Involvement of the TCL5 gene on human chromosome 1 in T-cell leukemia and melanoma. *Proceedings of the National Academy of Sciences of the United States of America, 86*(13), 5039–5043.
Fujiwara, Y., Browne, C. P., Cunniff, K., Goff, S. C., & Orkin, S. H. (1996). Arrested development of embryonic red cell precursors in mouse embryos lacking transcription factor GATA-1. *Proceedings of the National Academy of Sciences of the United States of America, 93*(22), 12355–12358.
Fujiwara, Y., Chang, A. N., Williams, A. M., & Orkin, S. H. (2004). Functional overlap of GATA-1 and GATA-2 in primitive hematopoietic development. *Blood, 103*(2), 583–585. http://dx.doi.org/10.1182/blood-2003-08-2870.
Fujiwara, T., Lee, H. Y., Sanalkumar, R., & Bresnick, E. H. (2010). Building multifunctionality into a complex containing master regulators of hematopoiesis. *Proceedings of the National Academy of Sciences of the United States of America, 107*(47), 20429–20434.
Gao, X., Johnson, K. D., Chang, Y. I., Boyer, M. E., Dewey, C. N., Zhang, J., & Bresnick, E. H. (2013). Gata2 cis-element is required for hematopoietic stem cell generation in the mammalian embryo. *The Journal of Experimental Medicine, 210*(13), 2833–2842.
Gerby, B., Tremblay, C. S., Tremblay, M., Rojas-Sutterlin, S., Herblot, S., Hebert, J., ... Hoang, T. (2014). SCL, LMO1 and Notch1 reprogram thymocytes into self-renewing cells. *PLoS Genetics, 10*(12), e1004768.
Gering, M., Rodaway, A. R., Gottgens, B., Patient, R. K., & Green, A. R. (1998). The SCL gene specifies haemangioblast development from early mesoderm. *The EMBO Journal, 17*(14), 4029–4045.
Gering, M., Yamada, Y., Rabbitts, T. H., & Patient, R. K. (2003). Lmo2 and Scl/Tal1 convert non-axial mesoderm into haemangioblasts which differentiate into endothelial cells in the absence of Gata1. *Development, 130*(25), 6187–6199.
Goardon, N., Lambert, J. A., Rodriguez, P., Nissaire, P., Herblot, S., Thibault, P., ... Hoang, T. (2006). ETO2 coordinates cellular proliferation and differentiation during erythropoiesis. *The EMBO Journal, 25*(2), 357–366.
Goardon, N., Schuh, A., Hajar, I., Ma, X., Jouault, H., Dzierzak, E., ... Maouche-Chretien, L. (2002). Ectopic expression of TAL-1 protein in Ly-6E.1-htal-1 transgenic mice induces defects in B- and T-lymphoid differentiation. *Blood, 100*(2), 491–500.
Goldfarb, A. N., Lewandowska, K., & Shoham, M. (1996). Determinants of helix-loop-helix dimerization affinity. Random mutational analysis of SCL/tal. *The Journal of Biological Chemistry, 271*(5), 2683–2688.
Goodings, C., Tripathi, R., Cleveland, S. M., Elliott, N., Guo, Y., Shyr, Y., & Dave, U. P. (2015). Enforced expression of E47 has differential effects on Lmo2-induced T-cell leukemias. *Leukemia Research, 39*(1), 100–109.
Greaves, M., & Maley, C. C. (2012). Clonal evolution in cancer. *Nature, 481*(7381), 306–313.
Green, A. R., Lints, T., Visvader, J., Harvey, R., & Begley, C. G. (1992). SCL is coexpressed with GATA-1 in hemopoietic cells but is also expressed in developing brain. *Oncogene, 7*(4), 653–660.
Green, A. R., Salvaris, E., & Begley, C. G. (1991). Erythroid expression of the 'helix-loop-helix' gene, SCL. *Oncogene, 6*(3), 475–479.

Hall, M. A., Curtis, D. J., Metcalf, D., Elefanty, A. G., Sourris, K., Robb, L., ... Begley, C. G. (2003). The critical regulator of embryonic hematopoiesis, SCL, is vital in the adult for megakaryopoiesis, erythropoiesis, and lineage choice in CFU-S12. *Proceedings of the National Academy of Sciences of the United States of America, 100*(3), 992–997.

Heng, T. S., Painter, M. W., & Immunological Genome Project Consortium. (2008). The Immunological Genome Project: networks of gene expression in immune cells. *Nature Immunology, 9*(10), 1091–1094.

Herblot, S., Aplan, P. D., & Hoang, T. (2002). Gradient of E2A activity in B-cell development. *Molecular and Cellular Biology, 22*(3), 886–900.

Herblot, S., Steff, A. M., Hugo, P., Aplan, P. D., & Hoang, T. (2000). SCL and LMO1 alter thymocyte differentiation: Inhibition of E2A-HEB function and pre-T alpha chain expression. *Nature Immunology, 1*(2), 138–144.

Hewitt, K. J., Kim, D. H., Devadas, P., Prathibha, R., Zuo, C., Sanalkumar, R., ... Bresnick, E. H. (2015). Hematopoietic signaling mechanism revealed from a stem/progenitor cell cistrome. *Molecular Cell, 59*(1), 62–74.

Hidaka, M., Stanford, W. L., & Bernstein, A. (1999). Conditional requirement for the Flk-1 receptor in the in vitro generation of early hematopoietic cells. *Proceedings of the National Academy of Sciences of the United States of America, 96*(13), 7370–7375.

Hoang, T. (2010). Of mice and men: How an oncogene transgresses the limits and predisposes to T cell acute lymphoblastic leukemia. *Science Translational Medicine, 2*(21), 21ps10.

Hoang, T., Paradis, E., Brady, G., Billia, F., Nakahara, K., Iscove, N. N., & Kirsch, I. R. (1996). Opposing effects of the basic helix-loop-helix transcription factor SCL on erythroid and monocytic differentiation. *Blood, 87*(1), 102–111.

Hsu, H. L., Cheng, J. T., Chen, Q., & Baer, R. (1991). Enhancer-binding activity of the tal-1 oncoprotein in association with the E47/E12 helix-loop-helix proteins. *Molecular and Cellular Biology, 11*(6), 3037–3042.

Hsu, H. L., Huang, L., Tsan, J. T., Funk, W., Wright, W. E., Hu, J. S., ... Baer, R. (1994). Preferred sequences for DNA recognition by the TAL1 helix-loop-helix proteins. *Molecular and Cellular Biology, 14*(2), 1256–1265.

Hsu, H. L., Wadman, I., & Baer, R. (1994). Formation of in vivo complexes between the TAL1 and E2A polypeptides of leukemic T cells. *Proceedings of the National Academy of Sciences of the United States of America, 91*(8), 3181–3185.

Hsu, H. L., Wadman, I., Tsan, J. T., & Baer, R. (1994). Positive and negative transcriptional control by the TAL1 helix-loop-helix protein. *Proceedings of the National Academy of Sciences of the United States of America, 91*(13), 5947–5951.

Hu, X., Li, X., Valverde, K., Fu, X., Noguchi, C., Qiu, Y., & Huang, S. (2009). LSD1-mediated epigenetic modification is required for TAL1 function and hematopoiesis. *Proceedings of the National Academy of Sciences of the United States of America, 106*(25), 10141–10146.

Huang, S., Qiu, Y., Shi, Y., Xu, Z., & Brandt, S. J. (2000). P/CAF-mediated acetylation regulates the function of the basic helix-loop-helix transcription factor TAL1/SCL. *The EMBO Journal, 19*(24), 6792–6803.

Huang, S., Qiu, Y., Stein, R. W., & Brandt, S. J. (1999). p300 functions as a transcriptional coactivator for the TAL1/SCL oncoprotein. *Oncogene, 18*(35), 4958–4967. http://dx.doi.org/10.1038/sj.onc.1202889.

Ismailoglu, I., Yeamans, G., Daley, G. Q., Perlingeiro, R. C., & Kyba, M. (2008). Mesodermal patterning activity of SCL. *Experimental Hematology, 36*(12), 1593–1603.

Johansson, B. M., & Wiles, M. V. (1995). Evidence for involvement of activin A and bone morphogenetic protein 4 in mammalian mesoderm and hematopoietic development. *Molecular and Cellular Biology, 15*(1), 141–151.

Jones, M. E., & Zhuang, Y. (2007). Acquisition of a functional T cell receptor during T lymphocyte development is enforced by HEB and E2A transcription factors. *Immunity*, *27*(6), 860–870.

Kabrun, N., Buhring, H. J., Choi, K., Ullrich, A., Risau, W., & Keller, G. (1997). Flk-1 expression defines a population of early embryonic hematopoietic precursors. *Development*, *124*(10), 2039–2048.

Kallianpur, A. R., Jordan, J. E., & Brandt, S. J. (1994). The SCL/TAL-1 gene is expressed in progenitors of both the hematopoietic and vascular systems during embryogenesis. *Blood*, *83*(5), 1200–1208.

Kanatsu, M., & Nishikawa, S. I. (1996). In vitro analysis of epiblast tissue potency for hematopoietic cell differentiation. *Development*, *122*(3), 823–830.

Kassouf, M. T., Chagraoui, H., Vyas, P., & Porcher, C. (2008). Differential use of SCL/TAL-1 DNA-binding domain in developmental hematopoiesis. *Blood*, *112*(4), 1056–1067.

Kassouf, M. T., Hughes, J. R., Taylor, S., McGowan, S. J., Soneji, S., Green, A. L., … Porcher, C. (2010). Genome-wide identification of TAL1's functional targets: Insights into its mechanisms of action in primary erythroid cells. *Genome Research*, *20*(8), 1064–1083.

Kattman, S. J., Huber, T. L., & Keller, G. M. (2006). Multipotent flk-1 + cardiovascular progenitor cells give rise to the cardiomyocyte, endothelial, and vascular smooth muscle lineages. *Developmental Cell*, *11*(5), 723–732.

Keller, G., Lacaud, G., & Robertson, S. (1999). Development of the hematopoietic system in the mouse. *Experimental Hematology*, *27*(5), 777–787.

Kennedy, M., Firpo, M., Choi, K., Wall, C., Robertson, S., Kabrun, N., & Keller, G. (1997). A common precursor for primitive erythropoiesis and definitive haematopoiesis. *Nature*, *386*(6624), 488–493.

Kerenyi, M. A., & Orkin, S. H. (2010). Networking erythropoiesis. *The Journal of Experimental Medicine*, *207*(12), 2537–2541.

Kiel, M. J., Yilmaz, O. H., Iwashita, T., Yilmaz, O. H., Terhorst, C., & Morrison, S. J. (2005). SLAM family receptors distinguish hematopoietic stem and progenitor cells and reveal endothelial niches for stem cells. *Cell*, *121*(7), 1109–1121.

Kim, P. G., Albacker, C. E., Lu, Y. F., Jang, I. H., Lim, Y., Heffner, G. C., … Daley, G. Q. (2013). Signaling axis involving Hedgehog, Notch, and Scl promotes the embryonic endothelial-to-hematopoietic transition. *Proceedings of the National Academy of Sciences of the United States of America*, *110*(2), E141–E150.

Kinder, S. J., Tsang, T. E., Quinlan, G. A., Hadjantonakis, A. K., Nagy, A., & Tam, P. P. (1999). The orderly allocation of mesodermal cells to the extraembryonic structures and the anteroposterior axis during gastrulation of the mouse embryo. *Development*, *126*(21), 4691–4701.

Knudsen, K. J., Rehn, M., Hasemann, M. S., Rapin, N., Bagger, F. O., Ohlsson, E., … Porse, B. T. (2015). ERG promotes the maintenance of hematopoietic stem cells by restricting their differentiation. *Genes & Development*, *29*(18), 1915–1929.

Kolodziej, S., Kuvardina, O. N., Oellerich, T., Herglotz, J., Backert, I., Kohrs, N., … Lausen, J. (2014). PADI4 acts as a coactivator of Tal1 by counteracting repressive histone arginine methylation. *Nature Communications*, *5*, 3995.

Krosl, G., He, G., Lefrancois, M., Charron, F., Romeo, P. H., Jolicoeur, P., … Hoang, T. (1998). Transcription factor SCL is required for c-kit expression and c-Kit function in hemopoietic cells. *The Journal of Experimental Medicine*, *188*(3), 439–450.

Kruse, E. A., Loughran, S. J., Baldwin, T. M., Josefsson, E. C., Ellis, S., Watson, D. K., … Kile, B. T. (2009). Dual requirement for the ETS transcription factors Fli-1 and Erg in hematopoietic stem cells and the megakaryocyte lineage. *Proceedings of the National Academy of Sciences of the United States of America*, *106*(33), 13814–13819.

Kunisato, A., Chiba, S., Saito, T., Kumano, K., Nakagami-Yamaguchi, E., Yamaguchi, T., & Hirai, H. (2004). Stem cell leukemia protein directs hematopoietic stem cell fate. *Blood, 103*(9), 3336–3341.

Labastie, M. C., Cortes, F., Romeo, P. H., Dulac, C., & Peault, B. (1998). Molecular identity of hematopoietic precursor cells emerging in the human embryo. *Blood, 92*(10), 3624–3635.

Lacombe, J., Herblot, S., Rojas-Sutterlin, S., Haman, A., Barakat, S., Iscove, N. N., … Hoang, T. (2010). Scl regulates the quiescence and the long-term competence of hematopoietic stem cells. *Blood, 115*(4), 792–803.

Lacombe, J., Krosl, G., Tremblay, M., Gerby, B., Martin, R., Aplan, P. D., … Hoang, T. (2013). Genetic interaction between Kit and Scl. *Blood, 122*(7), 1150–1161.

Lahlil, R., Lecuyer, E., Herblot, S., & Hoang, T. (2004). SCL assembles a multifactorial complex that determines glycophorin A expression. *Molecular and Cellular Biology, 24*(4), 1439–1452.

Lancrin, C., Sroczynska, P., Stephenson, C., Allen, T., Kouskoff, V., & Lacaud, G. (2009). The haemangioblast generates haematopoietic cells through a haemogenic endothelium stage. *Nature, 457*(7231), 892–895.

Lazrak, M., Deleuze, V., Noel, D., Haouzi, D., Chalhoub, E., Dohet, C., … Mathieu, D. (2004). The bHLH TAL-1/SCL regulates endothelial cell migration and morphogenesis. *Journal of Cell Science, 117*(Pt 7), 1161–1171.

Lecuyer, E., Herblot, S., Saint-Denis, M., Martin, R., Begley, C. G., Porcher, C., … Hoang, T. (2002). The SCL complex regulates c-kit expression in hematopoietic cells through functional interaction with Sp1. *Blood, 100*(7), 2430–2440.

Lecuyer, E., & Hoang, T. (2004). SCL: From the origin of hematopoiesis to stem cells and leukemia. *Experimental Hematology, 32*(1), 11–24.

Lecuyer, E., Lariviere, S., Sincennes, M. C., Haman, A., Lahlil, R., Todorova, M., … Hoang, T. (2007). Protein stability and transcription factor complex assembly determined by the SCL-LMO2 interaction. *The Journal of Biological Chemistry, 282*(46), 33649–33658.

Leroy-Viard, K., Vinit, M. A., Lecointe, N., Jouault, H., Hibner, U., Romeo, P. H., & Mathieu-Mahul, D. (1995). Loss of TAL-1 protein activity induces premature apoptosis of Jurkat leukemic T cells upon medium depletion. *The EMBO Journal, 14*(10), 2341–2349.

Lessard, J., & Sauvageau, G. (2003). Bmi-1 determines the proliferative capacity of normal and leukaemic stem cells. *Nature, 423*(6937), 255–260. http://dx.doi.org/10.1038/nature01572.

Leung, A., Ciau-Uitz, A., Pinheiro, P., Monteiro, R., Zuo, J., Vyas, P., … Porcher, C. (2013). Uncoupling VEGFA functions in arteriogenesis and hematopoietic stem cell specification. *Developmental Cell, 24*(2), 144–158.

Liao, W., Bisgrove, B. W., Sawyer, H., Hug, B., Bell, B., Peters, K., … Stainier, D. Y. (1997). The zebrafish gene cloche acts upstream of a flk-1 homologue to regulate endothelial cell differentiation. *Development, 124*(2), 381–389.

Liao, E. C., Paw, B. H., Oates, A. C., Pratt, S. J., Postlethwait, J. H., & Zon, L. I. (1998). SCL/Tal-1 transcription factor acts downstream of cloche to specify hematopoietic and vascular progenitors in zebrafish. *Genes & Development, 12*(5), 621–626.

Lin, Y. W., Nichols, R. A., Letterio, J. J., & Aplan, P. D. (2006). Notch1 mutations are important for leukemic transformation in murine models of precursor-T leukemia/lymphoma. *Blood, 107*(6), 2540–2543.

Liu, F., Li, D., Yu, Y. Y., Kang, I., Cha, M. J., Kim, J. Y., … Choi, K. (2015). Induction of hematopoietic and endothelial cell program orchestrated by ETS transcription factor ER71/ETV2. *EMBO Reports, 16*(5), 654–669.

Look, A. T. (1997). Oncogenic transcription factors in the human acute leukemias. *Science, 278*(5340), 1059–1064.
Love, P. E., Warzecha, C., & Li, L. (2014). Ldb1 complexes: The new master regulators of erythroid gene transcription. *Trends in Genetics, 30*(1), 1–9.
Martin, R., Lahlil, R., Damert, A., Miquerol, L., Nagy, A., Keller, G., & Hoang, T. (2004). SCL interacts with VEGF to suppress apoptosis at the onset of hematopoiesis. *Development, 131*(3), 693–702.
McCormack, M. P., Young, L. F., Vasudevan, S., de Graaf, C. A., Codrington, R., Rabbitts, T. H., ... Curtis, D. J. (2010). The Lmo2 oncogene initiates leukemia in mice by inducing thymocyte self-renewal. *Science, 327*(5967), 879–883.
McGrath, K. E., Frame, J. M., Fegan, K. H., Bowen, J. R., Conway, S. J., Catherman, S. C., ... Palis, J. (2015). Distinct sources of hematopoietic progenitors emerge before HSCs and provide functional Blood cells in the mammalian embryo. *Cell Reports, 11*(12), 1892–1904.
Mead, P. E., Deconinck, A. E., Huber, T. L., Orkin, S. H., & Zon, L. I. (2001). Primitive erythropoiesis in the xenopus embryo: The synergistic role of LMO-2, SCL and GATA-binding proteins. *Development, 128*(12), 2301–2308.
Mead, P. E., Kelley, C. M., Hahn, P. S., Piedad, O., & Zon, L. I. (1998). SCL specifies hematopoietic mesoderm in Xenopus embryos. *Development, 125*(14), 2611–2620.
Medvinsky, A., & Dzierzak, E. (1996). Definitive hematopoiesis is autonomously initiated by the AGM region. *Cell, 86*(6), 897–906.
Mikkola, H. K., Klintman, J., Yang, H., Hock, H., Schlaeger, T. M., Fujiwara, Y., & Orkin, S. H. (2003). Haematopoietic stem cells retain long-term repopulating activity and multipotency in the absence of stem-cell leukaemia SCL/tal-1 gene. *Nature, 421*(6922), 547–551.
Moignard, V., Macaulay, I. C., Swiers, G., Buettner, F., Schutte, J., Calero-Nieto, F. J., ... Gottgens, B. (2013). Characterization of transcriptional networks in blood stem and progenitor cells using high-throughput single-cell gene expression analysis. *Nature Cell Biology, 15*(4), 363–372.
Mouthon, M. A., Bernard, O., Mitjavila, M. T., Romeo, P. H., Vainchenker, W., & Mathieu-Mahul, D. (1993). Expression of tal-1 and GATA-binding proteins during human hematopoiesis. *Blood, 81*(3), 647–655.
Murre, C. (2000). Intertwining proteins in thymocyte development and cancer. *Nature Immunology, 1*(2), 97–98.
Nakamura-Ishizu, A., Takizawa, H., & Suda, T. (2014). The analysis, roles and regulation of quiescence in hematopoietic stem cells. *Development, 141*(24), 4656–4666.
Nakayama, N., Lee, J., & Chiu, L. (2000). Vascular endothelial growth factor synergistically enhances bone morphogenetic protein-4-dependent lymphohematopoietic cell generation from embryonic stem cells in vitro. *Blood, 95*(7), 2275–2283.
Nishikawa, S. I., Nishikawa, S., Hirashima, M., Matsuyoshi, N., & Kodama, H. (1998). Progressive lineage analysis by cell sorting and culture identifies FLK1+VE-cadherin+ cells at a diverging point of endothelial and hemopoietic lineages. *Development, 125*(9), 1747–1757.
Nowell, P. C. (1977). Preleukemia. Cytogenetic clues in some confusing disorders. *The American Journal of Pathology, 89*(2), 459–476.
Nutt, S. L., Heavey, B., Rolink, A. G., & Busslinger, M. (1999). Commitment to the B-lymphoid lineage depends on the transcription factor Pax5. *Nature, 401*(6753), 556–562.
Ogawa, M., Kizumoto, M., Nishikawa, S., Fujimoto, T., Kodama, H., & Nishikawa, S. I. (1999). Expression of alpha4-integrin defines the earliest precursor of hematopoietic cell lineage diverged from endothelial cells. *Blood, 93*(4), 1168–1177.

O'Neil, J., Shank, J., Cusson, N., Murre, C., & Kelliher, M. (2004). TAL1/SCL induces leukemia by inhibiting the transcriptional activity of E47/HEB. *Cancer Cell, 5*(6), 587–596.

Org, T., Duan, D., Ferrari, R., Montel-Hagen, A., Van Handel, B., Kerenyi, M. A., ... Mikkola, H. K. (2015). Scl binds to primed enhancers in mesoderm to regulate hematopoietic and cardiac fate divergence. *The EMBO Journal, 34*(6), 759–777.

Palii, C. G., Perez-Iratxeta, C., Yao, Z., Cao, Y., Dai, F., Davison, J., ... Brand, M. (2011). Differential genomic targeting of the transcription factor TAL1 in alternate haematopoietic lineages. *The EMBO Journal, 30*(3), 494–509.

Palis, J. (2014). Primitive and definitive erythropoiesis in mammals. *Frontiers in Physiology, 5*, 3.

Palomero, T., Odom, D. T., O'Neil, J., Ferrando, A. A., Margolin, A., Neuberg, D. S., ... Look, A. T. (2006). Transcriptional regulatory networks downstream of TAL1/SCL in T-cell acute lymphoblastic leukemia. *Blood, 108*(3), 986–992.

Pardanaud, L., & Dieterlen-Lievre, F. (1999). Manipulation of the angiopoietic/hemangiopoietic commitment in the avian embryo. *Development, 126*(4), 617–627.

Pardanaud, L., Luton, D., Prigent, M., Bourcheix, L. M., Catala, M., & Dieterlen-Lievre, F. (1996). Two distinct endothelial lineages in ontogeny, one of them related to hemopoiesis. *Development, 122*(5), 1363–1371.

Park, C., Afrikanova, I., Chung, Y. S., Zhang, W. J., Arentson, E., Fong Gh, G., ... Choi, K. (2004). A hierarchical order of factors in the generation of FLK1- and SCL-expressing hematopoietic and endothelial progenitors from embryonic stem cells. *Development, 131*(11), 2749–2762.

Park, I. K., Qian, D., Kiel, M., Becker, M. W., Pihalja, M., Weissman, I. L., ... Clarke, M. F. (2003). Bmi-1 is required for maintenance of adult self-renewing haematopoietic stem cells. *Nature, 423*(6937), 302–305.

Pereira, C. F., Chang, B., Qiu, J., Niu, X., Papatsenko, D., Hendry, C. E., ... Moore, K. (2013). Induction of a hemogenic program in mouse fibroblasts. *Cell Stem Cell, 13*(2), 205–218.

Petrie, H. T., & Zuniga-Pflucker, J. C. (2007). Zoned out: Functional mapping of stromal signaling microenvironments in the thymus. *Annual Review of Immunology, 25*, 649–679. http://dx.doi.org/10.1146/annurev.immunol.23.021704.115715.

Pevny, L., Simon, M. C., Robertson, E., Klein, W. H., Tsai, S. F., D'Agati, V., ... Costantini, F. (1991). Erythroid differentiation in chimaeric mice blocked by a targeted mutation in the gene for transcription factor GATA-1. *Nature, 349*(6306), 257–260.

Pimanda, J. E., Ottersbach, K., Knezevic, K., Kinston, S., Chan, W. Y., Wilson, N. K., ... Gottgens, B. (2007). Gata2, Fli1, and Scl form a recursively wired gene-regulatory circuit during early hematopoietic development. *Proceedings of the National Academy of Sciences of the United States of America, 104*(45), 17692–17697.

Pola, R., Ling, L. E., Silver, M., Corbley, M. J., Kearney, M., Blake Pepinsky, R., ... Isner, J. M. (2001). The morphogen Sonic hedgehog is an indirect angiogenic agent upregulating two families of angiogenic growth factors. *Nature Medicine*, 7(6), 706–711.

Porcher, C., Liao, E. C., Fujiwara, Y., Zon, L. I., & Orkin, S. H. (1999). Specification of hematopoietic and vascular development by the bHLH transcription factor SCL without direct DNA binding. *Development, 126*(20), 4603–4615.

Porcher, C., Swat, W., Rockwell, K., Fujiwara, Y., Alt, F. W., & Orkin, S. H. (1996). The T cell leukemia oncoprotein SCL/tal-1 is essential for development of all hematopoietic lineages. *Cell, 86*(1), 47–57.

Pulford, K., Lecointe, N., Leroy-Viard, K., Jones, M., Mathieu-Mahul, D., & Mason, D. Y. (1995). Expression of TAL-1 proteins in human tissues. *Blood, 85*(3), 675–684.

Ravet, E., Reynaud, D., Titeux, M., Izac, B., Fichelson, S., Romeo, P. H., ... Pflumio, F. (2004). Characterization of DNA-binding-dependent and -independent functions of SCL/TAL1 during human erythropoiesis. *Blood, 103*(9), 3326–3335.

Real, P. J., Ligero, G., Ayllon, V., Ramos-Mejia, V., Bueno, C., Gutierrez-Aranda, I., … Menendez, P. (2012). SCL/TAL1 regulates hematopoietic specification from human embryonic stem cells. *Molecular Therapy*, *20*(7), 1443–1453.

Reynaud, D., Ravet, E., Titeux, M., Mazurier, F., Renia, L., Dubart-Kupperschmitt, A., … Pflumio, F. (2005). SCL/TAL1 expression level regulates human hematopoietic stem cell self-renewal and engraftment. *Blood*, *106*(7), 2318–2328.

Riddell, J., Gazit, R., Garrison, B. S., Guo, G., Saadatpour, A., Mandal, P. K., … Rossi, D. J. (2014). Reprogramming committed murine blood cells to induced hematopoietic stem cells with defined factors. *Cell*, *157*(3), 549–564.

Robb, L., Elwood, N. J., Elefanty, A. G., Kontgen, F., Li, R., Barnett, L. D., & Begley, C. G. (1996). The scl gene product is required for the generation of all hematopoietic lineages in the adult mouse. *The EMBO Journal*, *15*(16), 4123–4129.

Robb, L., Lyons, I., Li, R., Hartley, L., Kontgen, F., Harvey, R. P., … Begley, C. G. (1995). Absence of yolk sac hematopoiesis from mice with a targeted disruption of the scl gene. *Proceedings of the National Academy of Sciences of the United States of America*, *92*(15), 7075–7079.

Robertson, S. M., Kennedy, M., Shannon, J. M., & Keller, G. (2000). A transitional stage in the commitment of mesoderm to hematopoiesis requiring the transcription factor SCL/tal-1. *Development*, *127*(11), 2447–2459.

Rodrigues, N. P., Janzen, V., Forkert, R., Dombkowski, D. M., Boyd, A. S., Orkin, S. H., … Scadden, D. T. (2005). Haploinsufficiency of GATA-2 perturbs adult hematopoietic stem-cell homeostasis. *Blood*, *106*(2), 477–484.

Rojas-Sutterlin, S., & Hoang, T. (2013). Hematopoietic stem cell quiescence and long term maintenance: Role of SCL/TAL1. In H. MA (Ed.), *Tumor dormancy and cellular quiescence: Vol. V.* (pp. 87–105). Netherlands: Springer Science Company.

Rojas-Sutterlin, S., Lecuyer, E., & Hoang, T. (2014). Kit and Scl regulation of hematopoietic stem cells. *Current Opinion in Hematology*, *21*(4), 256–264.

Rothenberg, E. V. (2014). Transcriptional control of early T and B cell developmental choices. *Annual Review of Immunology*, *32*, 283–321. http://dx.doi.org/10.1146/annurev-immunol-032712-100024.

Sanalkumar, R., Johnson, K. D., Gao, X., Boyer, M. E., Chang, Y. I., Hewitt, K. J., … Bresnick, E. H. (2014). Mechanism governing a stem cell-generating cis-regulatory element. *Proceedings of the National Academy of Sciences of the United States of America*, *111*(12), E1091–E1100. http://dx.doi.org/10.1073/pnas.1400065111.

Sanchez-Garcia, I., & Rabbitts, T. H. (1994). Transcriptional activation by TAL1 and FUS-CHOP proteins expressed in acute malignancies as a result of chromosomal abnormalities. *Proceedings of the National Academy of Sciences of the United States of America*, *91*(17), 7869–7873.

Sanda, T., Lawton, L. N., Barrasa, M. I., Fan, Z. P., Kohlhammer, H., Gutierrez, A., … Look, A. T. (2012). Core transcriptional regulatory circuit controlled by the TAL1 complex in human T cell acute lymphoblastic leukemia. *Cancer Cell*, *22*(2), 209–221.

Sauvageau, G., Thorsteinsdottir, U., Eaves, C. J., Lawrence, H. J., Largman, C., Lansdorp, P. M., & Humphries, R. K. (1995). Overexpression of HOXB4 in hematopoietic cells causes the selective expansion of more primitive populations in vitro and in vivo. *Genes & Development*, *9*(14), 1753–1765.

Schaniel, C., Bruno, L., Melchers, F., & Rolink, A. G. (2002). Multiple hematopoietic cell lineages develop in vivo from transplanted Pax5-deficient pre-B I-cell clones. *Blood*, *99*(2), 472–478.

Schaniel, C., Gottar, M., Roosnek, E., Melchers, F., & Rolink, A. G. (2002). Extensive in vivo self-renewal, long-term reconstitution capacity, and hematopoietic multipotency of Pax5-deficient precursor B-cell clones. *Blood*, *99*(8), 2760–2766.

Schlaeger, T. M., Schuh, A., Flitter, S., Fisher, A., Mikkola, H., Orkin, S. H., … Porcher, C. (2004). Decoding hematopoietic specificity in the helix-loop-helix domain of the transcription factor SCL/Tal-1. *Molecular and Cellular Biology*, *24*(17), 7491–7502.

Schoenebeck, J. J., Keegan, B. R., & Yelon, D. (2007). Vessel and blood specification override cardiac potential in anterior mesoderm. *Developmental Cell*, *13*(2), 254–267.

Schuh, A. C., Faloon, P., Hu, Q. L., Bhimani, M., & Choi, K. (1999). In vitro hematopoietic and endothelial potential of flk-1(−/−) embryonic stem cells and embryos. *Proceedings of the National Academy of Sciences of the United States of America*, *96*(5), 2159–2164.

Schuh, A. H., Tipping, A. J., Clark, A. J., Hamlett, I., Guyot, B., Iborra, F. J., … Porcher, C. (2005). ETO-2 associates with SCL in erythroid cells and megakaryocytes and provides repressor functions in erythropoiesis. *Molecular and Cellular Biology*, *25*(23), 10235–10250.

Semerad, C. L., Mercer, E. M., Inlay, M. A., Weissman, I. L., & Murre, C. (2009). E2A proteins maintain the hematopoietic stem cell pool and promote the maturation of myelolymphoid and myeloerythroid progenitors. *Proceedings of the National Academy of Sciences of the United States of America*, *106*(6), 1930–1935.

Shalaby, F., Ho, J., Stanford, W. L., Fischer, K. D., Schuh, A. C., Schwartz, L., … Rossant, J. (1997). A requirement for Flk1 in primitive and definitive hematopoiesis and vasculogenesis. *Cell*, *89*(6), 981–990.

Shalaby, F., Rossant, J., Yamaguchi, T. P., Gertsenstein, M., Wu, X. F., Breitman, M. L., & Schuh, A. C. (1995). Failure of blood-island formation and vasculogenesis in Flk-1-deficient mice. *Nature*, *376*(6535), 62–66. http://dx.doi.org/10.1038/376062a0.

Shivdasani, R. A., Mayer, E. L., & Orkin, S. H. (1995). Absence of blood formation in mice lacking the T-cell leukaemia oncoprotein tal-1/SCL. *Nature*, *373*(6513), 432–434. http://dx.doi.org/10.1038/373432a0.

Shlush, L. I., & Minden, M. D. (2015). Preleukemia: The normal side of cancer. *Current Opinion in Hematology*, *22*(2), 77–84.

Silver, L., & Palis, J. (1997). Initiation of murine embryonic erythropoiesis: A spatial analysis. *Blood*, *89*(4), 1154–1164.

Sinclair, A. M., Gottgens, B., Barton, L. M., Stanley, M. L., Pardanaud, L., Klaine, M., … Green, A. R. (1999). Distinct 5' SCL enhancers direct transcription to developing brain, spinal cord, and endothelium: Neural expression Is mediated by GATA factor binding sites. *Developmental Biology*, *209*(1), 128–142.

Singbrant, S., van Galen, P., Lucas, D., Challen, G., Rossi, D. J., & Daley, G. Q. (2015). Two new routes to make blood: Hematopoietic specification from pluripotent cell lines versus reprogramming of somatic cells. *Experimental Hematology*, *43*(9), 756–759.

Sive, J. I., & Gottgens, B. (2014). Transcriptional network control of normal and leukaemic haematopoiesis. *Experimental Cell Research*, *329*(2), 255–264.

Socolovsky, M., Lodish, H. F., & Daley, G. Q. (1998). Control of hematopoietic differentiation: Lack of specificity in signaling by cytokine receptors. *Proceedings of the National Academy of Sciences of the United States of America*, *95*(12), 6573–6575.

Sorrentino, B. P. (2010). Scl and stem cell quiescence. *Blood*, *115*(4), 751–752.

Souroullas, G. P., Salmon, J. M., Sablitzky, F., Curtis, D. J., & Goodell, M. A. (2009). Adult hematopoietic stem and progenitor cells require either Lyl1 or Scl for survival. *Cell Stem Cell*, *4*(2), 180–186.

Tanigawa, T., Elwood, N., Metcalf, D., Cary, D., DeLuca, E., Nicola, N. A., & Begley, C. G. (1993). The SCL gene product is regulated by and differentially regulates cytokine responses during myeloid leukemic cell differentiation. *Proceedings of the National Academy of Sciences of the United States of America*, *90*(16), 7864–7868.

Tanigawa, T., Nicola, N., McArthur, G. A., Strasser, A., & Begley, C. G. (1995). Differential regulation of macrophage differentiation in response to leukemia inhibitory factor/

oncostatin-M/interleukin-6: The effect of enforced expression of the SCL transcription factor. *Blood*, *85*(2), 379–390.

Taoudi, S., Bee, T., Hilton, A., Knezevic, K., Scott, J., Willson, T. A., ... Hilton, D. J. (2011). ERG dependence distinguishes developmental control of hematopoietic stem cell maintenance from hematopoietic specification. *Genes & Development*, *25*(3), 251–262.

Travnickova, J., Tran Chau, V., Julien, E., Mateos-Langerak, J., Gonzalez, C., Lelievre, E., ... Kissa, K. (2015). Primitive macrophages control HSPC mobilization and definitive haematopoiesis. *Nature Communications*, *6*, 6227.

Tremblay, M., Tremblay, C. S., Herblot, S., Aplan, P. D., Hebert, J., Perreault, C., & Hoang, T. (2010). Modeling T-cell acute lymphoblastic leukemia induced by the SCL and LMO1 oncogenes. *Genes & Development*, *24*(11), 1093–1105.

Trumpp, A., Essers, M., & Wilson, A. (2010). Awakening dormant haematopoietic stem cells. *Nature Reviews. Immunology*, *10*(3), 201–209.

Valtieri, M., Tocci, A., Gabbianelli, M., Luchetti, L., Masella, B., Vitelli, L., ... Peschle, C. (1998). Enforced TAL-1 expression stimulates primitive, erythroid and megakaryocytic progenitors but blocks the granulopoietic differentiation program. *Cancer Research*, *58*(3), 562–569.

van Eekelen, J. A., Bradley, C. K., Gothert, J. R., Robb, L., Elefanty, A. G., Begley, C. G., & Harvey, A. R. (2003). Expression pattern of the stem cell leukaemia gene in the CNS of the embryonic and adult mouse. *Neuroscience*, *122*(2), 421–436.

Van Handel, B., Montel-Hagen, A., Sasidharan, R., Nakano, H., Ferrari, R., Boogerd, C. J., ... Mikkola, H. K. (2012). Scl represses cardiomyogenesis in prospective hemogenic endothelium and endocardium. *Cell*, *150*(3), 590–605.

Van Vlierberghe, P., & Ferrando, A. (2012). The molecular basis of T cell acute lymphoblastic leukemia. *The Journal of Clinical Investigation*, *122*(10), 3398–3406.

Van Vlierberghe, P., Pieters, R., Beverloo, H. B., & Meijerink, J. P. (2008). Molecular-genetic insights in paediatric T-cell acute lymphoblastic leukaemia. *British Journal of Haematology*, *143*(2), 153–168.

Visvader, J., Begley, C. G., & Adams, J. M. (1991). Differential expression of the LYL, SCL and E2A helix-loop-helix genes within the hemopoietic system. *Oncogene*, *6*(2), 187–194.

Visvader, J. E., Fujiwara, Y., & Orkin, S. H. (1998). Unsuspected role for the T-cell leukemia protein SCL/tal-1 in vascular development. *Genes & Development*, *12*(4), 473–479.

Vo, L. T., & Daley, G. Q. (2015). De novo generation of HSCs from somatic and pluripotent stem cell sources. *Blood*, *125*(17), 2641–2648.

Voronova, A. F., & Lee, F. (1994). The E2A and tal-1 helix-loop-helix proteins associate in vivo and are modulated by Id proteins during interleukin 6-induced myeloid differentiation. *Proceedings of the National Academy of Sciences of the United States of America*, *91*(13), 5952–5956.

Vyas, P., McDevitt, M. A., Cantor, A. B., Katz, S. G., Fujiwara, Y., & Orkin, S. H. (1999). Different sequence requirements for expression in erythroid and megakaryocytic cells within a regulatory element upstream of the GATA-1 gene. *Development*, *126*(12), 2799–2811.

Wadman, I. A., Hsu, H. L., Cobb, M. H., & Baer, R. (1994). The MAP kinase phosphorylation site of TAL1 occurs within a transcriptional activation domain. *Oncogene*, *9*(12), 3713–3716.

Wadman, I. A., Osada, H., Grutz, G. G., Agulnick, A. D., Westphal, H., Forster, A., & Rabbitts, T. H. (1997). The LIM-only protein Lmo2 is a bridging molecule assembling an erythroid, DNA-binding complex which includes the TAL1, E47, GATA-1 and Ldb1/NLI proteins. *The EMBO Journal*, *16*(11), 3145–3157.

Wang, L., Li, L., Shojaei, F., Levac, K., Cerdan, C., Menendez, P., … Bhatia, M. (2004). Endothelial and hematopoietic cell fate of human embryonic stem cells originates from primitive endothelium with hemangioblastic properties. *Immunity*, *21*(1), 31–41.

Wareing, S., Eliades, A., Lacaud, G., & Kouskoff, V. (2012). ETV2 expression marks blood and endothelium precursors, including hemogenic endothelium, at the onset of blood development. *Developmental Dynamics*, *241*(9), 1454–1464.

Wareing, S., Mazan, A., Pearson, S., Gottgens, B., Lacaud, G., & Kouskoff, V. (2012). The Flk1-Cre-mediated deletion of ETV2 defines its narrow temporal requirement during embryonic hematopoietic development. *Stem Cells*, *30*(7), 1521–1531.

Wilson, N. K., Foster, S. D., Wang, X., Knezevic, K., Schutte, J., Kaimakis, P., … Gottgens, B. (2010). Combinatorial transcriptional control in Blood stem/progenitor cells: Genome-wide analysis of ten major transcriptional regulators. *Cell Stem Cell*, 7(4), 532–544.

Winnier, G., Blessing, M., Labosky, P. A., & Hogan, B. L. (1995). Bone morphogenetic protein-4 is required for mesoderm formation and patterning in the mouse. *Genes & Development*, *9*(17), 2105–2116.

Wu, H., Liu, X., Jaenisch, R., & Lodish, H. F. (1995). Generation of committed erythroid BFU-E and CFU-E progenitors does not require erythropoietin or the erythropoietin receptor. *Cell*, *83*(1), 59–67.

Xu, Z., Huang, S., Chang, L. S., Agulnick, A. D., & Brandt, S. J. (2003). Identification of a TAL1 target gene reveals a positive role for the LIM domain-binding protein Ldb1 in erythroid gene expression and differentiation. *Molecular and Cellular Biology*, *23*(21), 7585–7599.

Yamashita, J., Itoh, H., Hirashima, M., Ogawa, M., Nishikawa, S., Yurugi, T., … Nishikawa, S. (2000). Flk1-positive cells derived from embryonic stem cells serve as vascular progenitors. *Nature*, *408*(6808), 92–96.

Yang, Q., Jeremiah Bell, J., & Bhandoola, A. (2010). T-cell lineage determination. *Immunological Reviews*, *238*(1), 12–22.

Yun, W. J., Kim, Y. W., Kang, Y., Lee, J., Dean, A., & Kim, A. (2014). The hematopoietic regulator TAL1 is required for chromatin looping between the beta-globin LCR and human gamma-globin genes to activate transcription. *Nucleic Acids Research*, *42*(7), 4283–4293.

Zhang, J., Kalkum, M., Yamamura, S., Chait, B. T., & Roeder, R. G. (2004). E protein silencing by the leukemogenic AML1-ETO fusion protein. *Science*, *305*(5688), 1286–1289.

Zhang, J. A., Mortazavi, A., Williams, B. A., Wold, B. J., & Rothenberg, E. V. (2012). Dynamic transformations of genome-wide epigenetic marking and transcriptional control establish T cell identity. *Cell*, *149*(2), 467–482.

CHAPTER SEVEN

Navigating Transcriptional Coregulator Ensembles to Establish Genetic Networks: A GATA Factor Perspective

A.W. DeVilbiss*[,†], N. Tanimura*[,†], S.C. McIver*[,†], K.R. Katsumura*[,†], K.D. Johnson*[,†], E.H. Bresnick*[,†,1]

*UW-Carbone Cancer Center, University of Wisconsin School of Medicine and Public Health, Madison, WI, United States
[†]UW-Madison Blood Research Program, Madison, WI, United States
[1]Corresponding author: e-mail address: ehbresni@wisc.edu

Contents

Current Topics in Developmental Biology, Volume 118
ISSN 0070-2153
http://dx.doi.org/10.1016/bs.ctdb.2016.01.003

Abstract

Complex developmental programs require orchestration of intrinsic and extrinsic signals to control cell proliferation, differentiation, and survival. Master regulatory transcription factors are vital components of the machinery that transduce these stimuli into cellular responses. This is exemplified by the GATA family of transcription factors that establish cell type-specific genetic networks and control the development and homeostasis of systems including blood, vascular, adipose, and cardiac. Dysregulated GATA factor activity/expression underlies anemia, immunodeficiency, myelodysplastic syndrome, and leukemia. Parameters governing the capacity of a GATA factor expressed in multiple cell types to generate cell type-specific transcriptomes include selective coregulator usage and target gene-specific chromatin states. As knowledge of GATA-1 mechanisms in erythroid cells constitutes a solid foundation, we will focus predominantly on GATA-1, while highlighting principles that can be extrapolated to other master regulators. GATA-1 interacts with ubiquitous and lineage-restricted transcription factors, chromatin modifying/remodeling enzymes, and other coregulators to activate or repress transcription and to maintain preexisting transcriptional states. Major unresolved issues include: how does a GATA factor selectively utilize diverse coregulators; do distinct epigenetic landscapes and nuclear microenvironments of target genes dictate coregulator requirements; and do gene cohorts controlled by a common coregulator ensemble function in common pathways. This review will consider these issues in the context of GATA factor-regulated hematopoiesis and from a broader perspective.

1. INTRODUCTION

The combinatorial activities of transcription factors and coregulators establish and maintain genetic networks that control essentially all biological processes. Given the shear importance of the underlying mechanisms, it is highly instructive to consider how transcription factors navigate large coregulator ensembles in diverse contexts—in distinct cell and tissue types, development vs homeostasis, and physiological vs pathological states. In all of these contexts, the GATA family of dual zinc finger transcription factors have been demonstrated to play crucial roles. The GATA factor family consists of GATA-1–6 in mice and humans (Charron & Nemer, 1999; Molkentin, 2000; Orkin, 1992). The highly conserved zinc finger architecture contains two functionally distinct zinc fingers (Martin & Orkin, 1990). The zinc finger closest to the carboxyl-terminus (C-terminus) mediates sequence-specific DNA-binding and protein–protein interactions, while the amino-terminal (N-terminal) finger mediates protein/protein interactions, enhances DNA binding at palindromic sites on naked DNA, and facilitates chromatin occupancy (Ko & Engel, 1993; Martin & Orkin, 1990; Merika & Orkin, 1993; Newton, MacKay, & Crossley, 2001;

Omichinski et al., 1993; Trainor et al., 1996; Tsang et al., 1997). GATA factors exhibit tissue-specific expression and control the development and homeostasis of multiple tissue types. GATA-1, GATA-2, and GATA-3 are expressed in the hematopoietic compartment and select nonhematopoietic cells/tissues, while GATA-4, GATA-5, and GATA-6 are expressed predominantly, if not exclusively, in nonhematopoietic tissues (Bresnick, Katsumura, Lee, Johnson, & Perkins, 2012; Charron & Nemer, 1999; Molkentin, 2000; Orkin, 1992).

Principles of GATA factor function have emerged from studies on GATA-1 in erythroid cells (Bresnick et al., 2012). GATA-1 is also expressed in megakaryocytes, mast cells, eosinophils, basophils, dendritic cells, and Sertoli cells (Martin, Zon, Mutter, & Orkin, 1990; Nei et al., 2013; Pevny et al., 1991; Scheenstra et al., 2015; Yomogida et al., 1994; Yu et al., 2002). Targeted deletion of *Gata1* in the mouse yields lethality at day E10.5 with a gross failure of terminal erythroid and megakaryocyte maturation and defective production of mast cells and eosinophils (Fujiwara, Browne, Cunniff, Goff, & Orkin, 1996; Pevny et al., 1991, 1995; Simon et al., 1992; Yu et al., 2002). Prior to GATA-1 expression in erythroid cells, GATA-2 is expressed in erythroid precursors (Leonard, Brice, Engel, & Papayannopoulou, 1993; Weiss, Keller, & Orkin, 1994). As GATA-1 levels increase, GATA-1 replaces GATA-2 at many sites in the genome, a process termed GATA switching (Bresnick, Lee, Fujiwara, Johnson, & Keles, 2010; Bresnick, Martowicz, Pal, & Johnson, 2005; Bresnick et al., 2012; Grass et al., 2003). We restrict the use of the term GATA switching to the replacement of GATA-2 by GATA-1 at sites on chromatin, rather than the transition from GATA-2 to GATA-1 expression during erythroid maturation, which is commonly measured in heterogenous cell populations. GATA switching illustrates how GATA factors can elicit qualitatively distinct activities. GATA-1 and GATA-2 often induce distinct transcriptional outputs from the same chromatin occupancy site. A similar switch in expression from GATA-2 to GATA-1 occurs during monocytic precursor maturation into dendritic cells (Scheenstra et al., 2015) and during megakaryocyte maturation (Dore, Chlon, Brown, White, & Crispino, 2012; Pimkin et al., 2014).

By contrast to the GATA switch mechanism, GATA-1 and GATA-2 both confer survival to yolk sac-derived primitive erythroblasts (Fujiwara, Chang, Williams, & Orkin, 2004). GATA-2 is also expressed in hematopoietic stem and progenitor cells (HSPCs) and certain nonhematopoietic cells, such as endothelial cells (Dorfman, Wilson, Bruns, & Orkin, 1992; Wilson, Dorfman, & Orkin, 1990) and neurons (Craven et al., 2004; Nardelli, Thiesson, Fujiwara, Tsai, & Orkin, 1999; Yamamoto et al., 1990). Targeted deletion of *Gata2* yields embryonic lethality at day E10.5 and failure to produce all definitive blood lineages (Tsai & Orkin, 1997; Tsai et al., 1994). GATA-2 is required for hematopoietic stem cell (HSC) generation from hemogenic endothelium in the aorta gonad mesonephros (AGM) region of the developing embryo and the function of HSPCs (de Pater et al., 2013; Gao et al., 2013; Johnson et al., 2012; Ling et al., 2004; Rodrigues et al., 2005). Mutant mouse strains lacking *Gata2 cis*-elements exhibit embryonic lethality at E13.5–14.5 (Johnson et al., 2012) or after E15.5 (Johnson et al., 2015) and therefore provide unique conditional knockout models to dissect GATA-2 function at specific stages of embryogenesis. These models revealed an absolute GATA-2 requirement for the initial mechanistic step in HSC genesis, the hemogenic endothelial to HSC transition (Gao et al., 2013; Johnson et al., 2012), and for conferring myeloid progenitor activity to generate diverse myeloerythroid cell progeny (Johnson et al., 2015).

GATA-2 levels must be maintained within a stringent physiological window to ensure homeostasis. Insufficient GATA-2 causes a primary immunodeficiency syndrome that progresses to myelodysplastic syndrome and acute myeloid leukemia (AML) (Dickinson et al., 2011; Hahn et al., 2011; Hsu et al., 2011; Ostergaard et al., 2011), and high GATA-2 expression correlates with poor prognosis of AML (Luesink et al., 2012; Vicente et al., 2012). Given that GATA switching links GATA-1 and GATA-2 function (Bresnick et al., 2005, 2010), not surprisingly, altering GATA-1 function can be leukemogenic. Mutations that generate an ectopic GATA-1 translation initiation sequence and reduce GATA-1 activity cause megakaryoblastic leukemia (Rainis et al., 2003; Vyas & Crispino, 2007; Wechsler et al., 2002).

A common theme in GATA factor mechanisms is recognition of DNA containing a WGATAR motif in naked DNA (Evans, Reitman, & Felsenfeld, 1988) and a more complex permutation ((C/G)(A/T)GATAA (G/A/C)(G/A/C)) in chromatin (Fujiwara et al., 2009). GATA-1 was discovered as an erythroid-specific transcription factor that binds WGATAR at promoters and enhancers of the β-globin locus and other erythroid genes (Evans & Felsenfeld, 1989; Evans et al., 1988; Tsai et al., 1989). However, GATA-1 occupies <0.01% of genomic WGATAR motifs in erythroid cells, and GATA-2 occupancy has a similarly low frequency in erythroid precursors and endothelial cells (Fujiwara et al., 2009; Kang et al., 2012; Linnemann, O'Geen, Keles, Farnham, & Bresnick, 2011). GATA motifs frequently reside near motifs for other erythroid transcription factors, including Krüppel-like factor EKLF (KLF1), Scl/TAL1, and

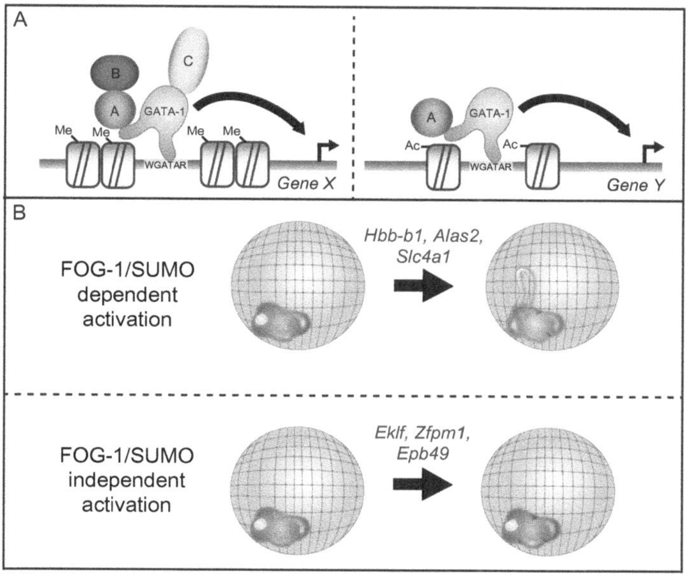

Fig. 1 Unique coregulator requirements at distinct GATA factor target genes. (A) GATA-1 requires different combinations of coregulators to activate and repress target genes. The molecular attributes of loci that mandate these requirements are predicted to include local chromatin environment, higher order chromatin structure, and subnuclear localization. (B) GATA-1-activated genes exhibit differential coregulator requirements for localization in specific nuclear neighborhoods. FOG-1/SUMO-sensitive, GATA-1-activated genes are expelled from the nuclear periphery upon activation; whereas FOG-1/SUMO-insensitive genes remain at the periphery (Lee et al., 2009; Lee, Johnson, Boyer, & Bresnick, 2011). (See the color plate.)

NF-E2. In principle, the WGATAR motif might need to be positioned near additional *cis*-elements to ensure chromatin occupancy. However, no simple formula has emerged in which GATA motif amalgamation with other motifs yields a high probability for occupancy. While the details of the mechanisms determining selectivity of GATA factor occupancy remain enigmatic, they almost certainly involve local and higher order target gene chromatin and subnuclear localization (Fig. 1A).

2. THE GATA FACTOR COREGULATOR REPERTOIRE

Chromatin modifying and remodeling coregulators control chromatin structure/function and target gene positioning in nuclear microenvironments. In addition to dissecting mechanisms governing the establishment/maintenance of GATA factor complexes on chromatin, there is a pressing need to elucidate how GATA factors navigate coregulator ensembles in a context (different chromatin sites and cellular environments)-dependent manner. Qualitatively distinct GATA-1 and GATA-2 activities and unique chromatin occupancy patterns by the same GATA factor in distinct cell types suggest a critical level of control involving selective coregulator usage.

Models to explain how a single transcription factor contends with a complex entourage of coregulators include sequential and simultaneous coregulator recruitment. This problem has largely been studied using artificial reporter genes in cell culture systems, which often do not recapitulate endogenous factor action at native chromosomal targets. GATA-1 and GATA-2 interplay to establish/maintain genetic networks during erythroid cell development/maturation makes the erythroid system attractive for elucidating principles of selective coregulator usage, and the respective mechanisms can be extrapolated to diverse contexts.

2.1 Friend of GATA-1 (FOG-1)

The erythroid-specific GATA-1-binding coregulator Friend of GATA-1 (FOG-1) was discovered using a yeast two-hybrid screen (Tsang, Fujiwara, Hom, & Orkin, 1998; Tsang et al., 1997). FOG-1 contains nine zinc fingers with four zinc fingers (1, 5, 6, and 9) capable of mediating binding to the GATA-1 N-terminal zinc finger (Cantor, Katz, & Orkin, 2002; Fox et al., 1999; Liew et al., 2005). FOG-1 and GATA-1 are coexpressed in yolk sac and fetal liver erythroid precursors, and targeted deletion of murine *Zfpm1* (encoding FOG-1) yields a phenotype mimicking the GATA-1 knockout, suggesting a common mode of GATA-1 and FOG-1 function (Tsang et al., 1997, 1998). Both mouse mutants exhibit maturation blockades in primitive and definitive proerythroblasts. Lethality of *Zfpm1*$^{-/-}$ mouse embryos later than *Gata1*$^{-}$ embryos (Fujiwara et al., 1996; Tsang et al., 1998) suggests that GATA-1 might have important FOG-1-independent modes. Evidence for FOG-1-independent, GATA-1-mediated activation and repression has emerged (Bresnick et al., 2012). *Zfpm1*$^{-/-}$ mice exhibit defective megakaryopoiesis at an earlier stage than *Gata1*$^{-}$ mouse, suggesting a potential GATA-1-independent function of FOG-1 (Tsang et al., 1998).

FOG-1 does not appear to bind DNA directly, and GATA-1 recruits FOG-1 to chromatin (Kim, Bultman, Jing, Blobel, & Bresnick, 2007) through protein–protein interactions (Crispino, Lodish, MacKay, & Orkin, 1999). GATA-1 mutations, including valine 205 to glycine (V205G) or methionine (V205M), reduce FOG-1 binding affinity, without impacting DNA-binding capacity and N-finger structural integrity

(Crispino et al., 1999; Nichols et al., 2000). These mutations impair GATA-1 competence to induce megakaryocyte differentiation and terminal erythroid maturation (Chang et al., 2002; Muntean & Crispino, 2005; Nichols et al., 2000). These activities are rescued with FOG-1 suppressor mutations that reestablish binding to mutant GATA-1, indicating that a GATA-1–FOG-1 physical interaction is critical (Crispino et al., 1999). GATA-1(V205M) causes familial dyserythropoietic anemia (Nichols et al., 2000). Structural analysis of the GATA-1 N-finger revealed that residues required for FOG-1 binding, but not for N-finger structural integrity, establish a contiguous surface on the N-finger (Kowalski, Czolij, King, Crossley, & Mackay, 1999; Liew et al., 2005). Although GATA-1 requires FOG-1 to activate and repress most of its target genes, some GATA-1 targets are regulated by GATA-1 mutants defective in FOG-1 binding, in *Zfpm1*$^{-/-}$ cells, and after small interfering RNA (siRNA)- or shRNA-mediated FOG-1 downregulation (Crispino et al., 1999; Johnson et al., 2007; Kim et al., 2007; Lee et al., 2009, 2011; Pal, Nemeth, et al., 2004). In aggregate, these findings indicate that one mode of GATA-1 function is FOG-1-independent.

Although GATA-1 was initially described as an activating factor (Evans & Felsenfeld, 1989; Evans et al., 1988; Tsai et al., 1989), GATA-1-mediated repression is common and required for erythroid and megakaryocyte maturation (Welch et al., 2004). GATA-1 represses target genes including *c-Kit* (Munugalavadla et al., 2005), *c-Myc* (Rylski et al., 2003), *Gata2* (Grass et al., 2003, 2006; Martowicz, Grass, Boyer, Guend, & Bresnick, 2005; Weiss et al., 1994), and *Lyl1* (Johnson et al., 2007) that control HSPC proliferation, survival, and function. As with activation, GATA-1-mediated repression requires FOG-1 at most target genes including *Gata2* and *c-Myc* (Crispino et al., 1999; DeVilbiss, Boyer, & Bresnick, 2013; Johnson et al., 2007). By contrast, the quantitatively large repression of *Lyl1* occurs without FOG-1 binding and in *Zfpm1*$^{-/-}$ cells (Johnson et al., 2007). GATA-1-mediated activation and repression can therefore be FOG-1-sensitive or -insensitive. FOG-1 occupancy detected by ChIP does not predict sensitivity, as FOG-1 can occupy "FOG-1-insensitive," GATA-1-regulated loci. This epitomizes a classic limitation of relying on ChIP for dissecting mechanisms; occupancy often does not translate into critical function (DeVilbiss, Sanalkumar, Johnson, Keles, & Bresnick, 2014). It seems likely that the sensitivity reflects postchromatin occupancy steps and the broader constellation of factors/coregulators at the locus. At certain loci, the FOG-1 contribution to GATA-1 activity might be redundant with other coregulators, and/or the chromatin attributes of the locus might negate the FOG-1 requirement, even though GATA-1 recruits FOG-1 to the locus. Finally, FOG-1-insensitivity might reflect the inability to measure FOG-1 function in certain contexts in the systems commonly used to dissect GATA-1/FOG-1 mechanisms. It will be intriguing to determine whether the mechanistic principles apply to complex scenarios involving microenvironment–hematopoietic precursor cell interactions, and whether intercellular signaling impacts selective coregulator usage.

GATA-1 directly induces *Zfpm1* transcription and elevates FOG-1 (Crispino et al., 1999; Kim et al., 2007), recruits FOG-1 to chromatin (Kim et al., 2007), and FOG-1 facilitates GATA-1 chromatin occupancy at certain sites (Letting, Chen, Rakowski, Reedy, & Blobel, 2004; Pal, Cantor, et al., 2004). The GATA switch that represses *Gata2* during erythropoiesis occurs at five dispersed *Gata2* locus sites (Bresnick et al., 2010; Grass et al., 2003, 2006; Martowicz et al., 2005). In cells lacking FOG-1, GATA-1 chromatin occupancy is reduced and unable to repress *Gata2* (Pal, Cantor, et al., 2004). Moreover, GATA-1(V205G) occupies certain, but not all, chromatin sites (Letting et al., 2004) and acquires ectopic sites differing from GATA-1-occupied sites (Chlon, Dore, & Crispino, 2012). Thus, FOG-1 suppresses occupancy at certain sites, while promoting occupancy at others. FOG-1 overexpression can antagonize GATA-1 in transient transfection assays, and this involves FOG-1 binding to the repressor C-terminal binding protein (CtBP) (Fox et al., 1999; Snow, Kim, Currie, Xu, & Orkin, 2010). However, mice homozygous for a knock-in *Zfpm1* allele encoding mutant FOG-1 lacking a CtBP binding sequence develop normally without erythroid or megakaryocytic phenotypes (Katz, Cantor, & Orkin, 2002). It is intriguing that FOG-1 orchestrates GATA-1 chromatin occupancy genome-wide (Chlon et al., 2012), and it will be important to establish whether other GATA-1-interacting coregulators cooperate with or antagonize FOG-1 in this mechanism.

2.2 Nucleosome Remodeling and Deacetylase Complex

Extensive chromatin remodeling and modification are hallmarks of mechanisms controlling transcriptional activation and repression. The repositioning of nucleosomes and posttranslational modification (PTM) (eg, acetylation or methylation) of histones regulate chromatin structure/accessibility. In addition, these chromatin transactions create docking sites for regulatory factors containing domains that recognize modified histones, thereby controlling transcription factor access to DNA. Active genes are commonly enriched in acetylated histone H3 and H4 at enhancers and promoters (Forsberg & Bresnick, 2001; Forsberg et al., 2000; Hebbes, Clayton, Thorne, & Crane-Robinson, 1994; Litt, Simpson, Recillas-Targa, Prioleau, & Felsenfeld, 2001; O'Neill & Turner, 1995). Specific histone modifications are often differentially enriched at genomic sites with distinct functions (eg, enhancers vs promoters) (Barski et al., 2007; Heintzman et al., 2009; Huang, Sabari, Garcia, Allis, & Zhao, 2014). Transcriptional repression can involve decreased histone acetylation locally or more broadly at a locus. Reduced histone acetylation at a locus can involve the targeted recruitment of histone deacetylase (HDAC) enzymes, enhanced HDAC activity, repulsion of histone acetyltransferases (HATs), or reduced HAT activity. The broadly expressed nucleosome remodeling and deacetylase (NuRD) complex contains ATPase (Mi2β or Mi2α) and histone deacetylase (HDAC1 and HDAC2) subunits (Tong, Hassig, Schnitzler, Kingston, & Schreiber, 1998; Wade, Jones, Vermaak, & Wolffe, 1998; Xue et al., 1998; Zhang, LeRoy, Seelig, Lane, & Reinberg, 1998; Zhang et al., 1999), and the subunit composition can be cell type-specific (Fujita et al., 2004; Le Guezennec et al., 2006). At GATA-1-repressed genes, GATA-1(V205G) fails to reduce histone acetylation near GATA-1-occupied sites, suggesting a FOG-1-requirement (Hong et al., 2005). A FOG-1 N-terminal repression domain binds the NuRD complex through a high-affinity interaction with multiple NuRD components (Gao et al., 2010; Lejon et al., 2011; Miccio et al., 2010). A conserved 12-amino acid motif within the repression domain is homologous to other repressors (Hong et al., 2005; Miccio et al., 2010; Roche et al., 2008). The FOG-1/NuRD interaction appears to be required for GATA-1 to repress certain FOG-1-sensitive target genes in erythroid cells (Miccio et al., 2010). GATA-1/FOG-1 recruits the NuRD complex to chromatin. FOG-1 mutations that reduce NuRD binding, including a lysine-5-to-alanine (K5A) mutation, impair GATA-1-mediated *Kit* repression (Hong et al., 2005). FOG-1(K5A) homozygous knock-in mutant mice exhibit ineffective erythropoiesis and defective megakaryopoiesis, underscoring the importance of the FOG-1/NuRD interaction (Gregory et al., 2010; Miccio et al., 2010). Although GATA-1/FOG-1/NuRD complexes are implicated in silencing the fetal gamma-globin genes in mouse embryos harboring a yeast artificial chromosome (YAC) containing the human globin locus (Harju-Baker, Costa, Fedosyuk, Neades, & Peterson, 2008), gamma-globin gene silencing was unaffected in mice containing the FOG-1(K5A) knock-in allele (Miccio & Blobel, 2010). The NuRD component Mi2β represses the endogenous murine fetal and embryonic globin genes *Hbb-y* and *Hbb-bh1* in G1E-ER-GATA-1 cells (DeVilbiss et al., 2013). G1E cells are a mouse ES cell-derived, GATA-1-null erythroid precursor cell line (Weiss, Yu, & Orkin, 1997). These cells permit a powerful genetic complementary assay. When a conditional *Gata1* allele, encoding an estrogen receptor ligand-binding domain to GATA-1 (ER-GATA-1), is stably expressed in G1E cells, estradiol-mediated ER-GATA-1 activation establishes a physiologically relevant erythroid gene expression program and recapitulates a normal window of erythroid maturation (Gregory et al., 1999; Weiss et al., 1997; Welch et al., 2004).

Consistent with chromatin remodeling and modification mechanisms generating active and repressive structures, NuRD mediates repression and activation. NuRD occupies chromatin at *Hbb-b1* and other GATA-1/FOG-1-activated loci (Miccio et al., 2010; Yu et al., 2009). A FOG-1/NuRD interaction is required for activation and repression of genes in megakaryocyte-erythroid precursors in mice harboring a homozygous FOG-1(K5A) knock-in allele. FOG-1 mutations that disrupt NuRD interaction or downregulating NuRD components impair GATA-1/FOG-1-mediated activation of a luciferase reporter when transiently coexpressed in 3T3 cells (Miccio et al., 2010). Loss-of-function analysis and expression profiling indicate that Mi2β and GATA-1 coactivate 205 genes in G1E-ER-GATA-1 cells (DeVilbiss et al., 2013). As is the case for FOG-1, NuRD complex recruitment to a GATA-1-occupied gene does not predict whether the gene will be activated or repressed.

GATA-1/FOG-1-repressed genes, insensitive to Mi2β downregulation, have been identified (DeVilbiss et al., 2013). The GATA-1/FOG-1 complex can therefore confer repression without this component of the NuRD complex. Alternatively, Mi2β may be in excess, and the downregulation may be insufficient to disrupt NuRD complex activity. Finally, the relative importance of another corepressor might increase when NuRD components are limiting. GATA-1 has been reported to bind HDAC5 via its N-finger (Watamoto et al., 2003), and GATA-1 requires the histone methyltransferase SetD8 to repress a subset of its direct target genes (DeVilbiss et al., 2013, 2015).

2.3 CREB-Binding Protein (CBP)/p300

GATA-1 regulates histone acetylation at GATA-1-repressed and -activated genes (Im et al., 2005; Johnson et al., 2002; Kim et al., 2007; Letting, Rakowski, Weiss, & Blobel, 2003). At the β-globin locus, genetic complementation analysis in G1E-ER-GATA-1 cells (Weiss et al., 1997) revealed ER-GATA-1-induced acetylation of histones H3 and H4 at the *βmajor* promoter and DNase hypersensitive sites of the β-globin locus control region (LCR) (Gregory et al. 1999; Im et al., 2005; Kiekhaefer, Grass, Johnson, Boyer, & Bresnick, 2002; Kim et al., 2007; Weiss et al., 1997; Welch et al., 2004) Acetylation patterns at the β-globin locus are developmentally regulated, and HDAC activity suppresses acetylation at the repressed fetal globin promoters in definitive murine erythroid cells (Forsberg et al., 2000; Im, Grass, Christensen, Perkins, & Bresnick, 2002). The broadly expressed HAT CBP and its paralog p300 facilitate transcriptional activation by many transcription factors through histone and nonhistone protein (eg, transcription factor) acetylation (Vo & Goodman, 2001). CBP binds GATA-1 directly, and the GATA-1 C-terminal finger suffices for this interaction (Blobel, Nakajima, Eckner, Montminy, & Orkin, 1998). It is likely that multiple coregulators establish and maintain the histone modification state of GATA-1 target genes, and much more work is required to gain a clear picture.

The GATA switch component of the *Gata2* repression mechanism is associated with decreased CBP occupancy at *cis*-elements and reduced histone H3 and H4 acetylation (Grass et al., 2003, 2006; Martowicz et al., 2005). CBP/p300 occupancy can precede GATA-1 occupancy in certain contexts. Since GATA-2 occupies certain chromatin sites prior to GATA-1, GATA-2 might recruit CBP/p300 prior to GATA-1 occupancy. GATA-2-mediated CBP/p300 recruitment could contribute to the transcriptional state of a locus prior to GATA-1 occupancy or create primed chromatin poised for subsequent GATA-1-mediated regulation.

2.4 Mediator Complex

The Mediator complex (Mediator) is a large 28 subunit protein complex that directly interacts with RNA polymerase II (Pol II) (Kornberg, 2005). Mediator can link enhancer-bound transcriptional activators with the basal transcriptional machinery and promotes activation of many Pol II-transcribed genes (Belakavadi & Fondell, 2006). Mice lacking the MED1 component exhibit defective cardiovascular development, are anemic, and die at ~E11.5 (Ito, Yuan, Okano, Darnell, & Roeder, 2000; Zhu et al., 2000). The Mx1-Cre-mediated conditional hematopoietic-specific *Med1* knockout in adult mice revealed defective erythroid maturation (Stumpf et al., 2006). Although lymphopoiesis was unaffected, and myeloid progenitor activity was normal, MED1 loss yielded an erythroid maturation blockade and dysregulated GATA-1 target genes. GATA-1 target genes in *Med1*$^{-/-}$ mice are not activated to the same extent as in wild-type mice (Stumpf et al., 2006). MED1 promotes GATA-1-mediated activation of a subset of its target genes (Pope & Bresnick, 2010, 2013). MED1 binds GATA-1 (Stumpf et al., 2006) and can be recruited to GATA-1-occupied sites, although its recruitment is independent of the GATA-1/FOG-1 interaction (Pope & Bresnick, 2010). Unlike CBP/p300, MED1 does not persist at GATA-1-repressed target genes (Pope & Bresnick, 2010), although a caveat of negative ChIP data is that nucleoprotein remodeling may occlude a factor and/or prevent crosslinking (DeVilbiss et al., 2014; Im et al., 2004).

Downregulating MED1 or the Mediator component MED25 with siRNAs in G1E-ER-GATA-1 cells reduces GATA-1-mediated target gene activation (Pope & Bresnick, 2010, 2013), but to a much lesser extent than that resulting from FOG-1 downregulation. Since the siRNA approach reduces, without eliminating, MED1, the quantitatively lower impact of the MED1 knockdown might reflect limitations of the system. Whereas Mediator is generally considered to be a coactivator, microarray analysis in erythroid cells revealed MED1 activation and repression of erythroid genes (Pope & Bresnick, 2013); whether the repressed genes are indirectly regulated is not known. A MED1/MED25 target gene in erythroid cells *Rrad* encodes a small G-protein that promotes erythroid cell survival in G1E-ER-GATA-1 cells, especially when cytokine levels are limiting (Pope & Bresnick, 2013). Given the numerous Mediator subunits (Bourbon et al., 2004), many lacking known functions, major work is required to understand how GATA-1 utilizes Mediator and to determine the applicability of the respective mechanisms to other GATA factors. Key issues include (i) establishing whether the modest contribution to GATA-1 function is physiologically significant; (ii) if this mechanism is physiologically significant, to elucidate how GATA-1 utilizes Mediator in the context of additional GATA-1 coregulators and other activators that also need to engage Mediator; and (iii) to determine if Mediator has cell type-specific functions, eg, in the maturing erythroblast, that are not evident from canonical mechanisms.

2.5 SetD8

SetD8 is a histone methyltransferase (Beck, Oda, Shen, & Reinberg, 2012) that functions as a corepressor at select GATA-1 target genes (DeVilbiss et al., 2013, 2015). SetD8 is the only enzyme known to catalyze histone H4 monomethylation at lysine 20 (H4K20me1) (Beck et al., 2012). Targeted deletion of *SetD8* in mice (Oda et al., 2009) and *Drosophila*

(Fang et al., 2002; Nishioka et al., 2002) is lethal early in embryonic development. The functional consequences of SetD8-dependent catalysis of H4K20me1 appear to differ in different systems. In *Drosophila*, H4K20me1 localizes to transcriptionally silent heterochromatic regions of polytene chromosomes (Nishioka et al., 2002). By contrast, in Mouse Erythroleukemia (MEL) cells, H4K20me1 correlates with transcriptionally active chromatin (Talasz, Lindner, Sarg, & Helliger, 2005). In CD4+ T-cells, H4K20me1 is enriched in actively transcribed gene bodies (Barski et al., 2007). Downregulating SetD8 reduces Wnt target gene activation in HEK293 cells (Li, Nie, Wang, & Li, 2011), providing evidence for a SetD8 function in activation. Gene expression profiling in SetD8-downregulated G1E-ER-GATA-1 cells indicates that SetD8 functions principally as a repressor in this system (DeVilbiss et al., 2013, 2015). ER-GATA-1 induces H4K20me1 at GATA-1-repressed, but not -activated, loci (DeVilbiss et al., 2013, 2015). At GATA-1/SetD8-repressed genes, GATA-1 activation induces H4K20me1 spreading upstream of the promoter and through the gene body. GATA-1 and SetD8 can occupy overlapping chromatin regions, and SetD8 occupancy precedes GATA-1 occupancy. It is unclear whether GATA-1 and SetD8 physically interact, as no proteomic analyses of GATA-1 complexes have identified SetD8. An important principle illustrated by the GATA-1/SetD8 corepression of genes in erythroid cells is that SetD8 facilitates GATA-1-mediated repression and can limit the extent to which GATA-1 activates target genes.

Given the early developmental function of SetD8 (Nishioka et al., 2002; Oda et al., 2009), conditional gene ablation approaches are required to analyze its contribution to GATA-1 function in vivo. Studies in primary cells indicate that SetD8 is required for erythroid cell maturation and survival (DeVilbiss et al., 2015). Downregulating SetD8 impairs fetal liver erythroid precursor cell maturation (DeVilbiss et al., 2015; Malik, Getman, & Steiner, 2015) and maturation of extensively self-renewing erythroblasts (ESREs) (Malik et al., 2015). In maturing ESREs, SetD8 promotes nuclear condensation, enucleation, and hemoglobinization (Malik et al., 2015). SetD8 downregulation induces apoptosis in fetal liver-derived erythroid precursor cells and increases c-H2Ax foci in ESREs (DeVilbiss et al., 2015; Malik et al., 2015), consistent with its established role in DNA damage repair. The erythroid maturation blockade does not result from increased cell death. In fetal liver proerythroblasts, SetD8 contributes to *Gata2* repression, and enforced GATA-2 expression induces a similar maturation blockade at the proerythroblast stage (DeVilbiss et al., 2015).

Downregulating GATA-2, in the context of SetD8 knockdown, does not negate the SetD8 knockdown-induced maturation blockade (DeVilbiss et al., 2015). *Gata2* repression may only be one of multiple SetD8-regulated maturation steps. SetD8 catalyzes H4K20me1 at an intronic *cis*-element +9.5 kb from the *Gata2* 1S promoter (+9.5 site), and restricts Scl/TAL1 occupancy at this site (DeVilbiss et al., 2015). As Scl/TAL1 activates *Gata2* through the +9.5 site (Sanalkumar et al., 2014; Wozniak, Boyer, Grass, Lee, & Bresnick, 2007; Wozniak et al., 2008), SetD8 catalysis of H4K20me1 at the +9.5 site (DeVilbiss et al., 2015) may create a chromatin environment incompatible with transcription factor complex assembly and/or stability.

2.6 Brahma-Related Gene 1 (BRG1)

SWI/SNF or BAF protein complexes are chromatin remodeling complexes that reposition and alter the structure of nucleosomes and facilitate nuclear factor access, thereby regulating nuclear processes (Kruger et al., 1995; Peterson, Dingwall, & Scott, 1994; Wang, Cote, et al., 1996). This process is dependent on ATP hydrolysis by a core ATPase subunit BRG1 (Bultman et al., 2000; Khavari, Peterson, Tamkun, Mendel, & Crabtree, 1993) or BRM1 (Bultman et al., 2000). At GATA-1-activated genes, eg, *Hbb-b1*, GATA-1 recruits BRG1 to chromatin (Im et al., 2005; Kim et al., 2007). EKLF/KLF1 utilizes SWI/SNF to maximally activate *Hbb-b1* during erythroid maturation (Armstrong, Bieker, & Emerson, 1998). BRG1 physically associates with a pentameric GATA-1 complex, containing Scl/TAL1, LMO2, LDB1, and E47, that binds an E-box-8 bp spacer-GATA composite element in erythroid cells (for detailed review, see chapter "The Hematopoietic Stem and Progenitor Cell Cistrome: GATA Factor-Dependent *cis*-Regulatory Mechanisms" by Hewitt et al.). Mice harboring a hypomorphic *Brg1* allele, encoding a mutant with normal ATPase function and SWI/SNF assembly but defective in establishing DNase I hypersensitive sites, exhibit drastically reduced *Hbb-b1* expression in E12.5 fetal liver cells (Bultman, Gebuhr, & Magnuson, 2005). The hypomorphic allele also reduces Pol II recruitment and serine 5 phosphorylation of the Pol II C-terminal domain (CTD) at *Hbb-b1* (Kim et al., 2007). In G1E-ER-GATA-1 cells, BRG1 recruitment to the β-globin LCR is one of the earliest mechanistic steps following ER-GATA-1 activation (Im et al., 2005; Kim et al., 2007). BRG1 induces chromatin looping at the *Hbb-b1* and α-globin loci (Kim, Bresnick, & Bultman, 2009; Kim, Bultman, Kiefer, Dean, & Bresnick, 2009). While BRG1 often mediates activation, enforced expression of BRG1 in maturing MEL cells represses the gene encoding the cytoskeletal protein 4.2 (Xu, Meng, Cai, Koury, & Brandt, 2006).

BRG1 also functions in the context of GATA-2 complexes at E-box-GATA composite elements (Sanalkumar et al., 2014). The *Gata2* +9.5 site, which binds GATA-2, Scl/TAL1 and BRG1 (Sanalkumar et al., 2014; Wozniak et al., 2007, 2008), strongly enhances *Gata2* transcription in HSPCs (Gao et al., 2013; Johnson et al., 2012; Sanalkumar et al., 2014). Downregulating BRG1 expression impairs conversion of the repressed to the activated *Gata2* locus in G1E-ER-GATA-1 cells (Sanalkumar et al., 2014). How BRG1 interfaces with FOG-1 and Mediator to orchestrate GATA-1 function remains unclear. FOG-1 also induces chromatin looping (Vakoc et al., 2005), which might or might not relate to its activity to promote GATA-1 chromatin occupancy.

2.7 Polycomb Repressive Complex

Polycomb proteins are conserved from *Drosophila* to mammals and mediate transcriptional silencing by promoting specific histone PTMs (Shilatifard, 2006, 2012; Simon & Kingston, 2013). Polycomb Repressive Complex 2 (PRC2) consists of the histone methyltransferases EZH1 or EZH2, the core subunit Suz12, and the WD-repeat domain-containing protein EED (Margueron & Reinberg, 2011). PRC2 facilitates transcriptional silencing by catalyzing histone H3 lysine 27 trimethylation (H3K27me3) (Kuzmichev, Nishioka, Erdjument-Bromage, Tempst, & Reinberg, 2002). EZH1 and EZH2 exhibit a switch during erythroid maturation, in which EZH1 expression is induced via a GATA switch-dependent mechanism, and EZH2 expression is downregulated (Xu et al., 2015). EZH1- and EZH2-containing PRC2 complexes are not exclusively involved in repression and can activate target genes in erythroid cells. Whereas EZH1/Suz12 complexes commonly occupy active genes, EED/EZH2 complexes occupy bivalent or repressed loci (Xu et al., 2015). Downregulating EZH2 induces global repositioning of EZH1 complexes, including to sites occupied previously by EZH2 (Xu et al., 2015). Suz12 localizes to GATA-1-repressed targets in G1E-ER-GATA-1 cells, and GATA-1, Suz12, and EZH2 physically associate (Xu et al., 2015). EZH2 promotes GATA-1-dependent *Hes1* repression (Ross, Mavoungou, Bresnick, & Milot, 2012). The erythroid-specific knockout of EED, which impacts EZH1 and EZH2 complexes, yields a partial erythroid maturation blockade at the proerythroblast stage (Yu et al., 2009). While the importance of PRC2 for GATA-1 function and erythroid maturation is established, given the complexity of the coregulator complexes involved, and the dearth of insights regarding how GATA-1 navigates the complexes, much work remains.

3. GATA FACTOR POSTTRANSLATIONAL MODIFICATIONS

Transcription factors, like most cellular proteins, are commonly regulated by PTMs, including phosphorylation, ubiquitination, acetylation, and methylation. PTMs can alter DNA-binding activity, protein–protein interactions, subcellular localization, and can target a protein for degradation. It is of particular interest to establish how microenvironment-induced PTMs of master regulators, such as GATA factors, influence their capacity to establish/maintain genetic networks. Despite transformative advances in analytical technologies, knowledge on how GATA factors integrate cellular signals via PTMs remains rudimentary. From the initial data accrued, this line of investigation will undoubtedly inform physiological and pathological mechanisms.

3.1 Phosphorylation

GATA-1 is extensively phosphorylated at multiple serine residues, and phosphorylation increases during MEL cell maturation (Crossley & Orkin, 1994). Phosphorylated GATA-1 enhances naked DNA binding (Partington & Patient, 1999). MAPK-mediated phosphorylation of GATA-1 at serine 26 promotes cell survival via *BCL-xL* induction in Ba/F cells (Yu et al., 2005). GATA-1 serine 310 is phosphorylated by Akt in fetal liver erythroid cells (Kadri et al., 2005; Zhao, Kitidis, Fleming, Lodish, & Ghaffari, 2006). GATA-1(S310A) has a reduced capacity to induce hemoglobinization in G1E cells. However, GATA-1 phosphorylation at conserved serines 72, 142, and 310 does not appear to regulate DNA binding or nuclear translocation, and mice bearing mutations in these residues develop normally without altered red blood cell numbers or changes in other hematopoietic parameters (Rooke & Orkin, 2006). Phosphorylation of these serines is therefore not critical, or redundant/compensatory mechanisms mask activities in vivo.

Multisite GATA-2 phosphorylation on serine residues is linked to the regulation of chromatin occupancy, transcription, and pathogenesis (Katsumura, Yang, Boyer, Li, & Bresnick, 2014; Menghini et al., 2005; Towatari et al., 1995). Serine 192 phosphorylation is required for multiple GATA-2 phosphorylation events, as a serine-192-to-alanine mutation prevents GATA-2 multisite phosphorylation (Katsumura et al., 2014). Several *GATA2* mutations have been linked to the primary immunodeficiency disorder MonoMAC syndrome, including mutations affecting threonine 354 in the C-terminal zinc finger (Bresnick et al., 2012; Dickinson et al., 2014; Spinner et al., 2013). In erythroid precursors and 293T cells, mutating T354 to M (a mutation found in MonoMAC patients) increases the amount of the hyperphosphorylated GATA-2 isoform (Katsumura et al., 2014). In erythroid cells, p38 signaling enhances S192-dependent GATA-2 phosphorylation (Katsumura et al., 2014). GATA-2 is a downstream target of RAS signaling, as expression of constitutively active RAS(G12V) in erythroid precursor cells elevates the amount of the hyperphosphorylated GATA-2 isoform and increases GATA-2-mediated transactivation (Katsumura et al., 2014). It is attractive to propose that microenvironment–HSPC interactions utilize this mechanism to sustain GATA-2 activity within a critical physiological window, which in turn would control HSPC function. In principle, disruption of such interactions could yield GATA-2 haploinsufficiency, thus causing immunodeficiency, MDS, and AML.

3.2 Acetylation

GATA-1 recruits CBP/p300 to certain chromatin sites and expels it from others (Grass et al., 2003, 2006; Letting et al., 2003; Martowicz et al., 2005). CBP/p300 directly acetylates GATA-1 on at least seven lysine residues (Hung, Lau, Kim, Weiss, & Blobel, 1999). One report indicated that acetylated GATA-1 has enhanced naked DNA-binding capacity (Boyes, Byfield, Nakatani, & Ogryzko, 1998), whereas another described no impact of acetylation on DNA binding (Hung et al., 1999). GATA-1 acetylation promotes GATA-1 chromatin occupancy (Lamonica, Vakoc, & Blobel, 2006). In addition, a mutant GATA-1 harboring mutations at conserved acetylation sites has a reduced capacity to induce G1E cell maturation (Hung et al., 1999). The leukemogenic protein AML1-ETO inhibits erythroid maturation, and this was suggested to involve blocking p300-mediated GATA-1 acetylation (Choi, Elagib, Delehanty, & Goldfarb, 2006). Acetylated GATA-1 can bind the bromodomain-containing protein BRD3 (Lamonica et al., 2011). BRD3 is a member of the BET family of bromodomain proteins that bind acetylated histone tails. BRD3 and GATA-1 co-occupy chromatin, and the GATA-1–BRD3 interaction promotes maximal GATA-1 chromatin occupancy (Lamonica et al., 2011). BRD3 is another example of a coactivator that can occupy GATA-1-repressed loci. GATA-1 directly represses *c-Kit* (Munugalavadla et al., 2005), and BRD3 persists at *c-Kit* during repression in G1E-ER-GATA-1 cells (Lamonica et al., 2011). Pharmacological inhibition of the GATA-1–BRD3 interaction impairs erythroid maturation (Lamonica et al., 2011). BRD2, BRD3, and BRD4 can all be recruited to GATA-1-occupied chromatin sites (Stonestrom et al., 2015). BRD2 and BRD4 function nonredundantly to regulate erythroid gene expression, while BRD3 has an overlapping function with BRD2 and rescues erythroid gene expression in the context of depleted BRD2 (Stonestrom et al., 2015).

p300 can acetylate GATA-2 in a biochemical assay and upon GATA-2 overexpression in a transient transfection assay (Hayakawa et al., 2004). GATA-2 acetylation facilitates DNA binding and transactivation in cell lines (Hayakawa et al., 2004). While GATA-2 expression is growth inhibitory in 32D myeloid cells, this growth inhibition is abrogated by mutations of the acetylated residues (Hayakawa et al., 2004). Morpholino-mediated GATA-2 downregulation revealed its importance for primitive erythropoiesis in *Xenopus* (Dalgin et al., 2007). Whereas GATA-2 rescues the primitive erythropoiesis defect, a GATA-2 mutant lacking four lysines near the zinc finger that can be acetylated does not rescue (Dalgin et al., 2007). In *Xenopus*, GATA-2 acetylation is controlled by the CaMKIV signaling pathway (Dalgin et al., 2007). GATA-3 is acetylated in T-cells, which appears to be involved in the control of T-cell homing and prolonged T-cell lifespan following T-cell activation (Yamagata et al., 2000). Thus, GATA factor acetylation is a common regulatory mechanism in diverse cell types.

3.3 Sumoylation

Proteins can be reversibly modified by the covalent attachment of the small ubiquitin-like peptide SUMO via an E3 ligase (Flotho & Melchior, 2013). GATA-1 can be sumoylated on lysine 137 by the ligase PIASy (Collavin et al., 2004), and sumoylation promotes activation of FOG-1-sensitive GATA-1 target genes in erythroid cells (Lee et al., 2009). Ablation of GATA-1 sumoylation does not affect GATA-1-mediated activation of FOG-1-insensitive target genes (Lee et al., 2009). Impaired target gene activation by a GATA-1 mutant lacking FOG-1 binding capacity can be bypassed by tethering SUMO to GATA-1 (Lee et al., 2009).

Establishing and maintaining transcription factor subnuclear localization represents a critical, but poorly understood, mode of transcriptional regulation (Feuerborn & Cook, 2015; Furlan-Magaril, Varnai, Nagano, & Fraser, 2015; Rajapakse & Groudine, 2011; Takizawa, Meaburn, & Misteli, 2008) (Fig. 1B). SUMO/FOG-1-dependent GATA-1 target genes are expelled from the nuclear periphery as a component of the mechanism by which

GATA-1 activates transcription (Lee et al., 2009, 2011). GATA-1 target genes insensitive to SUMO/FOG-1 persist at the periphery upon GATA-1-mediated activation. This differential behavior distinguishes FOG-1-sensitive from FOG-1-insensitive genes. Subnuclear localization of chromosomal regions can position a locus in functionally distinct microenvironments and dictate whether GATA-1 will require a given coregulator to activate or repress genes. This illustrates a key element of selective coregulator usage, as certain coregulators mediate establishment and/or maintenance of subnuclear positioning and may or may not be required for subsequent regulatory steps. Subnuclear positioning may dictate coregulator requirements for activation or repression. Achieving an integrated view of the dynamics of these events will be transformative for cell biology and much more broadly. For example, chromosomal segments in close proximity are more likely to form translocations (Roukos & Misteli, 2014; Roukos et al., 2013), which has critical implications for cancer, as locus subnuclear positioning can be aberrant in cancer cells.

PIASy interacts with and conjugates SUMO-2 to GATA-2 (Chun, Itoh, Subramanian, Ingiguez-Lluhi, & Nakao, 2003). In a transient transfection assay with an endothelin-1 (ET-1) promoter reporter construct in endothelial cells, PIASy suppresses GATA-2-mediated transactivation. Whereas the GATA-2/PIASy interaction appears to be important for PIASy-mediated suppression of GATA-2 activity, the RING-like domain, which is essential for conjugating SUMO, is dispensable (Chun et al., 2003). Either PIASy regulates GATA-2 activity by a sumoylation-independent mechanism, or the results may be unique to the specific system studied. Although GATA-2 bears two putative sumoylation sites (amino acids 221–224 and 388–391 in human GATA-2), the function of these sites is unknown.

3.4 Methylation

Although arginine methylation has not been described for GATA-1, GATA-2, and GATA-3 are methylated on arginine residues. GATA-3 is expressed in T-lineage lymphocytes and is required for T-cell development (Rothenberg, 2012; Tindemans, Serafini, Di Santo, & Hendriks, 2014). GATA-3 controls CD4 vs CD8 lineage selection and TH2 differentiation (Hernandez-Hoyos, Anderson, Wang, Rothenberg, & Alberola-Ila, 2003; Pai, Truitt, & Ho, 2004). In TH2 T-cells, GATA-3 is methylated on Arg256 and Arg261, which promotes GATA-3-mediated *Il5* activation (Hosokawa et al., 2015). Impaired arginine methylation results in strong association of GATA-3 with HSP60, which negatively regulates GATA-3-mediated transactivation (Hosokawa et al., 2015). Arginine 261 is conserved in mouse and human GATA-3, as well as in human GATA-1 and GATA-2. Thus, arginine methylation may constitute a general mode of controlling GATA factor function.

4. TRANSCRIPTION FACTOR COOPERATION

4.1 Ikaros

Ikaros is a hematopoietic transcription factor that regulates HSCs, lymphopoiesis, and erythroid cells (Bottardi, Mavoungou, & Milot, 2015; Ng, Yoshida, & Georgopoulos, 2007; Yoshida & Georgopoulos, 2014). Ikaros associates with SWI/SNF and NuRD complexes (Georgopoulos et al., 1994; Kim et al., 1999; O'Neill et al., 1999, 2000). Mutations that abolish Ikaros DNA-binding activity lead to lethality 1–3 weeks after birth and impair lymphopoiesis (Georgopoulos et al., 1994; Wang, Nichogiannopoulou, et al., 1996), and Ikaros mutations cause leukemia and lymphoma (Mullighan et al., 2009; Winandy, Wu, & Georgopoulos, 1995). *Ikaros*$^{-/-}$ mice harboring the human β-globin locus on a YAC exhibit delayed fetal to adult globin switching (Lopez, Schoetz, DeAngelis, O'Neill, & Bank, 2002). Ikaros has been reported to increase GATA-1 chromatin occupancy, as depletion of Ikaros impaired GATA-1 occupancy at the GATA-2 locus and elevated *Gata2* transcription (Bottardi et al., 2009). In yolk sac erythroid precursors, Ikaros facilitates GATA-1 recruitment to *cis*-elements to enhance human γ-globin expression (Bottardi et al., 2009). Ikaros can also occupy the human γ-globin promoter, along with GATA-1, HDAC1, and other NuRD components, in the fetal liver when γ-globin expression is repressed. Ikaros depletion reduces GATA-1 occupancy at the γ-globin promoter (Bottardi et al., 2009). Selective coregulator usage is apparent with the stage-specific control of γ-globin by Ikaros, as Ikaros recruits BRG1 in yolk sac, but interacts with NuRD components at the γ-globin locus in fetal liver (Bottardi et al., 2009, 2011, 2013). Thus, GATA-1 and Ikaros differentially regulate γ-globin in a developmental stage-specific manner.

4.2 FoxO3

The forkhead family of transcription factors is conserved from *C. elegans* to *D. melanogaster* to humans (Tran, Brunet, Griffith, & Greenberg, 2003;

Zhang, Rielland, Yalcin, & Ghaffari, 2011). In erythroid cells, FOXO3a (human) and FoxO3 (mice) counteract the extreme oxidative stress in which erythrocytes are subjected (Marinkovic et al., 2007). FoxO3 promotes GATA-1 activity to induce transcription of genes mediating autophagy (Kang et al., 2012; McIver et al., 2014), a cellular process critical for erythrocyte formation. Whereas autophagy has many physiological and pathological roles, in the context of erythroid maturation, it promotes organelle and macromolecule degradation and recycling to clear obsolete or damaged cellular constituents (Kundu et al., 2008; Li-Harms et al., 2014; Mortensen et al., 2010; Sandoval et al., 2008; Schweers et al., 2007; Zhang & Ney, 2009). FoxO3 and GATA-1 co-occupy overlapping chromatin sites at autophagy genes, and FoxO3 promotes GATA-1-mediated autophagy gene activation (Kang et al., 2012). GATA-1 and FoxO3 also corepress genes involved in RNA metabolism and processing, including exosome complex components (McIver et al., 2014). The exosome complex mediates degradation and processing of coding and non-coding RNAs in a highly selective manner (Makino, Halbach, & Conti, 2013; Makino et al., 2015). Downregulating exosome complex components in erythroid cells increases erythroid maturation (McIver et al., 2014). The exosome complex is therefore a barricade to erythroid maturation, and GATA-1/FoxO3 repress genes encoding exosome complex components, thereby breaking down the barricade and promoting erythroid maturation. The relationship between GATA-1 and FoxO3 in erythroid cells can be described as a type I coherent feed-forward loop, in which GATA-1 activates Foxo3, and GATA-1 and FoxO3 upregulate autophagy gene transcription and repress exosome complex components (Kang et al., 2012; McIver et al., 2014).

4.3 EKLF/KLF1 and Sp1

EKLF (KLF1) is an erythroid-specific transcription factor that binds CCNCNCCCN (CACC box) (Bieker, 2001; Miller & Bieker, 1993) and is a critical regulator of erythropoiesis (Hodge et al., 2005; Perkins, Sharpe, & Orkin, 1995). GATA-2 upregulates *Klf1* transcription in progenitor cells prior to erythroid lineage commitment (Lohmann & Bieker, 2008). Subsequently, GATA-1 directly upregulates *Klf1* via a FOG-1-independent mechanism (Crispino et al., 1999; Im et al., 2005; Kim et al., 2007). The ubiquitous transcription factor Sp1 binds similar motifs (Kadonaga, Carner, Masiarz, & Tjian, 1987). In vitro studies indicate that GATA-1 binds KLF1 and Sp1 and these factors synergize in transient transfection assays. Whether these components stably interact in cells is unknown (Merika & Orkin, 1995). The Sp1/KLF1 interaction with GATA-1 occurs through the C-terminal zinc finger, and since the C-finger engages DNA, it is unclear whether chromatin-bound GATA-1 would be competent to interact with Sp1/KLF1.

KLF1 and GATA-1-regulated genes exhibit considerable overlap. In one ChIP-seq study, ~1000 KLF1 occupancy peaks were detected with unfractionated fetal liver cells (Tallack et al., 2010). 45% of the peaks reside within 1 kb of a GATA-1 peak. In another study, ~14,000 KLF1 peaks were identified in primary erythroblasts (Pilon et al., 2011; Tallack et al., 2010), but these overlap with only 6% of GATA-1 peaks in induced G1E-ER-GATA1 cells. A high degree of overlap of KLF1 peaks was observed between the two studies, and the distinct peak numbers may reflect in the sequencing depth. Although only 6% of GATA-1 peaks overlapped with KLF1 in this study (Pilon et al., 2011), many of the genes sharing GATA-1 and KLF1 binding are erythroid genes encoding globin subunits, heme biosynthetic machinery, iron import machinery, and erythroid-specific cytoskeletal components. KLF1 and GATA-1 commonly co-occupy and coregulate many erythroid genes, but also regulate distinct components of the erythroid transcriptome (Tallack et al., 2012). With respect to interpreting ChIP studies, *trans*-acting factor occupancy does not readily translate into nonredundant function (DeVilbiss et al., 2014). There are numerous examples of occupancy without a measurable impact on expression of nearest-neighbor genes. This might be explained by occupancy without critical function, "occupancy" as an artifact of crosslinking, or occupancy that regulates a gene in three-dimensional space distinct from the nearest-neighbors. Removal of transcription factor-occupied *cis*-elements from the genome using traditional homologous recombination (Capecchi, 1989), TALENs (Sung et al., 2013), or CRISPR/Cas9 (Cho, Kim, Kim, & Kim, 2013) allows one to definitively assess function. It will be exciting to determine whether individual and ensembles of *cis*-elements generate and/or contribute to three-dimensional chromatin organization (including locus positioning in specific nuclear microenvironments) and therefore gene function that cannot be predicted from *cis*-element positioning relative to nearest-neighbor genes in two-dimensions.

4.4 PU.1

PU.1 is an ETS family transcription factor expressed in myeloid and B-lineage cells and required for HSC function as well as monocyte/macrophage,

B cell, neutrophil, and erythroid cell development (Klemsz, McKercher, Celada, Van Beveren, & Maki, 1990; McKercher et al., 1996; Pop et al., 2010; Scott, Simon, Anastasi, & Singh, 1994; Scott et al., 1997; Staber et al., 2013). PU.1 downregulation is necessary for megakaryocyte and erythroid differentiation (Pop et al., 2010; Stopka, Amanatullah, Papetti, & Skoultchi, 2005). PU.1 can bind GATA-1 and GATA-2 C-fingers (Liew et al., 2006; Rekhtman, Radparvar, Evans, & Skoultchi, 1999; Zhang et al., 2000). Although GATA-1 and GATA-2 do not inhibit PU.1 binding to naked DNA, GATA-1 can inhibit coactivation of genes by PU.1 and its partner c-Jun (Zhang et al., 2000). PU.1 inhibits transcription of GATA-1 target genes in hematopoietic progenitors by recruiting a repressive protein complex containing Rb, Suv39, and HP1α to promoters of those genes (Stopka et al., 2005). PU.1 downregulation results in dissociation of repressive proteins from GATA-1 activated genes, and increased CBP/p300 and histone acetylation at these sites (Stopka et al., 2005). GATA-1 also represses PU.1 at the transcriptional level (Chou et al., 2009).

The GATA-2–PU.1 interaction has context-dependent influences on hematopoietic cell differentiation. Using cells derived from *Spi1*$^{-/-}$ mice, PU.1 negatively regulates *Gata2* expression and promotes differentiation into macrophages (Walsh et al., 2002). Both PU.1 and GATA-2 promote differentiation into mast cells (Kitajima et al., 2006; Walsh et al., 2002). Using a mouse embryonic stem cell differentiation system, when GATA-2 is overexpressed at an early stage, GATA-2 suppresses PU.1 expression by inhibiting H4 acetylation at the PU.1 promoter (Kitajima et al., 2006). This prevents differentiation into macrophages and induces megakaryocytic differentiation. Given the multitude of mechanisms governing GATA factor and PU.1 expression and function, understanding the context-dependence of the cross talk (Monteiro, Pouget, & Patient, 2011) will require a deep systems-level perspective.

5. ASSEMBLING AND DECIPHERING A COREGULATOR MATRIX GOVERNING GATA FACTOR FUNCTION

Despite extensive insights into GATA factor mechanisms and function in diverse cell and tissue types, an integrative view of how GATA factors selectively utilize coregulators has not emerged. GATA-1 target genes are differentially sensitive to changes in the levels of coregulators, which raises the question: why are the coregulator requirements for activation and repression locus-specific? It seems reasonable to assume that local chromatin environment, including nucleosome density and DNA accessibility, histone PTMs, and locus subnuclear positioning create or negate coregulator requirements.

GATA-1-mediated transcriptional repression can require diverse coregulators including ETO2, LMO2, SetD8, NuRD, FOG-1 (Bresnick et al., 2012; Katsumura, DeVilbiss, Pope, Johnson, & Bresnick, 2013). Since FOG-1 binds NuRD (Hong et al., 2005), GATA-1-repressed genes that require FOG-1 would be expected to require the NuRD complex. However, loss-of-function analysis in G1E-ER-GATA-1 cells revealed that FOG-1-sensitive GATA-1 target genes are differentially sensitive to Mi2β downregulation (DeVilbiss et al., 2013). Thus, FOG-1 sensitivity does not predict NuRD sensitivity. LMO2 and ETO2 are components of a pentameric complex consisting of GATA-1, Scl/TAL1 and its heterodimeric partner E2A, and LDB1 (Love, Warzecha, & Li, 2014; Wadman et al., 1997). While ETO2 and LMO2, components of a multimeric GATA factor/Scl/TAL1 complex, coregulate a small gene cohort, they exhibit major differences in their target gene ensembles and regulate certain distinct biological processes (Fujiwara, Lee, Sanalkumar, & Bresnick, 2010). These results provide evidence for multifunctionality of the multimeric GATA-1 complex and/or diverse GATA-1 complexes with distinct functions.

Different GATA-1-regulated loci have unique coregulator requirements, and a simple way to conceptualize these mechanisms genome-wide is to assemble a matrix illustrating the relationship between target genes and requisite coregulators (DeVilbiss et al., 2013). It will be highly instructive to expand this matrix to the full complement of GATA-1 target genes and GATA-1-interacting coregulators (Fig. 2). GATA factor occupancy, coregulator occupancy, overlapping transcription factors, chromatin marks, DNase hypersensitive sites, etc., are insufficient to ascribe function. Many ChIP-seq datasets were derived using cells under static conditions, in which genes are either active or repressed. A more dynamic perspective will increase clarity of the respective mechanisms. In G1E-ER-GATA-1 cells, RNA interference or genome editing technology have been used to reduce endogenous coregulator expression levels, to excise prospective *cis*-elements, and to titrate the active GATA-1 level (DeVilbiss et al., 2013; Hewitt et al., 2015; Johnson, Kim, & Bresnick, 2006; Lee et al., 2011). Combined with genome-wide occupancy profiling and transcriptomics, a coregulator matrix in this system will provide a rich foundation for extrapolation to different physiological and pathological contexts. Locus regulation may reflect direct or indirect regulation, and clearly, there are

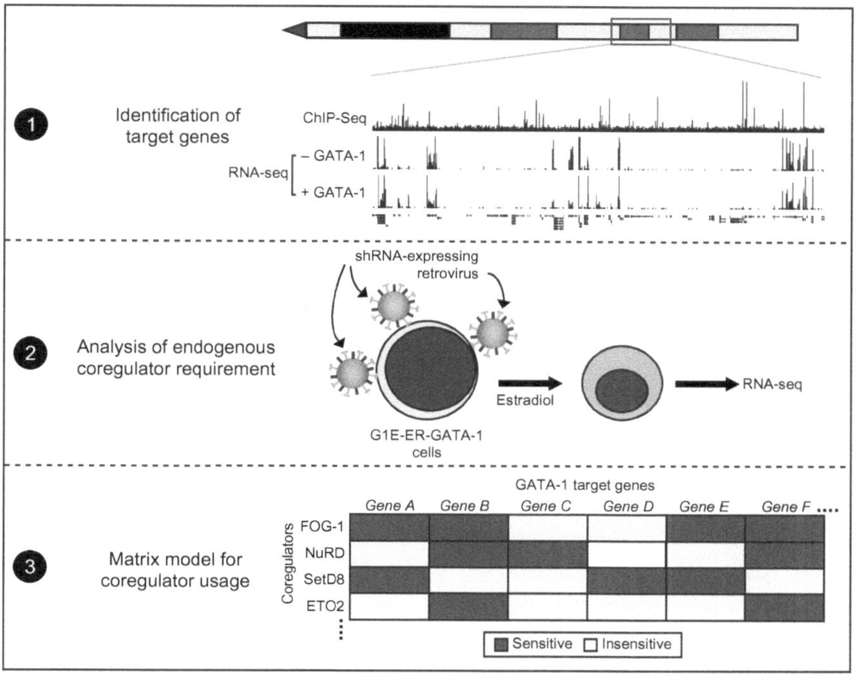

Fig. 2 Coregulator matrix model of selective coregulator usage. Developing a coregulator matrix involves: (1) Identifying the target gene ensemble using ChIP-seq combined with genome-wide RNA profiling; (2) Using RNA interference or genome editing, coupled with expression profiling, to determine the requirement of endogenous coregulators for expression of each target gene; (3) Combine regulatory information for each coregulator at each gene into a matrix. This type of matrix constitutes an invaluable resource for identifying locus attributes linked to regulation by a particular cohort of coregulators. (See the color plate.)

merits to restricting the matrix constituents to those involving direct regulation. Incorporating direct and indirect regulatory modes will be informative as well, since the actions of certain coregulators are interlinked in pathways and not necessarily by functioning at a common locus.

Selective coregulator usage at GATA factor-occupied sites in distinct cell types may endow different *cis*-elements with unique developmental stage-specific activities. The *Gata2* +9.5 site is required for HSC generation in vivo (Gao et al., 2013; Johnson et al., 2012), while sites −1.8, −2.8, and −3.9 kb from the *Gata2* 1S promoter are not (Sanalkumar et al., 2014; Snow, Trowbridge, et al., 2010; Snow et al., 2011). These additional sites are dispensable for hematopoiesis and development. The −77 site (Grass et al., 2006), however, has an embryonic lethal phenotype later than

E15.5 distinct from that of the +9.5 knockout in which lethality occurs at ~E13.5 (Johnson et al., 2015). Furthermore, a chromosomal rearrangement in AML repositions the −77 site ~40 megabases away next to the *MECOM* oncogene, thus activating expression and contributing to leukemogenesis (Groschel et al., 2014; Yamazaki et al., 2014). $-77^{-/-}$ mice exhibit normal HSC generation from the AGM and normal *Gata2* expression in lineage-negative, SCA1-positive, c-KIT-positive (LSK) cells (Johnson et al., 2015). However, *Gata2* expression is greatly reduced in common myeloid progenitors (CMPs) and granulocyte macrophage progenitors (GMPs) from $-77^{-/-}$ fetal livers (Johnson et al., 2015). These mutant progenitors fail to produce multiple lineages of myeloid and erythroid cells. The distinct activities of GATA factor-bound *cis*-elements at *Gata2* almost certainly reflect a GATA factor complex on the specific *cis*-element that utilizes a unique constellation of coregulators, which translates into cell type-specific activity. Elucidating the full ensemble of coregulators that function through these sites is an essential step towards understanding how GATA factors selectively utilize coregulators in distinct context to achieve unique transcriptional outputs and cellular phenotypes.

REFERENCES

Armstrong, J. A., Bieker, J. J., & Emerson, B. M. (1998). A SWI/SNF-related chromatin remodeling complex, E-RC1, is required for tissue-specific transcriptional regulation by EKLF in vitro. *Cell, 95*, 93–104.

Barski, A., Cuddapah, S., Cui, K., Roh, T. Y., Schones, D. E., Wang, Z., et al. (2007). High-resolution profiling of histone methylations in the human genome. *Cell, 129*, 823–837.

Beck, D. B., Oda, H., Shen, S. S., & Reinberg, D. (2012). PR-Set7 and H4K20me1: At the crossroads of genome integrity, cell cycle, chromosome condensation, and transcription. *Genes & Development, 26*, 325–337.

Belakavadi, M., & Fondell, J. D. (2006). Role of the mediator complex in nuclear hormone receptor signaling. *Reviews of Physiology, Biochemistry and Pharmacology, 156*, 23–43.

Bieker, J. J. (2001). Kruppel-like factors: Three fingers in many pies. *Journal of Biological Chemistry, 276*, 34355–34358.

Blobel, G. A., Nakajima, T., Eckner, R., Montminy, M., & Orkin, S. H. (1998). CREB-binding protein cooperates with transcription factor GATA-1 and is required for erythroid differentiation. *Proceedings of the National academy of Sciences of the United States of America, 95*, 2061–2066.

Bottardi, S., Mavoungou, L., Bourgoin, V., Mashtalir, N., Affar el, B., & Milot, E. (2013). Direct protein interactions are responsible for Ikaros-GATA and Ikaros-Cdk9 cooperativeness in hematopoietic cells. *Molecular and Cellular Biology, 33*, 3064–3076.

Bottardi, S., Mavoungou, L., & Milot, E. (2015). IKAROS: A multifunctional regulator of the polymerase II transcription cycle. *Trends in Genetics, 31*, 500–508.

Bottardi, S., Ross, J., Bourgoin, V., Fotouhi-Ardakani, N., Affar el, B., Trudel, M., et al. (2009). Ikaros and GATA-1 combinatorial effect is required for silencing of human gamma-globin genes. *Molecular and Cellular Biology, 29*, 1526–1537.

Bottardi, S., Zmiri, F. A., Bourgoin, V., Ross, J., Mavoungou, L., & Milot, E. (2011). Ikaros interacts with P-TEFb and cooperates with GATA-1 to enhance transcription elongation. *Nucleic Acids Research*, *39*, 3505–3519.

Bourbon, H. M., Aguilera, A., Ansari, A. Z., Asturias, F. J., Berk, A. J., Bjorklund, S., et al. (2004). A unified nomenclature for protein subunits of mediator complexes linking transcriptional regulators to RNA polymerase II. *Molecular Cell*, *14*, 553–557.

Boyes, J., Byfield, P., Nakatani, Y., & Ogryzko, V. (1998). Regulation of activity of the transcription factor GATA-1 by acetylation. *Nature*, *396*, 594–598.

Bresnick, E. H., Katsumura, K. R., Lee, H. Y., Johnson, K. D., & Perkins, A. S. (2012). Master regulatory GATA transcription factors: Mechanistic principles and emerging links to hematologic malignancies. *Nucleic Acids Research*, *40*, 5819–5831.

Bresnick, E. H., Lee, H. Y., Fujiwara, T., Johnson, K. D., & Keles, S. (2010). GATA switches as developmental drivers. *Journal of Biological Chemistry*, *285*, 31087–31093.

Bresnick, E. H., Martowicz, M. L., Pal, S., & Johnson, K. D. (2005). Developmental control via GATA factor interplay at chromatin domains. *Journal of Cellular Physiology*, *205*, 1–9.

Bultman, S. J., Gebuhr, T. C., & Magnuson, T. (2005). A Brg1 mutation that uncouples ATPase activity from chromatin remodeling reveals an essential role for SWI/SNF complexes in beta-globin expression and erythroid development. *Genes & Development*, *19*, 2849–2861.

Bultman, S. J., Gebuhr, T. C., Yee, D., La Mantia, C., Nicholson, J., Gilliam, A., et al. (2000). A Brg1 null mutation in the mouse reveals functional differences among mammalian SWI/SNF complexes. *Molecular Cell*, *6*, 1287–1295.

Cantor, A. B., Katz, S. G., & Orkin, S. H. (2002). Distinct domains of the GATA-1 cofactor FOG-1 differentially influence erythroid versus megakaryocytic maturation. *Molecular and Cellular Biology*, *22*, 4268–4279.

Capecchi, M. R. (1989). Altering the genome by homologous recombination. *Science*, *244*, 1288–1292.

Chang, A. N., Cantor, A. B., Fujiwara, Y., Lodish, M. B., Droho, S., Crispino, J. D., et al. (2002). GATA-factor dependence of the multitype zinc-finger protein FOG-1 for its essential role in megakaryopoiesis. *Proceedings of the National Academy of Sciences of the United States of America*, *99*, 9237–9242.

Charron, F., & Nemer, M. (1999). GATA transcription factors and cardiac development. *Seminars in Cell & Developmental Biology*, *10*, 85–91.

Chlon, T. M., Dore, L. C., & Crispino, J. D. (2012). Cofactor-mediated restriction of GATA-1 chromatin occupancy coordinates lineage-specific gene expression. *Molecular Cell*, *47*, 608–621.

Cho, S. W., Kim, S., Kim, J. M., & Kim, J. S. (2013). Targeted genome engineering in human cells with the Cas9 RNA-guided endonuclease. *Nature Biotechnology*, *31*, 230–232.

Choi, Y., Elagib, K. E., Delehanty, L. L., & Goldfarb, A. N. (2006). Erythroid inhibition by the leukemic fusion AML1-ETO is associated with impaired acetylation of the major erythroid transcription factor GATA-1. *Cancer Research*, *66*, 2990–2996.

Chou, S. T., Khandros, E., Bailey, L. C., Nichols, K. E., Vakoc, C. R., Yao, Y., et al. (2009). Graded repression of PU.1/Sfpi1 gene transcription by GATA factors regulates hematopoietic cell fate. *Blood*, *114*, 983–994.

Chun, T. H., Itoh, H., Subramanian, L., Ingiguez-Lluhi, J. A., & Nakao, K. (2003). Modification of GATA-2 transcriptional activity in endothelial cells by the SUMO E3 ligase PIASgamma. *Circulation Research*, *92*, 1201–1208.

Collavin, L., Gostissa, M., Avolio, F., Secco, P., Ronchi, A., Santoro, C., et al. (2004). Modification of the erythroid transcription factor GATA-1 by SUMO-1. *Proceedings of the National Academy of Sciences of the United States of America*, *101*, 8870–8875.

Craven, S. E., Lim, K. C., Ye, W., Engel, J. D., de Sauvage, F., & Rosenthal, A. (2004). Gata2 specifies serotonergic neurons downstream of sonic hedgehog. *Development*, *131*, 1165–1173.

Crispino, J. D., Lodish, M. B., MacKay, J. P., & Orkin, S. H. (1999). Use of altered specificity mutants to probe a specific protein-protein interaction in differentiation: The GATA-1: FOG complex. *Molecular Cell, 3*, 219–228.
Crossley, M., & Orkin, S. H. (1994). Phosphorylation of the erythroid transcription factor GATA-1. *Journal of Biological Chemistry, 269*, 16589–16596.
Dalgin, G., Goldman, D. C., Donley, N., Ahmed, R., Eide, C. A., & Christian, J. L. (2007). GATA-2 functions downstream of BMPs and CaM KIV in ectodermal cells during primitive hematopoiesis. *Developmental Biology, 310*, 454–469.
de Pater, E., Kaimakis, P., Vink, C. S., Yokomizo, T., Yamada-Inagawa, T., van der Linden, R., et al. (2013). Gata2 is required for HSC generation and survival. *Journal of Experimental Medicine, 210*, 2843–2850.
DeVilbiss, A. W., Boyer, M. E., & Bresnick, E. H. (2013). Establishing a hematopoietic genetic network through locus-specific integration of chromatin regulators. *Proceedings of the National Academy of Sciences of the United States of America, 110*, E3398–E3407.
DeVilbiss, A. W., Sanalkumar, R., Hall, B. D., Katsumura, K. R., de Andrade, I. F., & Bresnick, E. H. (2015). Epigenetic determinants of erythropoiesis: Role of the histone methyltransferase SetD8 in promoting erythroid cell maturation and survival. *Molecular and Cellular Biology, 35*, 2073–2087.
DeVilbiss, A. W., Sanalkumar, R., Johnson, K. D., Keles, S., & Bresnick, E. H. (2014). Hematopoietic transcriptional mechanisms: From locus-specific to genome-wide vantage points. *Experimental Hematology, 42*, 618–629.
Dickinson, R. E., Griffin, H., Bigley, V., Reynard, L. N., Hussain, R., Haniffa, M., et al. (2011). Exome sequencing identifies GATA-2 mutation as the cause of dendritic cell, monocyte, B and NK lymphoid deficiency. *Blood, 118*, 2656–2658.
Dickinson, R. E., Milne, P., Jardine, L., Zandi, S., Swierczek, S. I., McGovern, N., et al. (2014). The evolution of cellular deficiency in GATA2 mutation. *Blood, 123*, 863–874.
Dore, L. C., Chlon, T. M., Brown, C. D., White, K. P., & Crispino, J. D. (2012). Chromatin occupancy analysis reveals genome-wide GATA factor switching during hematopoiesis. *Blood, 119*, 3724–3733.
Dorfman, D. M., Wilson, D. B., Bruns, G. A., & Orkin, S. H. (1992). Human transcription factor GATA-2. Evidence for regulation of preproendothelin-1 gene expression in endothelial cells. *Journal of Biological Chemistry, 267*, 1279–1285.
Evans, T., & Felsenfeld, G. (1989). The erythroid-specific transcription factor Eryf1: A new finger protein. *Cell, 58*, 877–885.
Evans, T., Reitman, M., & Felsenfeld, G. (1988). An erythrocyte-specific DNA-binding factor recognizes a regulatory sequence common to all chicken globin genes. *Proceedings of the National Academy of Sciences of the United States of America, 85*, 5976–5980.
Fang, J., Feng, Q., Ketel, C. S., Wang, H., Cao, R., Xia, L., et al. (2002). Purification and functional characterization of SET8, a nucleosomal histone H4-lysine 20-specific methyltransferase. *Current Biology, 12*, 1086–1099.
Feuerborn, A., & Cook, P. R. (2015). Why the activity of a gene depends on its neighbors. *Trends in Genetics, 31*, 483–490.
Flotho, A., & Melchior, F. (2013). Sumoylation: A regulatory protein modification in health and disease. *Annual Review of Biochemistry, 82*, 357–385.
Forsberg, E. C., & Bresnick, E. H. (2001). Histone acetylation beyond promoters: Long-range acetylation patterns in the chromatin world. *Bioessays, 23*, 820–830.
Forsberg, E. C., Downs, K. M., Christensen, H. M., Im, H., Nuzzi, P. A., & Bresnick, E. H. (2000). Developmentally dynamic histone acetylation pattern of a tissue-specific chromatin domain. *Proceedings of the National Academy of Sciences of the United States of America, 97*, 14494–14499.
Fox, A. H., Liew, C., Holmes, M., Kowalski, K., MacKay, J. P., & Crossley, M. (1999). Transcriptional cofactors of the FOG family interact with GATA proteins bymeans of multiple zinc fingers. *EMBO Journal, 18*, 2812–2822.

Fujita, N., Jaye, D. L., Geigerman, C., Akyildiz, A., Mooney, M. R., Boss, J. M., et al. (2004). MTA3 and the Mi-2/NuRD complex regulate cell fate during B lymphocyte differentiation. *Cell*, *119*, 75–86.

Fujiwara, Y., Browne, C. P., Cunniff, K., Goff, S. C., & Orkin, S. H. (1996). Arrested development of embryonic red cell precursors in mouse embryos lacking transcription factor GATA-1. *Proceedings of the National Academy of Sciences of the United States of America*, *93*, 12355–12358.

Fujiwara, Y., Chang, A. N., Williams, A. M., & Orkin, S. H. (2004). Functional overlap of GATA-1 and GATA-2 in primitive hematopoietic development. *Blood*, *103*, 583–585.

Fujiwara, T., Lee, H. Y., Sanalkumar, R., & Bresnick, E. H. (2010). Building multifunctionality into a complex containing master regulators of hematopoiesis. *Proceedings of the National Academy of Sciences of the United States of America*, *107*, 20429–20434.

Fujiwara, T., O'Geen, H., Keles, S., Blahnik, K., Linnemann, A. K., Kang, Y. A., et al. (2009). Discovering hematopoietic mechanisms through genome-wide analysis of GATA factor chromatin occupancy. *Molecular Cell*, *36*, 667–681.

Furlan-Magaril, M., Varnai, C., Nagano, T., & Fraser, P. (2015). 3D genome architecture from populations to single cells. *Current Opinion in Genetics & Development*, *31*, 36–41.

Gao, Z., Huang, Z., Olivey, H. E., Gurbuxani, S., Crispino, J. D., & Svensson, E. C. (2010). FOG-1-mediated recruitment of NuRD is required for cell lineage re-enforcement during haematopoiesis. *EMBO Journal*, *29*, 457–468.

Gao, X., Johnson, K. D., Chang, Y. I., Boyer, M. E., Dewey, C. N., Zhang, J., et al. (2013). Gata2 cis-element is required for hematopoietic stem cell generation in the mammalian embryo. *Journal of Experimental Medicine*, *210*, 2833–2842.

Georgopoulos, K., Bigby, M., Wang, J. H., Molnar, A., Wu, P., Winandy, S., et al. (1994). The Ikaros gene is required for the development of all lymphoid lineages. *Cell*, *79*, 143–156.

Grass, J. A., Boyer, M. E., Pal, S., Wu, J., Weiss, M. J., & Bresnick, E. H. (2003). GATA-1-dependent transcriptional repression of GATA-2 via disruption of positive autoregulation and domain-wide chromatin remodeling. *Proceedings of the National Academy of Sciences of the United States of America*, *100*, 8811–8816.

Grass, J. A., Jing, H., Kim, S.-I., Martowicz, M. L., Pal, S., Blobel, G. A., et al. (2006). Distinct functions of dispersed GATA factor complexes at an endogenous gene locus. *Molecular and Cellular Biology*, *26*, 7056–7067.

Gregory, G. D., Miccio, A., Bersenev, A., Wang, Y., Hong, W., Zhang, Z., et al. (2010). FOG1 requires NuRD to promote hematopoiesis and maintain lineage fidelity within the megakaryocytic-erythroid compartment. *Blood*, *115*, 2156–2166.

Gregory, T., Yu, C., Ma, A., Orkin, S. H., Blobel, G. A., & Weiss, M. J. (1999). GATA-1 and erythropoietin cooperate to promoter erythroid cell survival by regulating bcl-xl expression. *Blood*, *94*, 87–96.

Groschel, S., Sanders, M. A., Hoogenboezem, R., de Wit, E., Bouwman, B. A., Erpelinck, C., et al. (2014). A single oncogenic enhancer rearrangement causes concomitant EVI1 and GATA2 deregulation in leukemia. *Cell*, *157*, 369–381.

Hahn, C. N., Chong, C. E., Carmichael, C. L., Wilkins, E. J., Brautigan, P. J., Li, X. C., et al. (2011). Heritable GATA2 mutations associated with familial myelodysplastic syndrome and acute myeloid leukemia. *Nature Genetics*, *43*, 1012–1017.

Harju-Baker, S., Costa, F. C., Fedosyuk, H., Neades, R., & Peterson, K. R. (2008). Silencing of Agamma-globin gene expression during adult definitive erythropoiesis mediated by GATA-1-FOG-1-Mi2 complex binding at the −566 GATA site. *Molecular and Cellular Biology*, *28*, 3101–3113.

Hayakawa, F., Towatari, M., Ozawa, Y., Tomita, A., Privalsky, M. L., & Saito, H. (2004). Functional regulation of GATA-2 by acetylation. *Journal of Leukocyte Biology*, *75*, 529–540.

Hebbes, T. R., Clayton, A. L., Thorne, A. W., & Crane-Robinson, C. (1994). Core histone hyperacetylation co-maps with generalized DNase I sensitivity in the chicken beta-globin chromosomal domain. *EMBO Journal, 13*, 1823–1830.

Heintzman, N. D., Hon, G. C., Hawkins, R. D., Kheradpour, P., Stark, A., Harp, L. F., et al. (2009). Histone modifications at human enhancers reflect global cell-type-specific gene expression. *Nature, 459*, 108–112.

Hernandez-Hoyos, G., Anderson, M. K., Wang, C., Rothenberg, E. V., & Alberola-Ila, J. (2003). GATA-3 expression is controlled by TCR signals and regulates CD4/CD8 differentiation. *Immunity, 19*, 83–94.

Hewitt, K. J., Kim, D. H., Devadas, P., Prathibha, R., Zuo, C., Sanalkumar, R., et al. (2015). Hematopoietic signaling mechanism revealed from a stem/progenitor cell cistrome. *Molecular Cell, 59*(1), 62–74.

Hodge, D., Coghill, E., Keys, J., Maguire, T., Hartmann, B., McDowall, A., et al. (2005). A global role for EKLF in definitive and primitive erythropoiesis. *Blood, 107*(8), 3359–3370.

Hong, W., Nakazawa, M., Chen, Y. Y., Kori, R., Vakoc, C. R., Rakowski, C., et al. (2005). FOG-1 recruits the NuRD repressor complex to mediate transcriptional repression by GATA-1. *EMBO Journal, 24*, 67–78.

Hosokawa, H., Kato, M., Tohyama, H., Tamaki, Y., Endo, Y., Kimura, M. Y., et al. (2015). Methylation of Gata3 protein at Arg-261 regulates transactivation of the Il5 gene in T helper 2 cells. *Journal of Biological Chemistry, 290*, 13095–13103.

Hsu, A. P., Sampaio, E. P., Khan, J., Calvo, K. R., Lemieux, J. E., Patel, S. Y., et al. (2011). Mutations in GATA2 are associated with the autosomal dominant and sporadic monocytopenia and mycobacterial infection (MonoMAC) syndrome. *Blood, 118*, 2653–2655.

Huang, H., Sabari, B. R., Garcia, B. A., Allis, C. D., & Zhao, Y. (2014). SnapShot: Histone modifications. *Cell, 159*. 458–458 e451.

Hung, H. L., Lau, J., Kim, A. Y., Weiss, M. J., & Blobel, G. A. (1999). CREB-binding protein acetylates hematopoietic transcription factor GATA-1 at functionally important sites. *Molecular and Cellular Biology, 19*, 3496–3505.

Im, H., Grass, J. A., Christensen, H. M., Perkins, A., & Bresnick, E. H. (2002). Histone deacetylase-dependent establishment and maintenance of broad low-level histone acetylation within a tissue-specific chromatin domain. *Biochemistry, 41*, 15152–15160.

Im, H., Grass, J. A., Johnson, K. D., Boyer, M. E., Wu, J., & Bresnick, E. H. (2004). Measurement of protein-DNA interactions in vivo by chromatin immunoprecipitation. *Methods in Molecular Biology, 284*, 129–146.

Im, H., Grass, J. A., Johnson, K. D., Kim, S.-I., Boyer, M. E., Imbalzano, A. N., et al. (2005). Chromatin domain activation via GATA-1 utilization of a small subset of dispersed GATA motifs within a broad chromosomal region. *Proceedings of the National Academy of Sciences of the United States of America, 102*, 17065–17070.

Ito, M., Yuan, C. X., Okano, H. J., Darnell, R. B., & Roeder, R. G. (2000). Involvement of the TRAP220 component of the TRAP/SMCC coactivator complex in embryonic development and thyroid hormone action. *Molecular Cell, 5*, 683–693.

Johnson, K. D., Boyer, M. E., Kang, J. A., Wickrema, A., Cantor, A. B., & Bresnick, E. H. (2007). Friend of GATA-1-independent transcriptional repression: A novel mode of GATA-1 function. *Blood, 109*, 5230–5233.

Johnson, K. D., Grass, J. A., Boyer, M. E., Kiekhaefer, C. M., Blobel, G. A., Weiss, M. J., et al. (2002). Cooperative activities of hematopoietic regulators recruit RNA polymerase II to a tissue-specific chromatin domain. *Proceedings of the National Academy of Sciences of the United States of America, 99*, 11760–11765.

Johnson, K. D., Hsu, A. P., Ryu, M. J., Wang, J., Gao, X., Boyer, M. E., et al. (2012). Cis-element mutated in GATA2-dependent immunodeficiency governs hematopoiesis and vascular integrity. *Journal of Clinical Investigation, 122*, 3692–3704.

Johnson, K. D., Kim, S.-I., & Bresnick, E. H. (2006). Differential sensitivities of transcription factor target genes underlie cell type-specific gene expression patterns. *Proceedings of the National Academy of Sciences of the United States of America*, *103*, 15939–15944.

Johnson, K. D., Kong, G., Gao, X., Chang, Y.-I., Hewitt, K. J., Sanalkumar, R., et al. (2015). Cis-regulatory mechanisms governing stem and progenitor cell transitions. *Science Advances*, *1*(8), e1500503.

Kadonaga, J. T., Carner, K. R., Masiarz, F. R., & Tjian, R. (1987). Isolation of cDNA encoding transcription factor Sp1 and functional analysis of the DNA binding domain. *Cell*, *51*, 1079–1090.

Kadri, Z., Maouche-Chretien, L., Rooke, H. M., Orkin, S. H., Romeo, P. H., Mayeux, P., et al. (2005). Phosphatidylinositol 3-kinase/Akt induced by erythropoietin renders the erythroid differentiation factor GATA-1 competent for TIMP-1 gene transactivation. *Molecular and Cellular Biology*, *25*, 7412–7422.

Kang, Y. A., Sanalkumar, R., O'Geen, H., Linnemann, A. K., Chang, C. J., Bouhassira, E. E., et al. (2012). Autophagy driven by a master regulator of hematopoiesis. *Molecular and Cellular Biology*, *32*, 226–239.

Katsumura, K. R., DeVilbiss, A. W., Pope, N. J., Johnson, K. D., & Bresnick, E. H. (2013). Transcriptional mechanisms underlying hemoglobin synthesis. *Cold Spring Harbor Perspectives in Medicine*, *3*, a015412.

Katsumura, K. R., Yang, C., Boyer, M. E., Li, L., & Bresnick, E. H. (2014). Molecular basis of crosstalk between oncogenic Ras and the master regulator of hematopoiesis GATA-2. *EMBO Reports*, *15*, 938–947.

Katz, S. G., Cantor, A. B., & Orkin, S. H. (2002). Interaction between FOG-1 and the corepressor C-terminal bindng protein is dispensible for normal erythropoiesis in vivo. *Molecular and Cellular Biology*, *22*, 3121–3128.

Khavari, P. A., Peterson, C. L., Tamkun, J. W., Mendel, D. B., & Crabtree, G. R. (1993). BRG1 contains a conserved domain of the SWI2/SNF2 family necessary for normal mitotic growth and transcription. *Nature*, *366*, 170–174.

Kiekhaefer, C. M., Grass, J. A., Johnson, K. D., Boyer, M. E., & Bresnick, E. H. (2002). Hematopoietic-specific activators establish an overlapping pattern of histone acetylation and methylation within a mammalian chromatin domain. *Proceedings of the National Academy of Sciences of the United States of America*, *99*, 14309–14314.

Kim, S. I., Bresnick, E. H., & Bultman, S. J. (2009). BRG1 directly regulates nucleosome structure and chromatin looping of the alpha globin locus to activate transcription. *Nucleic Acids Research*, *37*, 6019–6027.

Kim, S.-I., Bultman, S. J., Jing, H., Blobel, G. A., & Bresnick, E. B. (2007). Dissecting molecular steps in chromatin domain activation during hematopoietic differentiation. *Molecular and Cellular Biology*, *27*, 4551–4565.

Kim, S.-I., Bultman, S. J., Kiefer, C. M., Dean, A., & Bresnick, E. H. (2009). BRG1 requirement for long-range interaction of a locus control region with a downstream promoter. *Proceedings of the National Academy of Sciences of the United States of America*, *106*, 2259–2264.

Kim, J., Sif, S., Jones, B., Jackson, A., Koipally, J., Heller, E., et al. (1999). Ikaros DNA-binding proteins direct formation of chromatin remodeling complexes in lymphocytes. *Immunity*, *10*, 345–355.

Kitajima, K., Tanaka, M., Zheng, J., Yen, H., Sato, A., Sugiyama, D., et al. (2006). Redirecting differentiation of hematopoietic progenitors by a transcription factor, GATA-2. *Blood*, *107*, 1857–1863.

Klemsz, M. J., McKercher, S. R., Celada, A., Van Beveren, C., & Maki, R. A. (1990). The macrophage and B cell-specific transcription factor PU.1 is related to the ets oncogene. *Cell*, *61*, 113–124.

Ko, L. J., & Engel, J. D. (1993). DNA-binding specificities of the GATA transcription factor family. *Molecular and Cellular Biology*, *13*, 4011–4022.

Kornberg, R. D. (2005). Mediator and the mechanism of transcriptional activation. *Trends in Biochemical Sciences, 30*, 235–239.

Kowalski, K., Czolij, R., King, G. F., Crossley, M., & Mackay, J. P. (1999). The solution structure of the N-terminal zinc finger of GATA-1 reveals a specific binding face for the transcriptional co-factor FOG. *Journal of Biomolecular NMR, 13*, 249–262.

Kruger, W., Peterson, C. L., Sil, A., Coburn, C., Arents, G., Moudrianakis, E. N., et al. (1995). Amino acid substitutions in the structured domains of histones H3 and H4 partially relieve the requirement of the yeast SWI/SNF complex for transcription. *Genes & Development, 9*, 2770–2779.

Kundu, M., Lindsten, T., Yang, C. Y., Wu, J., Zhao, F., Zhang, J., et al. (2008). Ulk1 plays a critical role in the autophagic clearance of mitochondria and ribosomes during reticulocyte maturation. *Blood, 112*, 1493–1502.

Kuzmichev, A., Nishioka, K., Erdjument-Bromage, H., Tempst, P., & Reinberg, D. (2002). Histone methyltransferase activity associated with a human multiprotein complex containing the enhancer of zeste protein. *Genes & Development, 16*, 2893–2905.

Lamonica, J. M., Deng, W., Kadauke, S., Campbell, A. E., Gamsjaeger, R., Wang, H., et al. (2011). Bromodomain protein Brd3 associates with acetylated GATA1 to promote its chromatin occupancy at erythroid target genes. *Proceedings of the National Academy of Sciences of the United States of America, 108*, E159–E168.

Lamonica, J. M., Vakoc, C. R., & Blobel, G. A. (2006). Acetylation of GATA-1 is required for chromatin occupancy. *Blood, 108*, 3736–3738.

Le Guezennec, X., Vermeulen, M., Brinkman, A. B., Hoeijmakers, W. A., Cohen, A., Lasonder, E., et al. (2006). MBD2/NuRD and MBD3/NuRD, two distinct complexes with different biochemical and functional properties. *Molecular and Cellular Biology, 26*, 843–851.

Lee, H. Y., Johnson, K. D., Boyer, M. E., & Bresnick, E. H. (2011). Relocalizing genetic loci into specific subnuclear neighborhoods. *Journal of Biological Chemistry, 286*, 18834–18844.

Lee, H.-Y., Johnson, K. D., Fujiwara, T., Boyer, M. E., Kim, S.-I., & Bresnick, E. H. (2009). Controlling hematopoiesis through sumoylation-dependent regulation of a GATA factor. *Molecular Cell, 36*, 984–995.

Lejon, S., Thong, S. Y., Murthy, A., AlQarni, S., Murzina, N. V., Blobel, G. A., et al. (2011). Insights into association of the NuRD complex with FOG-1 from the crystal structure of an RbAp48.FOG-1 complex. *Journal of Biological Chemistry, 286*, 1196–1203.

Leonard, M., Brice, M., Engel, J. D., & Papayannopoulou, T. (1993). Dynamics of GATA-1 transcription factor expression during erythroid differentiation. *Blood, 82*, 1–9.

Letting, D. L., Chen, Y. Y., Rakowski, C., Reedy, S., & Blobel, G. A. (2004). Context-dependent regulation of GATA-1 by friend of GATA-1. *Proceedings of the National Academy of Sciences of the United States of America, 101*, 476–481.

Letting, D. L., Rakowski, C., Weiss, M. J., & Blobel, G. A. (2003). Formation of a tissue-specific histone acetylation pattern by the hematopoietic transcription factor GATA-1. *Molecular and Cellular Biology, 23*, 1334–1340.

Li, Z., Nie, F., Wang, S., & Li, L. (2011). Histone H4 Lys 20 monomethylation by histone methylase SET8 mediates Wnt target gene activation. *Proceedings of the National Academy of Sciences of the United States of America, 108*, 3116–3123.

Liew, C. W., Rand, K. D., Simpson, R. J., Yung, W. W., Mansfield, R. E., Crossley, M., et al. (2006). Molecular analysis of the interaction between the hematopoietic master transcription factors GATA-1 and PU.1. *Journal of Biological Chemistry, 281*, 28296–28306.

Liew, C. K., Simpson, R. J. Y., Kwan, A. H. Y., Crofts, L. A., Loughlin, F. E., Matthews, J. M., et al. (2005). Zinc fingers as protein recognition motifs: Structural basis for the GATA-1/friend of GATA interaction. *Proceedings of the National Academy of Sciences of the United States of America, 102*, 583–588.

Li-Harms, X., Milasta, S., Lynch, J., Wright, C., Joshi, A., Iyengar, R., et al. (2014). Mito-protective autophagy is impaired in erythroid cells of aged mtDNA-mutator mice. *Blood, 125*(1), 162–174.

Ling, K. W., Ottersbach, K., van Hamburg, J. P., Oziemlak, A., Tsai, F. Y., Orkin, S. H., et al. (2004). GATA-2 plays two functionally distinct roles during the ontogeny of hematopoietic stem cells. *Journal of Experimental Medicine, 200*, 871–882.

Linnemann, A. K., O'Geen, H., Keles, S., Farnham, P. J., & Bresnick, E. H. (2011). Genetic framework for GATA factor function in vascular biology. *Proceedings of the National Academy of Sciences of the United States of America, 108*, 13641–13646.

Litt, M. D., Simpson, M., Recillas-Targa, F., Prioleau, M. N., & Felsenfeld, G. (2001). Transitions in histone acetylation reveal boundaries of three separately regulated neighboring loci. *EMBO Journal, 20*, 2224–2235.

Lohmann, F., & Bieker, J. J. (2008). Activation of Eklf expression during hematopoiesis by Gata2 and Smad5 prior to erythroid commitment. *Development, 135*, 2071–2082.

Lopez, R. A., Schoetz, S., DeAngelis, K., O'Neill, D., & Bank, A. (2002). Multiple hematopoietic defects and delayed globin switching in Ikaros null mice. *Proceedings of the National Academy of Sciences of the United States of America, 99*, 602–607.

Love, P. E., Warzecha, C., & Li, L. (2014). Ldb1 complexes: The new master regulators of erythroid gene transcription. *Trends in Genetics, 30*, 1–9.

Luesink, M., Hollink, I. H., van der Velden, V. H., Knops, R. H., Boezeman, J. B., de Haas, V., et al. (2012). High GATA2 expression is a poor prognostic marker in pediatric acute myeloid leukemia. *Blood, 120*, 2064–2075.

Makino, D. L., Halbach, F., & Conti, E. (2013). The RNA exosome and proteasome: Common principles of degradation control. *Nature Reviews. Molecular Cell Biology, 14*, 654–660.

Makino, D. L., Schuch, B., Stegmann, E., Baumgartner, M., Basquin, C., & Conti, E. (2015). RNA degradation paths in a 12-subunit nuclear exosome complex. *Nature, 524*, 54–58.

Malik, J., Getman, M., & Steiner, L. A. (2015). Histone methyltransferase Setd8 represses Gata2 expression and regulates erythroid maturation. *Molecular and Cellular Biology, 35*, 2059–2072.

Margueron, R., & Reinberg, D. (2011). The polycomb complex PRC2 and its mark in life. *Nature, 469*, 343–349.

Marinkovic, D., Zhang, X., Yalcin, S., Luciano, J. P., Brugnara, C., Huber, T., et al. (2007). Foxo3 is required for the regulation of oxidative stress in erythropoiesis. *Journal of Clinical Investigation, 117*, 2133–2144.

Martin, D. I., & Orkin, S. H. (1990). Transcriptional activation and DNA binding by the erythroid factor GF- 1/NF-E1/Eryf 1. *Genes & Development, 4*, 1886–1898.

Martin, D. I., Zon, L. I., Mutter, G., & Orkin, S. H. (1990). Expression of an erythroid transcription factor in megakaryocytic and mast cell lineages. *Nature, 344*, 444–447.

Martowicz, M. L., Grass, J. A., Boyer, M. E., Guend, H., & Bresnick, E. H. (2005). Dynamic GATA factor interplay at a multi-component regulatory region of the GATA-2 locus. *Journal of Biological Chemistry, 280*, 1724–1732.

McIver, S. C., Kang, Y. A., DeVilbiss, A. W., O'Driscoll, C. A., Ouellette, J. N., Pope, N. J., et al. (2014). The exosome complex establishes a barricade to erythroid maturation. *Blood, 124*, 2285–2297.

McKercher, S. R., Torbett, B. E., Anderson, K. L., Henkel, G. W., Vestal, D. J., Baribault, H., et al. (1996). Targeted disruption of the PU.1 gene results in multiple hematopoietic abnormalities. *EMBO Journal, 15*, 5647–5658.

Menghini, R., Marchetti, V., Cardellini, M., Hribal, M. L., Mauriello, A., Lauro, D., et al. (2005). Phosphorylation of GATA2 by Akt increases adipose tissue differentiation and reduces adipose tissue-related inflammation: A novel pathway linking obesity to atherosclerosis. *Circulation, 111*, 1946–1953.

Merika, M., & Orkin, S. H. (1993). DNA-binding specificity of GATA family transcription factors. *Molecular and Cellular Biology, 13*, 3999–4010.

Merika, M., & Orkin, S. H. (1995). Functional synergy and physical interactions of the erythroid transcription factor GATA-1 with the Kruppel family proteins Sp1 and EKLF. *Molecular and Cellular Biology, 15*, 2437–2447.

Miccio, A., & Blobel, G. A. (2010). Role of the GATA-1/FOG-1/NuRD pathway in the expression of human beta-like globin genes. *Molecular and Cellular Biology, 30*, 3460–3470.

Miccio, A., Wang, Y., Hong, W., Gregory, G. D., Wang, H., Yu, X., et al. (2010). NuRD mediates activating and repressive functions of GATA-1 and FOG-1 during blood development. *EMBO Journal, 29*, 442–456.

Miller, I. J., & Bieker, J. J. (1993). A novel, erythroid cell-specific murine transcription factor that binds to the CACCC element and is related to the Kruppel family of nuclear proteins. *Molecular and Cellular Biology, 13*, 2776–2786.

Molkentin, J. D. (2000). The zinc finger-containing transcription factors GATA-4, −5, and −6. Ubiquitously expressed regulators of tissue-specific gene expression. *Journal of Biological Chemistry, 275*, 38949–38952.

Monteiro, R., Pouget, C., & Patient, R. (2011). The gata1/pu.1 lineage fate paradigm varies between blood populations and is modulated by tif1 gamma. *EMBO Journal, 30*, 1093–1103.

Mortensen, M., Ferguson, D. J., Edelmann, M., Kessler, B., Morten, K. J., Komatsu, M., et al. (2010). Loss of autophagy in erythroid cells leads to defective removal of mitochondria and severe anemia in vivo. *Proceedings of the National Academy of Sciences of the United States of America, 107*, 832–837.

Mullighan, C. G., Su, X., Zhang, J., Radtke, I., Phillips, L. A., Miller, C. B., et al. (2009). Deletion of IKZF1 and prognosis in acute lymphoblastic leukemia. *New England Journal of Medicine, 360*, 470–480.

Muntean, A. G., & Crispino, J. D. (2005). Differential requirements for the activation domain and FOG-interaction surface of GATA-1 in megakaryocyte gene expression and development. *Blood, 106*, 1223–1231.

Munugalavadla, V., Dore, L. C., Tan, B. L., Hong, L., Vishnu, M., Weiss, M. J., et al. (2005). Repression of c-kit and its downstream substrates by GATA-1 inhibits cell proliferation during erythroid maturation. *Molecular and Cellular Biology, 25*, 6747–6759.

Nardelli, J., Thiesson, D., Fujiwara, Y., Tsai, F.-Y., & Orkin, S. H. (1999). Expression and genetic interaction of transcription factors GATA-2 and GATA-3 during development of the mouse central nervous system. *Developmental Biology, 210*, 305–321.

Nei, Y., Obata-Ninomiya, K., Tsutsui, H., Ishiwata, K., Miyasaka, M., Matsumoto, K., et al. (2013). GATA-1 regulates the generation and function of basophils. *Proceedings of the National Academy of Sciences of the United States of America, 110*, 18620–18625.

Newton, A., MacKay, J. P., & Crossley, M. (2001). The N-terminal zinc finger of the erythroid transcription factor GATA-1 binds GATC motifs in DNA. *Journal of Biological Chemistry, 276*, 35794–35801.

Ng, S. Y., Yoshida, T., & Georgopoulos, K. (2007). Ikaros and chromatin regulation in early hematopoiesis. *Current Opinion in Immunology, 19*, 116–122.

Nichols, K. E., Crispino, J. D., Poncz, M., White, J. G., Orkin, S. H., Maris, J. M., et al. (2000). Familial dyserythropoietic anaemia and thrombocytopenia due to an inherited mutation in GATA-1. *Nature Genetics, 24*, 266–270.

Nishioka, K., Rice, J. C., Sarma, K., Erdjument-Bromage, H., Werner, J., Wang, Y., et al. (2002). PR-Set7 is a nucleosome-specific methyltransferase that modifies lysine 20 of histone H4 and is associated with silent chromatin. *Molecular Cell, 9*, 1201–1213.

Oda, H., Okamoto, I., Murphy, N., Chu, J., Price, S. M., Shen, M. M., et al. (2009). Monomethylation of histone H4-lysine 20 is involved in chromosome structure and stability and is essential for mouse development. *Molecular and Cellular Biology, 29*, 2278–2295.

Omichinski, J. G., Trainor, C., Evans, T., Gronenborn, A. M., Clore, G. M., & Felsenfeld, G. (1993). A small single-"finger" peptide from the erythroid transcription factor GATA-1 binds specifically to DNA as a zinc or iron complex. *Proceedings of the National Academy of Sciences of the United States of America, 90*, 1676–1680.
O'Neill, D. W., Schoetz, S. S., Lopez, R. A., Castle, M., Rabinowitz, L., Shor, E., et al. (2000). An ikaros-containing chromatin-remodeling complex in adult-type erythroid cells. *Molecular and Cellular Biology, 20*, 7572–7582.
O'Neill, L. P., & Turner, B. M. (1995). Histone H4 acetylation distinguishes coding regions of the human genome from heterochromatin in a differentiation-dependent but transcription-independent manner. *EMBO Journal, 14*, 3946–3957.
O'Neill, D., Yang, J., Erdjument-Bromage, H., Bornschlegel, K., Tempst, P., & Bank, A. (1999). Tissue-specific and developmental stage-specific DNA binding by a mammalian SWI/SNF complex associated with human fetal-to-adult globin gene switching. *Proceedings of the National Academy of Sciences of the United States of America, 96*, 349–354.
Orkin, S. H. (1992). GATA-binding transcription factors in hematopoietic cells. *Blood, 80*, 575–581.
Ostergaard, P., Simpson, M. A., Connell, F. C., Steward, C. G., Brice, G., Woollard, W. J., et al. (2011). Mutations in GATA2 cause primary lymphedema associated with a predisposition to acute myeloid leukemia (Emberger syndrome). *Nature Genetics, 43*, 929–931.
Pai, S. Y., Truitt, M. L., & Ho, I. C. (2004). GATA-3 deficiency abrogates the development and maintenance of T helper type 2 cells. *Proceedings of the National Academy of Sciences of the United States of America, 101*, 1993–1998.
Pal, S., Cantor, A. B., Johnson, K. D., Moran, T., Boyer, M. E., Orkin, S. H., et al. (2004). Coregulator-dependent facilitation of chromatin occupancy by GATA-1. *Proceedings of the National Academy of Sciences of the United States of America, 101*, 980–985.
Pal, S., Nemeth, M. J., Bodine, D., Miller, J. L., Svaren, J., Thein, S. L., et al. (2004). Neurokinin-B transcription in erythroid cells: Direct activation by the hematopoietic transcription factor GATA-1. *Journal of Biological Chemistry, 279*, 31348–31356.
Partington, G. A., & Patient, R. K. (1999). Phosphorylation of GATA-1 increases its DNA-binding affinity and is correlated with induction of human K562 erythroleukaemia cells. *Nucleic Acids Research, 27*, 1168–1175.
Perkins, A. C., Sharpe, A. H., & Orkin, S. H. (1995). Lethal beta-thalassaemia in mice lacking the erythroid CACCC- transcription factor EKLF. *Nature, 375*, 318–322.
Peterson, C. L., Dingwall, A., & Scott, M. P. (1994). Five SWI/SNF gene products are components of a large multisubunit complex required for transcriptional enhancement. *Proceedings of the National academy of Sciences of the United States of America, 91*, 2905–2908.
Pevny, L., Lin, C. S., D'Agati, V., Simon, M. C., Orkin, S. H., & Costantini, F. (1995). Development of hematopoietic cells lacking transcription factor GATA-1. *Development, 121*, 163–172.
Pevny, L., Simon, M. C., Robertson, E., Klein, W. H., Tsai, S. F., D'Agati, V., et al. (1991). Erythroid differentiation in chimaeric mice blocked by a targeted mutation in the gene for transcription factor GATA-1. *Nature, 349*, 257–260.
Pilon, A. M., Ajay, S. S., Kumar, S. A., Steiner, L. A., Cherukuri, P. F., Wincovitch, S., et al. (2011). Genome-wide ChIP-Seq reveals a dramatic shift in the binding of the transcription factor erythroid Kruppel-like factor during erythrocyte differentiation. *Blood, 118*, e139–e148.
Pimkin, M., Kossenkov, A. V., Mishra, T., Morrissey, C. S., Wu, W., Keller, C. A., et al. (2014). Divergent functions of hematopoietic transcription factors in lineage priming and differentiation during erythro-megakaryopoiesis. *Genome Research, 24*, 1932–1944.
Pop, R., Shearstone, J. R., Shen, Q., Liu, Y., Hallstrom, K., Koulnis, M., et al. (2010). A key commitment step in erythropoiesis is synchronized with the cell cycle clock through mutual inhibition between PU.1 and S-phase progression. *PLoS Biology, 8*, e1000484.

Pope, N. J., & Bresnick, E. H. (2010). Differential coregulator requirements for function of the hematopoietic transcription factor GATA-1 at endogenous loci. *Nucleic Acids Research*, *38*(7), 2190–2200.

Pope, N. J., & Bresnick, E. H. (2013). Establishment of a cell-type-specific genetic network by the mediator complex component Med1. *Molecular and Cellular Biology*, *33*, 1938–1955.

Rainis, L., Bercovich, D., Strehl, S., Teigler-Schlegel, A., Stark, B., Trka, J., et al. (2003). Mutations in exon 2 of GATA1 are early events in megakaryocytic malignancies associated with trisomy 21. *Blood*, *102*, 981–986.

Rajapakse, I., & Groudine, M. (2011). On emerging nuclear order. *Journal of Cell Biology*, *192*, 711–721.

Rekhtman, N., Radparvar, F., Evans, T., & Skoultchi, A. I. (1999). Direct interaction of hematopoietic transcription factors PU.1 and GATA-1: Functional antagonism in erythroid cells. *Genes & Development*, *13*, 1398–1411.

Roche, A. E., Bassett, B. J., Samant, S. A., Hong, W., Blobel, G. A., & Svensson, E. C. (2008). The zinc finger and C-terminal domains of MTA proteins are required for FOG-2-mediated transcriptional repression via the NuRD complex. *Journal of Molecular and Cellular Cardiology*, *44*, 352–360.

Rodrigues, N. P., Janzen, V., Forkert, R., Dombkowski, D. M., Boyd, A. S., Orkin, S. H., et al. (2005). Haploinsufficiency of GATA-2 perturbs adult hematopoietic stem cell homeostasis. *Blood*, *106*, 477–484.

Rooke, H. M., & Orkin, S. H. (2006). Phosphorylation of Gata1 at serine residues 72, 142, and 310 is not essential for hematopoiesis in vivo. *Blood*, *107*, 3527–3530.

Ross, J., Mavoungou, L., Bresnick, E. H., & Milot, E. (2012). GATA-1 utilizes Ikaros and polycomb repressive complex 2 to suppress Hes1 and to promote erythropoiesis. *Molecular and Cellular Biology*, *32*, 3624–3638.

Rothenberg, E. V. (2012). Transcriptional drivers of the T-cell lineage program. *Current Opinion in Immunology*, *24*, 132–138.

Roukos, V., & Misteli, T. (2014). The biogenesis of chromosome translocations. *Nature Cell Biology*, *16*, 293–300.

Roukos, V., Voss, T. C., Schmidt, C. K., Lee, S., Wangsa, D., & Misteli, T. (2013). Spatial dynamics of chromosome translocations in living cells. *Science*, *341*, 660–664.

Rylski, M., Welch, J. J., Chen, Y. Y., Letting, D. L., Diehl, J. A., Chodosh, L. A., et al. (2003). GATA-1-mediated proliferation arrest during erythroid maturation. *Molecular and Cellular Biology*, *23*, 5031–5042.

Sanalkumar, R., Johnson, K. D., Gao, X., Boyer, M. E., Chang, Y. I., Hewitt, K. J., et al. (2014). Mechanism governing a stem cell-generating cis-regulatory element. *Proceedings of the National Academy of Sciences of the United States of America*, *111*(12), E1091–E1100.

Sandoval, H., Thiagarajan, P., Dasgupta, S. K., Schumacher, A., Prchal, J. T., Chen, M., et al. (2008). Essential role for nix in autophagic maturation of erythroid cells. *Nature*, *454*, 232–235.

Scheenstra, M. R., Salunkhe, V., De Cuyper, I. M., Hoogenboezem, M., Li, E., Kuijpers, T. W., et al. (2015). Characterization of hematopoietic GATA transcription factor expression in mouse and human dendritic cells. *Blood Cells, Molecules & Diseases*, *55*, 293–303.

Schweers, R. L., Zhang, J., Randall, M. S., Loyd, M. R., Li, W., Dorsey, F. C., et al. (2007). NIX is required for programmed mitochondrial clearance during reticulocyte maturation. *Proceedings of the National Academy of Sciences of the United States of America*, *104*, 19500–19505.

Scott, E. W., Fisher, R. C., Olson, M. C., Kehrli, E. W., Simon, M. C., & Singh, H. (1997). PU.1 functions in a cell-autonomous manner to control the differentiation of multipotential lymphoid-myeloid progenitors. *Immunity*, *6*, 437–447.

Scott, E. W., Simon, M. C., Anastasi, J., & Singh, H. (1994). Requirement of transcription factor PU.1 in the development of multiple hematopoietic lineages. *Science*, *265*, 1573–1577.

Shilatifard, A. (2006). Chromatin modifications by methylation and ubiquitination: Implications in the regulation of gene expression. *Annual Review of Biochemistry*, *75*, 243–269.

Shilatifard, A. (2012). The COMPASS family of histone H3K4 methylases: Mechanisms of regulation in development and disease pathogenesis. *Annual Review of Biochemistry*, *81*, 65–95.

Simon, J. A., & Kingston, R. E. (2013). Occupying chromatin: Polycomb mechanisms for getting to genomic targets, stopping transcriptional traffic, and staying put. *Molecular Cell*, *49*, 808–824.

Simon, M. C., Pevny, L., Wiles, M. V., Keller, G., Costantini, F., & Orkin, S. H. (1992). Rescue of erythroid development in gene targeted GATA-1- mouse embryonic stem cells. *Nature Genetics*, *1*, 92–98.

Snow, J. W., Kim, J., Currie, C. R., Xu, J., & Orkin, S. H. (2010). Sumoylation regulates interaction of FOG1 with C-terminal-binding protein (CTBP). *Journal of Biological Chemistry*, *285*, 28064–28075.

Snow, J. W., Trowbridge, J. J., Fujiwara, T., Emambokus, N. E., Grass, J. A., Orkin, S. H., et al. (2010). A single cis element maintains repression of the key developmental regulator Gata2. *PLoS Genetics*, *6*, e1001103.

Snow, J. W., Trowbridge, J. J., Johnson, K. D., Fujiwara, T., Emambokus, N. E., Grass, J. A., et al. (2011). Context-dependent function of "GATA switch" sites in vivo. *Blood*, *117*, 4769–4772.

Spinner, M. A., Sanchez, L. A., Hsu, A. P., Shaw, P. A., Zerbe, C. S., Calvo, K. R., et al. (2013). GATA2 deficiency: A protean disorder of hematopoiesis, lymphatics and immunity. *Blood*, *123*(6), 809–821.

Staber, P. B., Zhang, P., Ye, M., Welner, R. S., Nombela-Arrieta, C., Bach, C., et al. (2013). Sustained PU.1 levels balance cell-cycle regulators to prevent exhaustion of adult hematopoietic stem cells. *Molecular Cell*, *49*, 934–946.

Stonestrom, A. J., Hsu, S. C., Jahn, K. S., Huang, P., Keller, C. A., Giardine, B. M., et al. (2015). Functions of BET proteins in erythroid gene expression. *Blood*, *125*, 2825–2834.

Stopka, T., Amanatullah, D. F., Papetti, M., & Skoultchi, A. I. (2005). PU.1 inhibits the erythroid program by binding to GATA-1 on DNA and creating a repressive chromatin structure. *EMBO Journal*, *24*, 3712–3723.

Stumpf, M., Waskow, C., Krotschel, M., van Essen, D., Rodriguez, P., Zhang, X., et al. (2006). The mediator complex functions as a coactivator for GATA-1 in erythropoiesis via subunit Med1/TRAP220. *Proceedings of the National Academy of Sciences of the United States of America*, *103*, 18504–18509.

Sung, Y. H., Baek, I. J., Kim, D. H., Jeon, J., Lee, J., Lee, K., et al. (2013). Knockout mice created by TALEN-mediated gene targeting. *Nature Biotechnology*, *31*, 23–24.

Takizawa, T., Meaburn, K. J., & Misteli, T. (2008). The meaning of gene positioning. *Cell*, *135*, 9–13.

Talasz, H., Lindner, H. H., Sarg, B., & Helliger, W. (2005). Histone H4-lysine 20 monomethylation is increased in promoter and coding regions of active genes and correlates with hyperacetylation. *Journal of Biological Chemistry*, *280*, 38814–38822.

Tallack, M. R., Magor, G. W., Dartigues, B., Sun, L., Huang, S., Fittock, J. M., et al. (2012). Novel roles for KLF1 in erythropoiesis revealed by mRNA-seq. *Genome Research*, *22*, 2385–2398.

Tallack, M. R., Whitington, T., Yuen, W. S., Wainwright, E. N., Keys, J. R., Gardiner, B. B., et al. (2010). A global role for KLF1 in erythropoiesis revealed by ChIP-seq in primary erythroid cells. *Genome Research*, *20*, 1052–1063.

Tindemans, I., Serafini, N., Di Santo, J. P., & Hendriks, R. W. (2014). GATA-3 function in innate and adaptive immunity. *Immunity, 41*, 191–206.
Tong, J. K., Hassig, C. A., Schnitzler, G. R., Kingston, R. E., & Schreiber, S. L. (1998). Chromatin deacetylation by an ATP-dependent nucleosome remodelling complex. *Nature, 395*, 917–921.
Towatari, M., May, G. E., Marais, R., Perkins, G. R., Marshall, C. J., Cowley, S., et al. (1995). Regulation of GATA-2 phosphorylation by mitogen-activated protein kinase and interleukin-3. *Journal of Biological Chemistry, 270*, 4101–4107.
Trainor, C. D., Omichinski, J. G., Vandergon, T. L., Gronenborn, A. M., Clore, G. M., & Felsenfeld, G. (1996). A palindromic regulatory site within vertebrate GATA-1 promoters requires both zinc fingers of the GATA-1 DNA-binding domain for high- affinity interaction. *Molecular and Cellular Biology, 16*, 2238–2247.
Tran, H., Brunet, A., Griffith, E. C., & Greenberg, M. E. (2003). The many forks in FOXO's road. *Science's STKE, 2003*, RE5.
Tsai, F. Y., Keller, G., Kuo, F. C., Weiss, M., Chen, J., Rosenblatt, M., et al. (1994). An early haematopoietic defect in mice lacking the transcription factor GATA-2. *Nature, 371*, 221–226.
Tsai, S. F., Martin, D. I., Zon, L. I., D'Andrea, A. D., Wong, G. G., & Orkin, S. H. (1989). Cloning of cDNA for the major DNA-binding protein of the erythroid lineage through expression in mammalian cells. *Nature, 339*, 446–451.
Tsai, F.-Y., & Orkin, S. H. (1997). Transcription factor GATA-2 is required for proliferation/survival of early hematopoietic cells and mast cell formation, but not for erythroid and myeloid terminal differentiation. *Blood, 89*, 3636–3643.
Tsang, A. P., Fujiwara, Y., Hom, D. B., & Orkin, S. H. (1998). Failure of megakaryopoiesis and arrested erythropoiesis in mice lacking the GATA-1 transcriptional cofactor FOG. *Genes & Development, 12*, 1176–1188.
Tsang, A. P., Visvader, J. E., Turner, C. A., Fujiwara, Y., Yu, C., Weiss, M. J., et al. (1997). FOG, a multitype zinc finger protein, acts as a cofactor for transcription factor GATA-1 in erythroid and megakaryocytic differentiation. *Cell, 90*, 109–119.
Vakoc, C. R., Letting, D. L., Gheldof, N., Sawado, T., Bender, M. A., Groudine, M., et al. (2005). Proximity among distant regulatory elements at the beta globin locus requires GATA-1 and FOG-1. *Molecular Cell, 17*, 453–462.
Vicente, C., Vazquez, I., Conchillo, A., Garcia-Sanchez, M. A., Marcotegui, N., Fuster, O., et al. (2012). Overexpression of GATA2 predicts an adverse prognosis for patients with acute myeloid leukemia and it is associated with distinct molecular abnormalities. *Leukemia, 26*, 550–554.
Vo, N., & Goodman, R. H. (2001). CREB-binding protein and p300 in transcriptional regulation. *Journal of Biological Chemistry, 276*, 13505–13508.
Vyas, P., & Crispino, J. D. (2007). Molecular insights into Down syndrome-associated leukemia. *Current Opinion in Pediatrics, 19*, 9–14.
Wade, P. A., Jones, P. L., Vermaak, D., & Wolffe, A. P. (1998). A multiple subunit Mi-2 histone deacetylase from Xenopus laevis cofractionates with an associated Snf2 superfamily ATPase. *Current Biology, 8*, 843–846.
Wadman, I. A., Osada, H., Grutz, G. G., Agulnick, A. D., Westphal, H., Forster, A., et al. (1997). The LIM-only protein Lmo2 is a bridging molecule assembling an erythroid, DNA-binding complex which includes the TAL1, E47, GATA-1 and Ldb1/NLI proteins. *EMBO Journal, 16*, 3145–3157.
Walsh, J. C., DeKoter, R. P., Lee, H.-J., Smith, E. D., Lancki, D. W., Gurish, M. F., et al. (2002). Cooperative and antagonistic interplay between PU.1 and GATA-2 in the specification of myeloid cell fates. *Immunity, 17*, 665–676.
Wang, W., Cote, J., Xue, Y., Zhou, S., Khavari, P. A., Biggar, S. R., et al. (1996). Purification and biochemical heterogeneity of the mammalian SWI/SNF complex. *EMBO Journal, 15*, 5370–5382.

Wang, J. H., Nichogiannopoulou, A., Wu, L., Sun, L., Sharpe, A. H., Bigby, M., et al. (1996). Selective defects in the development of the fetal and adult lymphoid system in mice with an Ikaros null mutation. *Immunity, 5*, 537–549.

Watamoto, K., Towatari, M., Ozawa, Y., Miyata, Y., Okamoto, M., Abe, A., et al. (2003). Altered interaction of HDAC5 with GATA-1 during MEL cell differentiation. *Oncogene, 22*, 9176–9184.

Wechsler, J., Greene, M., McDevitt, M. A., Anastasi, J., Karp, J. E., Le Beau, M. M., et al. (2002). Acquired mutations in GATA-1 in the megakaryoblastic leukemia of Down syndrome. *Nature Genetics, 32*, 148–152.

Weiss, M. J., Keller, G., & Orkin, S. H. (1994). Novel insights into erythroid development revealed through in vitro differentiation of GATA-1 embryonic stem cells. *Genes & Development, 8*, 1184–1197.

Weiss, M. J., Yu, C., & Orkin, S. H. (1997). Erythroid-cell-specific properties of transcription factor GATA-1 revealed by phenotypic rescue of a gene-targeted cell line. *Molecular and Cellular Biology, 17*, 1642–1651.

Welch, J. J., Watts, J. A., Vakoc, C. R., Yao, Y., Wang, H., Hardison, R. C., et al. (2004). Global regulation of erythroid gene expression by transcription factor GATA-1. *Blood, 104*, 3136–3147.

Wilson, D. B., Dorfman, D. M., & Orkin, S. H. (1990). A nonerythroid GATA-binding protein is required for function of the human preproendothelin-1 promoter in endothelial cells. *Molecular and Cellular Biology, 10*, 4854–4862.

Winandy, S., Wu, P., & Georgopoulos, K. (1995). A dominant mutation in the Ikaros gene leads to rapid development of leukemia and lymphoma. *Cell, 83*, 289–299.

Wozniak, R. J., Boyer, M. E., Grass, J. A., Lee, Y., & Bresnick, E. H. (2007). Context-dependent GATA factor function: Combinatorial requirements for transcriptional control in hematopoietic and endothelial cells. *Journal of Biological Chemistry, 282*, 14665–14674.

Wozniak, R. J., Keles, S., Lugus, J. J., Young, K., Boyer, M. E., Tran, T. T., et al. (2008). Molecular hallmarks of endogenous chromatin complexes containing master regulators of hematopoiesis. *Molecular and Cellular Biology, 28*, 6681–6694.

Xu, Z., Meng, X., Cai, Y., Koury, M. J., & Brandt, S. J. (2006). Recruitment of the SWI/SNF protein Brg1 by a multiprotein complex effects transcriptional repression in murine erythroid progenitors. *Biochemistry Journal, 399*, 297–304.

Xu, J., Shao, Z., Li, D., Xie, H., Kim, W., Huang, J., et al. (2015). Developmental control of polycomb subunit composition by GATA factors mediates a switch to non-canonical functions. *Molecular Cell, 57*, 304–316.

Xue, Y., Wong, J., Moreno, G. T., Young, M. K., Cote, J., & Wang, W. (1998). NURD, a novel complex with both ATP-dependent chromatin-remodeling and histone deacetylase activities. *Molecular Cell, 2*, 851–861.

Yamagata, T., Mitani, K., Oda, H., Suzuki, T., Honda, H., Asai, T., et al. (2000). Acetylation of GATA-3 affects T-cell survival and homing to secondary lymphoid organs. *EMBO Journal, 19*, 4676–4687.

Yamamoto, M., Ko, L. J., Leonard, M. W., Beug, H., Orkin, S. H., & Engel, J. D. (1990). Activity and tissue-specific expression of the transcription factor NF- E1 multigene family. *Genes & Development, 4*, 1650–1662.

Yamazaki, H., Suzuki, M., Otsuki, A., Shimizu, R., Bresnick, E. H., Engel, J. D., et al. (2014). A remote GATA2 hematopoietic enhancer drives leukemogenesis in inv(3)(q21;q26) by activating EVI1 expression. *Cancer Cell, 25*, 415–427.

Yomogida, K., Ohtani, H., Harigae, H., Ito, E., Nishimune, Y., Engel, J. D., et al. (1994). Developmental stage- and spermatogenic cycle-specific expression of transcription factor GATA-1 in mouse Sertoli cells. *Development, 120*, 1759–1766.

Yoshida, T., & Georgopoulos, K. (2014). Ikaros fingers on lymphocyte differentiation. *International Journal of Hematology, 100*, 220–229.

Yu, C., Cantor, A. B., Yang, H., Browne, C., Wells, R. A., Fujiwara, Y., et al. (2002). Targeted deletion of a high-affinity GATA-binding site in the GATA-1 promoter leads to selective loss of the eosinophil lineage in vivo. *Journal of Experimental Medicine, 195*, 1387–1395.

Yu, Y. L., Chiang, Y. J., Chen, Y. C., Papetti, M., Juo, C. G., Skoultchi, A. I., et al. (2005). MAPK-mediated phosphorylation of GATA-1 promotes Bcl-XL expression and cell survival. *Journal of Biological Chemistry, 280*, 29533–29542.

Yu, M., Riva, L., Xie, H., Schindler, Y., Moran, T. B., Cheng, Y., et al. (2009). Insights into GATA-1-mediated gene activation versus repression via genome-wide chromatin occupancy analysis. *Molecular Cell, 36*, 682–695.

Zhang, Y., LeRoy, G., Seelig, H. P., Lane, W. S., & Reinberg, D. (1998). The dermatomyositis-specific autoantigen Mi2 is a component of a complex containing histone deacetylase and nucleosome remodeling activities. *Cell, 95*, 279–289.

Zhang, J., & Ney, P. A. (2009). Role of BNIP3 and NIX in cell death, autophagy, and mitophagy. *Cell Death and Differentiation, 16*, 939–946.

Zhang, Y., Ng, H. H., Erdjument-Bromage, H., Tempst, P., Bird, A., & Reinberg, D. (1999). Analysis of the NuRD subunits reveals a histone deacetylase core complex and a connection with DNA methylation. *Genes & Development, 13*, 1924–1935.

Zhang, X., Rielland, M., Yalcin, S., & Ghaffari, S. (2011). Regulation and function of FoxO transcription factors in normal and cancer stem cells: What have we learned? *Current Drug Targets, 12*, 1267–1283.

Zhang, P., Zhang, X., Iwama, A., Yu, C., Smith, K. A., Mueller, B. U., et al. (2000). PU.1 inhibits GATA-1 function and erythroid differentiation by blocking GATA-1 DNA binding. *Blood, 96*, 2641–2648.

Zhao, W., Kitidis, C., Fleming, M. D., Lodish, H. F., & Ghaffari, S. (2006). Erythropoietin stimulates phosphorylation and activation of GATA-1 via the PI3-kinase/AKT signaling pathway. *Blood, 107*, 907–915.

Zhu, Y., Qi, C., Jia, Y., Nye, J. S., Rao, M. S., & Reddy, J. K. (2000). Deletion of PBP/PPARBP, the gene for nuclear receptor coactivator peroxisome proliferator-activated receptor-binding protein, results in embryonic lethality. *Journal of Biological Chemistry, 275*, 14779–14782.

CHAPTER EIGHT

Noncoding Regulatory RNAs in Hematopoiesis

M. Jeong, M.A. Goodell[1]

Stem Cells and Regenerative Medicine Center, Baylor College of Medicine, Houston, TX, United States

[1]Corresponding author: e-mail address: goodell@bcm.edu

Contents

Abstract

Hematopoiesis is a dynamic process in which blood cells are continuously generated from hematopoietic stem cells (HSCs). The regulatory mechanisms controlling HSC fate have been studied extensively over the past several decades. Although many protein-coding genes have been shown to regulate hematopoietic differentiation, additional levels of HSC regulation are not well studied. Advances in deep sequencing have revealed many new classes of regulatory noncoding RNAs (ncRNAs), such as enhancer RNAs and antisense ncRNAs. Functional analysis of some of these ncRNAs has provided insights into the molecular mechanisms that regulate hematopoietic development and disease. In this review, we summarize recent advances in our understanding of functional regulatory ncRNAs associated with hematopoietic self-renewal and differentiation, as well as those dysregulated ncRNAs involved in hematologic malignancies.

Current Topics in Developmental Biology, Volume 118
ISSN 0070-2153
http://dx.doi.org/10.1016/bs.ctdb.2016.01.006

1. INTRODUCTION

Hematopoiesis is a hierarchical developmental system that depends upon a rare stem cell population, hematopoietic stem cells (HSCs), to ensure production of blood cells throughout the lifetime of an individual (Orkin & Zon, 2008). HSCs play an important role in maintaining blood homeostasis through both differentiation into lineage cells and self-renewal (Luo et al., 2015; Sun, Luo, et al., 2014). As a result of several decades of study focusing on hematopoietic hierarchy and HSC renewal and differentiation, our current understanding of the stages of hematopoiesis is well developed (Akashi, Traver, Miyamoto, & Weissman, 2000; Kondo et al., 2000; Rossi et al., 2012; Sun, Luo, et al., 2014; Weissman, 2000). HSCs are pluripotent cells with high self-renewal capacity that balance self-renewal and commitment toward specific differentiation pathways. During differentiation, HSCs progress through various intermediate maturational stages, generating multipotent and lineage-committed progenitor cells prior to producing mature cells, such as myelocytes, lymphocytes, erythrocytes, and platelets. Mature blood cells mediate inflammatory reactions, fight against infections and cancer cells, provide oxygen throughout the body, and prevent bleeding by clotting. A complex network of extrinsic and intrinsic regulatory factors orchestrates HSC fate decisions.

Transcription factors (TFs) play a central role in hematopoietic development, from HSC commitment to terminal differentiation and death. Knocking-out TFs such as Gata2, Prdm16, and Nurr1 leads to severe hematopoietic defects (Aguilo et al., 2011; Ling et al., 2004; Rodrigues et al., 2005; Sirin, Lukov, Mao, Conneely, & Goodell, 2010). Even a 50% reduction in the expression of some genes, such as PU.1, can impact cell lineage decisions (Dahl et al., 2003). Many growth factors and cytokines also activate cell-signaling pathways and protein–protein interactions and control the stability of proteins, posttranslational modifications, and epigenetic regulations during hematopoietic development.

While specific regulatory proteins play key roles in the regulation of gene expression during hematopoiesis, advances in genomic discoveries over the past 10 years have uncovered an additional level of regulation involving RNA. RNA is an important intermediate molecule between DNA and protein, and recently noncoding RNAs (ncRNAs) were discovered to play important regulatory roles in a wide range of cellular processes. Whole transcriptome sequencing identified many long RNA that are located in intergenic or intronic sites and on the antisense strand but do not encode proteins (Carninci et al., 2005; Cheng et al., 2005). The function of some of these ncRNAs has been established in a few biological processes, including cell-cycle regulation and embryonic stem cell function (Guttman et al., 2011; Hung et al., 2011; Klattenhoff et al., 2013; Prensner et al., 2011), but the vast majority have not yet been analyzed. Here, we present current data about noncoding regulatory RNAs and discuss their potential and confirmed roles in HSC pluripotency and in normal and pathological hematopoiesis.

2. EMERGING IMPORTANCE OF REGULATORY RNAs IN HEMATOPOIESIS

2.1 Landscape of Noncoding Regulatory RNAs

Genomic regions are separated into major functional units such as enhancer regions, promoter regions, the gene body, and gene terminal regions. The ENCODE project revealed that transcripts covering almost all regions of the eukaryotic genome can be detected under some conditions, yet only 2% of genome encodes proteins. Thus, the vast majority of cellular transcripts are noncoding, suggesting that many noncoding DNA regions generate ncRNAs that may have cellular functions (Carninci et al., 2005; Djebali et al., 2012; International Human Genome Sequencing Consortium, 2004; Mattick, 2007). Advances in the quality of deep sequencing and bioinformatics have enabled classification of ncRNAs into different groups, such as long noncoding RNAs (lncRNAs), which are classified based on transcript length; intronic ncRNAs; antisense ncRNAs classified based on their association with annotated protein-coding genes; and enhancer RNAs (eRNAs) classified through their association with other DNA elements of known function (Batista & Chang, 2013; Djebali et al., 2012; St Laurent, Wahlestedt, & Kapranov, 2015). The genomic and transcriptional landscape of regulatory ncRNAs is summarized in Fig. 1.

2.2 Enhancer RNAs

eRNAs are RNA molecules that are transcribed from genomic enhancer regions. Genomic enhancer regions are generally considered TF binding regulatory regions distal to gene transcription start sites. Enhancers are thought to bind a variety of proteins and interact with mediator complexes, which recruit Polymerase II and form loops that bring the promoter into

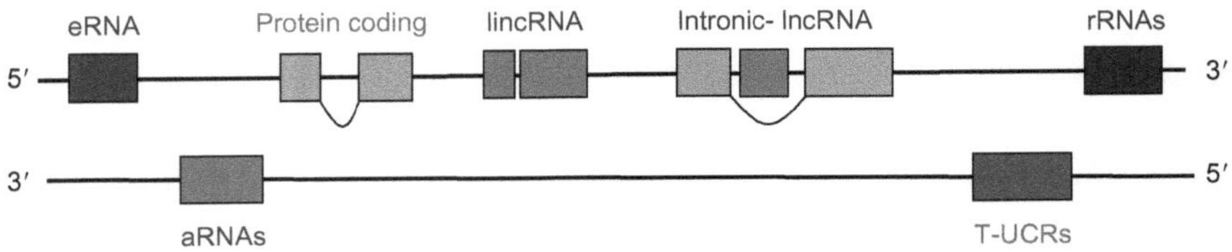

Fig. 1 Genome-wide landscape of regulatory lncRNAs. The various types of regulatory noncoding RNAs are transcribed from enhancers, promoters, introns of protein-coding or pseudogenes, antisense, or small RNAs within their transcription units. *aRNAs,* antisense RNAs; *eRNAs,* enhancer RNAs; *T-UCRs,* transcribed ultraconserved regions; *lincRNAs,* large intergenic ncRNAs; *rRNAs,* ribosomal RNAs. (See the color plate.)

proximity in order to positively impact gene expression from a distance. Enhancers are cell-type specific units that regulate tissue-specific expression of genes. A genome-wide association study showed that disease-associated single nucleotide polymorphisms are significantly enriched in enhancer regions, suggesting that differences in enhancer activity can contribute to phenotypes and disease outcome. Nevertheless, except in a few cases (eg, Hsu et al., 2013), data definitively demonstrating enhancer malfunction as causative in disease are lacking.

Enhancers are generally defined currently by histone modifications such as H3K27-acetylation (H3K27ac) and H3K4me1. By mapping transcripts that are associated with these marks, eRNAs have been identified. Using ChIP-seq for these marks, Kim et al. identified more than 12,000 enhancer-derived ncRNAs, specifically eRNAs, linked with H3K4me1, H3K27ac, and p300 binding, in murine neuronal cells. These eRNA transcripts are often bidirectional, lack a polyA tail, and have low copy number. Depletion of these eRNAs leads to decreased expression of their neighboring protein-coding genes, including master regulators of cellular differentiation (Kim, Hemberg, & Gray, 2015; Kim et al., 2010; Natoli & Andrau, 2012). Large-scale analyses confirmed that the expression level of eRNAs is positively correlated with the expression levels of nearby genes, as well as target genes of the corresponding enhancers (Andersson et al., 2014; Cheng, Pan, Tsai, & Tsai, 2015; Sanyal, Lajoie, Jain, & Dekker, 2012).

One example that has been studied in depth is the eRNA called HOTTIP. HOTTIP is encoded on the distal 5′-end of the HOXA gene cluster. HOTTIP forms a chromosomal loop that brings it into contact with multiple promoter-proximal target regions resulting in increased gene activation (Wang et al., 2011). Thus, eRNAs can serve as markers of activation for transcription, although their precise molecular mechanism is not fully defined, and it is not always possible to identify *a priori* the specific promoter region with which they interact. Further exploration of enhancer-associated transcription and the functional roles played by eRNAs will provide a better understanding of how gene expression is regulated during development.

2.3 Antisense ncRNAs

Antisense RNAs, which were initially considered transcriptional noise, are transcribed from complimentary DNA strands. Antisense transcripts are widespread throughout the genome and occur in 50–70% of all protein-coding genes (Katayama et al., 2005). However, the expression of antisense transcripts is ~10-fold lower than their cognate protein-coding genes, and these RNAs may degrade more rapidly than their sense strand transcripts (Ozsolak et al., 2010; Pelechano & Steinmetz, 2013).

Antisense lncRNAs have been studied in the context of maintenance of genomic imprinting (Mohammad, Mondal, & Kanduri, 2009), and their special mechanistic roles in genomic imprinting and X-chromosome inactivation has served as a paradigm for understanding lncRNA-dependent epigenetic-based gene regulation. Antisense lncRNAs are thought to epigenetically regulate expression of multiple genes in imprinted clusters by helping to recruit chromatin-modifying machinery.

One of the best-characterized antisense lncRNAs that mediates gene regulation is X-inactive specific transcript (XIST), which has a key role in mammalian X-chromosome inactivation (Pontier & Gribnau, 2011). During female development, one X chromosome produces XIST, which recruits the polycomb repressive complex 2 (PRC2) across the X chromosome by recruiting SHARP and activating HDAC3, which in turn leads to condensation and silencing of that X chromosome (Lee, 2009; McHugh et al., 2015; Zhao, Sun, Erwin, Song, & Lee, 2008). On the other active X allele, XIST is repressed by its antisense partner ncRNA, TSIX.

Another well-known antisense lncRNA is antisense PU.1. PU.1 is a master hematopoietic transcriptional factor that is necessary to maintain normal hematopoietic fate choice and suppress leukemia. Antisense PU.1 is thought to negatively regulate PU.1 expression by translational inhibition and has a lower expression level but longer half-life than sense PU.1. Antisense PU.1 may bind to the sense transcript and impact ribosome binding and thus specific steps in protein translation (Ebralidze et al., 2008). Further characterization of antisense transcription will shed light on the functional importance of antisense RNAs in gene regulation, which may play a critical role in development of diseases such as leukemia.

2.4 Diverse Mechanisms of lncRNA Function

Studies have elucidated the various mechanisms through which lncRNAs function. Unlike small RNAs (ie, miRNA), which have well-established mechanisms of action, lncRNAs can interact with proteins as well as with RNA or DNA. LncRNA–RNA interaction may lead to alterations in translational or in splicing. LncRNA–DNA interaction can lead to sequence-specific transcriptional regulation. When lncRNAs directly bind to DNA, they may regulate transcription by recruiting regulatory factors such as TFs and histone modifiers. LncRNAs can also function either *in cis*, within the same genomic locus, or *in trans*, affecting gene regulation in a different locus or even on different chromosomes (Morceau, Chateauvieux, Gaigneaux, Dicato, & Diederich, 2013).

3. REGULATORY RNAs IN NORMAL HEMATOPOIESIS

Over the past few years, there have been significant advances in identifying and understanding the diverse regulatory mechanisms of ncRNAs involved in hematopoiesis. A class of shorter ncRNAs and their important roles in hematopoiesis were previously well characterized (Wightman, Ha, & Ruvkun, 1993); recently, studies showed longer ncRNAs are also involved in hematopoietic development, including maintenance of HSCs, progenitors, and multilineage mature blood cells, including erythrocytes, megakaryocytes, myelocytes (monocytes, macrophages, and neutrophils), and lymphocytes (B cells and T cells). The functional roles of lncRNAs are currently being explored in mouse models using newer genetic engineering techniques. Unlike shorter ncRNAs, lncRNAs are poorly conserved among species, making it difficult to gauge their function in other organisms and translate results between species. However, some human lncRNAs, such as XIST and H19, are highly conserved and have similar tissue-specific expression patterns and functions in mice (Guttman et al., 2009). Representative examples of microRNAs and lncRNAs involved in the hematopoietic system are discussed in the following paragraphs and summarized in Fig. 2 and Table 1.

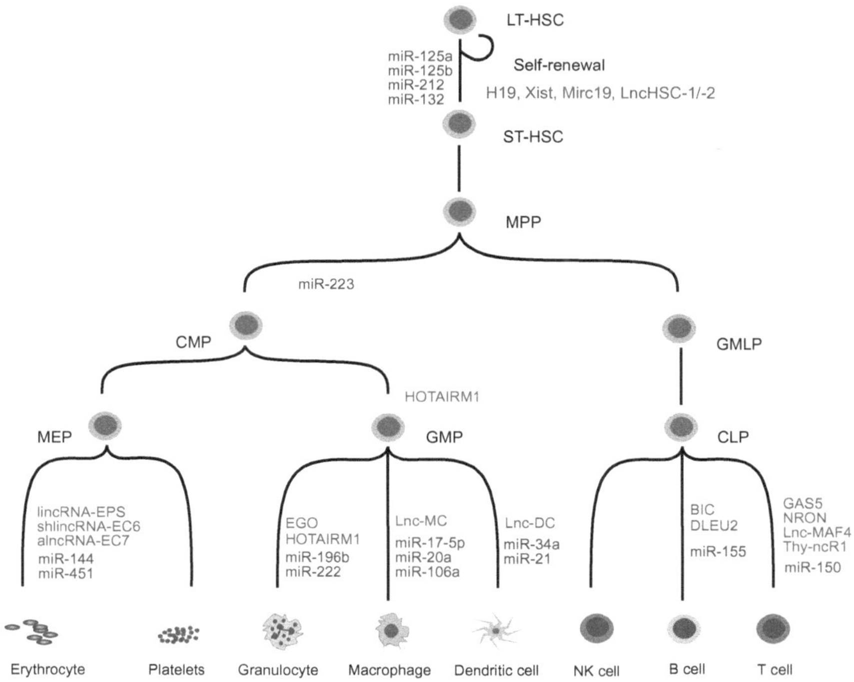

Fig. 2 The regulation of lncRNAs in hematopoietic development. LncRNAs are involved in regulating the different steps in hematopoiesis, and immune system development and activation. For instance, H19, XIST, and lncHSC-1/2 maintain long-term HSC quiescence and self-renewal. Lnc-DC controls DC differentiation by binding STAT3 (signal transducer and activator of transcription 3). NRON (noncoding RNA, repressor of NFAT) regulates the adaptive immune responses by restricting inappropriate activation of CD4+ T cells in the cytoplasm. LincRNA-Cox2 acts in the innate immune response by regulating TLR activation in macrophages and DCs. *LT-HSC,* long-term hematopoietic stem cell; *ST-HSC,* short-term hematopoietic stem cell; *MPP,* multipotent progenitors; *CMP,* common myeloid progenitor; *CMLP,* common myelo-lymphoid progenitor; *MEP,* megakaryocyte-erythroid progenitor; *GMP,* granulocyte–macrophage progenitor; *CLP,* common lymphoid progenitor. (See the color plate.)

3.1 MicroRNAs in Hematopoiesis

MicroRNAs are small ncRNAs, about 21–25 base pairs in length, present in a wide variety of organisms and their main function is to regulate gene expression by binding 3′ untranslated regions (UTRs) of the target messenger RNAs (mRNA). Currently, 2588 different human miRNAs and 1193 mouse miRNAs are listed in the miRBase hosted by Sanger Institute (miRBase v21; http://www.mirbase.org). Each miRNA exhibits complementarity to

Table 1 Long Noncoding RNAs Related to Hematopoiesis and Malignancies

LncRNA	Function and Disease	References
alncRNA-EC7	Regulating differentiation of MEP	Alvarez-Dominguez et al. (2014)
ANRIL, p15AS	AML/ALL by regulating p15	Yu et al. (2008)
CRNDE	Oncogene in AML, MM, T-cell leukemia	Ellis, Molloy, and Graham (2012)
DLEU1/DLEU2	Tumor suppressor in CCL, MM, and lymphoma	Garding et al. (2013)
EGO	Hematopoietic regulator in eosinophil development	Wagner et al. (2007)
EPS	Regulator of erythropoiesis	Hu, Yuan, Flygare, and Lodish (2011)
GAS5	Tumor suppressor in B-cell lymphoma and leukemia	Kino, Hurt, Ichijo, Nader, and Chrousos (2010)
H19	Regulating early hematopoiesis, oncogenic, or tumor suppressor in AML, CML, MPN, T-cell leukemia	Takeuchi et al. (2007)
HOTAIRM1	Regulator of myelopoiesis	Zhang et al. (2009)
HOXA-AS2	APL	Yang et al. (2008)
lincRNA-a7	Hematopoietic regulator by SCL/TAL1	Orom et al. (2010)
lincRNA-Cox2	Regulators of inflammatory response by TLR (Toll-like receptor) pathway	Carpenter et al. (2013)
lincRNA-EPS	Blocking differentiation of mouse erythroid progenitors	Hu et al. (2011)
lincRNA-p21	Potentially chronic lymphocytic leukemia	Blume et al. (2015)
lnc-DC	Regulating differentiation of dendritic cells	Wang, Xue, et al. (2014)
lncHSC-1/2	Regulating differentiation to short-term HSC	Luo et al. (2015)
MEG3	MDS/AML by regulating p53	Benetatos et al. (2010)
MIR100HG	Oncogenic in AML	Emmrich et al. (2014)

Table 1 Long Noncoding RNAs Related to Hematopoiesis and Malignancies—cont'd

LncRNA	Function and Disease	References
MIR155HG BIC	Host of miR-155 and regulating MYB and NFKB in CLL	Vargova et al. (2011)
MIR17HG	Host of miR-17–92 and targeting MYC in B-cell lymphoma	Ji et al. (2011)
MONC	Oncogenic in AML	Emmrich et al. (2014)
NRON	Regulating the activity of nuclear factor of activated T cells (NFATs)	Willingham et al. (2005)
PU.1-AS	Hematopoietic regulator of PU.1	Ebralidze et al. (2008)
RMRP	Non-Hodgkin lymphoma	Taskinen et al. (2008)
SNHG5	B-cell lymphoma	Braconi et al. (2011)
TCL-6	T-cell leukemia	Saitou, Sugimoto, Hatakeyama, Russo, and Isobe (2000)
THRIL	Regulators of innate activation of THP1 macrophages by TLR (Toll-like receptor) pathway	Li, Chao, et al. (2014)
ThyncR1	Regulator of T-cell selection and maturation	Aoki et al. (2010)
vtRNA2-1	Inhibiting PKR in AML	Kunkeaw et al. (2013)
XIST	Aberrant hematopoiesis and blood cell cancer	Yildirim et al. (2013)

multiple mRNAs, thus having the potential to control hundreds of different mRNAs in various ways, including mRNA degradation, deadenylation, and translational repression. In the hematopoietic system, more than 100 different miRNAs are expressed in different cell types (O'Connell et al., 2010), many of which play critical regulatory roles in hematopoiesis, including in HSC self-renewal and terminal differentiation.

The importance of miRNAs in regulating HSCs was shown by genetic ablation of the miRNA processing enzyme Dicer specifically in the hematopoietic stem/progenitor compartment. Dicer knock-out resulted in loss of functional HSCs and bone marrow failure due to increased apoptosis, suggesting that miRNAs are critical for HSC maintenance and normal hematopoiesis (Guo et al., 2010). Specific miRNAs have also been examined. miR-125a and miR-125b regulate the functions of long-term HSCs (LT-HSCs), and miR-125a specifically controls the size of the stem cell population by regulating hematopoietic stem/progenitor cell apoptosis (Guo et al., 2010). Overexpression of miR-126b in murine HSCs increases the number of HSCs and induces lymphoid-biased cell subsets (Ooi et al., 2010).

One of the most extensively studied miRNAs, miR-223, plays a pivotal role during myeloid differentiation by inhibiting *NFI-A* and *C/EBPalpha* expression (Fazi et al., 2005). Similarly, GATA-1, which plays a central role in erythroid differentiation, may induce expression of miR-144 and miR-451 (Dore et al., 2008), which in turn contribute to erythroid differentiation (Papapetrou, Korkola, & Sadelain, 2010). In the lymphoid lineages, miR-150 has been implicated. It is selectively expressed in mature B and T cells and contributes to B-cell development by targeting the transcription factor *c-Myb* (Xiao et al., 2007). miR-150 also regulates T-cell development by reducing NOTCH3 (Ghisi et al., 2011).

MicroRNA clusters have also been shown to impact HSC function. Mirc19 is a microRNA cluster (of miRNA-212/132) region located in chromosome 11, and while it is highly expressed in HSC, its expression increases in aging HSCs. Mirc19−/− mice have increased total HSC numbers and decreased progenitor numbers with age and LPS (lipopolysaccharide) treatment. Loss of Mirc19 reduces HSC cycling and functions through FOXO3, a TF associated with aging. Overexpression of miR-132 led to rapid HSC cycling and enlarged spleens, a hallmark of extramedullary hematopoiesis, demonstrating the important role Mirc19 plays in maintaining HSC quiescence and function through its interaction with FOXO3 (Mehta et al., 2015).

While miRNA expression plays numerous critical roles in the maintenance of normal hematopoiesis, they are also dysregulated in a variety of different human leukemias and lymphomas, which is a topic of intense research. miRNA dysregulation is in some cases sufficient to cause cancer (Bousquet, Harris, Zhou, & Lodish, 2010; Garzon et al., 2008; O'Connell et al., 2010). The roles of microRNAs in hematopoiesis are summarized in Fig. 1.

3.2 LncRNA in HSC Self-renewal and Proliferation

In the HSC compartment, there is a highly regulated balance between generation of committed progenitor cells and maintenance of the stem cell pool. The importance of this balance is demonstrated by numerous examples of aberrant HSC development linked to severe disease. Over the past several years, a large amount of RNA-seq data has accumulated, allowing for the identification of thousands of lncRNAs expressed in HSCs, progenitors, and lineage-specific cells, shedding light on the mechanisms of HSC maintenance and differentiation.

H19 was one of the first identified lncRNAs. H19 is well conserved across mammalian species and is an imprinted gene that is only transcribed from the maternally inherited allele on chromosome 11 (Goshen et al., 1993; Nunez et al., 2000). H19 is highly expressed in LT-HSCs and down-regulated in differentiated blood progenitors (Berg et al., 2011). Functional studies of H19 using conditional knock-out mice, in which differentially methylated regions (DMRs) controlling imprinting were conditionally deleted, showed that when the H19 DMR region is removed from the maternally derived allele, H19 was no longer expressed and LT-HSC numbers were depleted. HSCs from maternal-allele knock-out mice had a significantly increased short-term (ST)-HSC population. Bone marrow from these mice showed significantly reduced reconstitution ability after bone marrow transplantation (Venkatraman et al., 2013).

Our group extensively profiled the lncRNA of LT-HSCs as well as B cells and granulocytes. We uncovered many novel, HSC-specific lncRNAs (lncHSCs), including intergenic and enhancer transcripts. We further characterized two such lncRNAs, lncHSC-1 and lncHSC-2, and showed their nuclear localization and expression patterns in hematopoietic systems. We reported that lncHSC-1 and lncHSC-2 are important to HSC homeostasis, regulating HSC self-renewal and lineage commitment. LncHSC-1/-2 are highly expressed in stem and progenitor populations, but are either not expressed or are expressed at very low levels in terminally differentiated blood lineages. Knockdown of lncHSC-1 significantly increased colony numbers *in vitro* and myeloid progenitor cell numbers and myeloid differentiation *in vivo*. Knockdown of lncHSC-2 reduced HSC and progenitor cell numbers and increased T-cell differentiation. By using shRNA knockdown, lncHSC-1 was shown to contribute to myeloid skewing at the expense of B cells, and lncHSC-2 to promote self-renewal and T lymphopoiesis by regulating E2A (Luo et al., 2015).

While there are a few examples of lncRNAs that regulate HSC function, our overall understanding of their relative importance, compared to protein-coding genes and other regulatory mechanisms, is poor. Moreover, the mechanisms through which they act are not well understood, in part due to the technical challenges of identifying their binding partners, whether

RNA, DNA, or protein. Thus, there is a significant opportunity to better understand how these molecules impact HSC function.

3.3 LncRNA in Myeloid Differentiation

HSCs can develop into common myeloid progenitors (CMPs) and common lymphoid progenitors (CLPs). CMPs are the precursors of myeloid lineage cells including granulocytes, monocytes, and macrophages, which are critical effectors in innate immunity and regulators of adaptive immunity. Myeloid cells are constantly produced in the bone marrow in large numbers to maintain homeostatic immune responses. Furthermore, myeloid production can be rapidly modulated during infection or inflammation to adapt to the changing needs of the organism. Recent studies have shown that inflammatory cytokines can also directly act on HSCs to instruct their fate, either increasing proliferation or directing differentiation toward myeloid lineages (Mossadegh-Keller et al., 2013) at the expense of HSC self-renewal and/or maintenance in the bone marrow (Baldridge, King, Boles, Weksberg, & Goodell, 2010). In the same way, erythroid cell production is rapidly modulated during hematopoietic stresses such as hemorrhage and anemia. This rapid shift is likely driven by expression of successive combinations of TFs and other regulatory elements including lncRNAs.

An intronic lncRNA called eosinophil granule ontogeny lncRNA (EGO), located in the gene encoding inositol triphosphate receptor type 1 (ITPR1), was among the first hematopoietic lncRNAs characterized (Wagner et al., 2007). EGO, a conserved lncRNA, was identified in eosinophil differentiation of CD34+ hematopoietic progenitor cells. Eosinophils are myeloid cells that play a key role in parasitic immunity and allergic diseases, including asthma. EGO is highly expressed in human bone marrow and mature eosinophils. EGO levels have been shown to increase in response to Il-5, an important cytokine released during eosinophil differentiation. Knockdown of EGO in CD34+ progenitors resulted in decreased expression of eosinophil-derived neurotoxin, a gene critical to eosinophil differentiation, supporting the concept that EGO plays an important role in eosinophil development.

Another ncRNA thought to be involved in myeloid differentiation derives from the HOXA locus. HOX antisense intergenic RNA myeloid 1 (HOTAIRM1) is a long intergenic ncRNA (lincRNA) located between the HOXA1 and HOXA2 genes. HOTAIRM1 is expressed specifically in myeloid cells and is upregulated during retinoic acid (RA)-induced granulocytic differentiation of myeloid progenitors (Zhang et al., 2009). Silencing of HOTAIRM1 by shRNA inhibited RA-induced 3′ HOXA cluster genes including HOXA1, HOXA4, and HOXA5 and significantly inhibited differentiation of myeloid cells, as indicated by induction of cell surface markers such as CD11b and CD18 during differentiation, establishing a regulatory role for this lincRNA regulates myeloid cell production.

An additional dendritic cell-specific lncRNA (lnc-DC) was identified from extensive profiling of lncRNA expression during differentiation of monocytes into dendritic cells (DCs) (Wang, Xue, et al., 2014). Lnc-DC is a cytoplasmic lncRNA exclusively expressed in conventional human DCs. Knockdown of lnc-DC impaired DC differentiation by down-regulating DC-associated genes and reducing T-cell activation. Mechanistic studies suggested that lnc-DC contributes to preventing STAT3 dephosphorylation by Src homology region 2 domain-containing phosphatase-1 (SHP1) by directly binding to STAT3 in the cytoplasm.

Finally, Chen et al. identified an lncRNA involved in monocyte differentiation, lnc-MC. This lncRNA is involved in human myeloid differentiation through its interaction with PU.1 and miR199a-5p. Previous work demonstrated that miR-199a-5p negatively regulates monocyte/macrophage differentiation by targeting activin A receptor type 1B (ACVR1B). Knockdown and rescue of lnc-MC assay showed that lnc-MC is activated by PU.1 and negatively regulated by miR-119a-5p. Overexpression of lnc-MC reduced miR-199a-5p expression leading to significant increase of ACVR1B, an important regulator of myeloid differentiation (Chen et al., 2015).

Thus, several lncRNAs have been shown to interact in specific steps of myeloid cell differentiation. Again, the mechanisms are not entirely clear in most cases, and the general lack of conservation between mouse and human lncRNAs means that there are not straightforward animal models to definitively show the role of these transcripts, in most cases, in hematopoiesis.

3.4 LncRNA in Lymphoid Differentiation and Function

Lymphoid differentiation is the process by which lymphocytes (B cells, T cells, and NK cells) develop from progenitor cells. While CMPs generate myeloid cells, CLPs generate lymphocytes. B-cell lymphoid differentiation occurs in the bone marrow and T-cell lymphopoiesis occurs in the thymus.

Nuclear factor of activated T cells (NFATs) is a key regulator of T-cell development and function. NRON is an lncRNA that is thought to serve as a repressor of NFATs. NRON has composed three alternatively spliced exons which show distinct tissue-specific distribution (Willingham et al., 2005). In resting T cells, NFATs are phosphorylated and retained in cytoplasm by a complex that contains NRON (Tsao et al., 2013). In activated T cells, NRON expression is reduced and NFAT is dephosphorylated and transported into nucleus. This translocation activates subsequent gene transcription, and in patients with HIV-1, it promotes viral replication (Imam, Bano, Patel, Holla, & Jameel, 2015). NRON expression therefore contributes not only to normal hematopoiesis, but its downregulation also contributes to HIV progression.

Another example of a lymphoid lncRNA is growth arrest-specific 5 (GAS5). This human lncRNA is located on chromosome 1 and encodes multiple small nucleolar RNAs (snoRNAs) within its introns. In a leukemic T-cell line and in healthy T cells, GAS5 was shown to induce growth arrest and apoptosis, while downregulation of GAS5 using RNAi delayed growth arrest (Mourtada-Maarabouni, Hedge, Kirkham, Farzaneh, & Williams, 2008) and reduced the inhibitory impact of rapamycin (Mourtada-Maarabouni, Hasan, Farzaneh, & Williams, 2010). GAS5 may work through the formation of secondary structures which can interact with glucocorticoid receptor response elements (GREs) to inhibit gene expression and trigger apoptosis (Kino et al., 2010; Mourtada-Maarabouni et al., 2008; Nakamura et al., 2008; Williams, Mourtada-Maarabouni, & Farzaneh, 2011).

In addition, a thymus-associated lncRNA, Thy-ncR1, was identified as a 2.6-kb-long thymus-specific ncRNA. This transcript is specifically expressed in immature T cells. Thy-ncR1 produces several alternatively spliced isoforms and each isoform has cell-type specificity; two major Thy-ncR1 isoforms are localized in the cytoplasm and degraded via the nonsense-mediated decay pathway, suggesting that Thy-ncR1 is a possible target of NMD. Thus, the role of Thy-ncR1 in immature T cells remains to be clarified (Aoki et al., 2010).

In B cells, the lncRNA B-cell integration cluster (BIC) was identified as the host gene of miR-155-5p and miR-155-3p. BIC is expressed at a low level in most hematopoietic tissues, but miR-155-5p is highly expressed in both activated B and T cells and myeloid cells (Elton, Selemon, Elton, & Parinandi, 2013). Using BIC/miR-155-deficient mice, Rodriguez et al. demonstrated that BIC/miR-155 is required in order for B and

T lymphocytes to function properly (Rodriguez et al., 2007; van den Berg et al., 2003). Yin et al. demonstrated that the AP-1 site located upstream of BIC plays a major role in the BIC/miR-155 activation of B-cell activation (Yin, Wang, McBride, Fewell, & Flemington, 2008). These studies suggest that BIC/miR-155 plays a key role in the B- and T-cell homeostasis and function.

Ranzani et al. have generated the first comprehensive catalog of lincRNAs in 13 human T- and B-cell subtypes (Ranzani et al., 2015). Among the 500 subtype-specific lincRNAs discovered, linc-MAF-4 was specifically expressed in Th1 cells and plays a role during differentiation of Th1 cells. linc-MAF-4 is critical to maintaining the Th1 phenotype, as its downregulation leads to overexpression of MAF, an important Th2-associated TF, leading to a transition to Th2 phenotype. linc-MAF-4 is thought to repress MAF transcription by forming a chromatin loop close to the MAF promoter and recruiting PRC2-associated histone methyltransferase EZH2 and LSC1 (Ranzani et al., 2015).

A more comprehensive analysis of lncRNA expression during T-cell development was performed in 42 different T-cell types at various developmental and differentiation stages using RNA-seq. This study identified 1524 lincRNAs from mice and revealed dynamic and stage-specific expression patterns of lincRNAs during T-cell differentiation (Hu et al., 2013).

Thus, there are a number of lncRNAs shown to have a specific role in regulation of lymphoid development or function, but there are a much larger number that have been cataloged but not functionally examined. Thus, further exploration of the functions of these transcripts may lend new insights into lymphoid development.

3.5 LncRNAs in Erythroid Differentiation

Erythropoiesis is regulated by multiple mechanisms to ensure the proper generation of mature RBCs under multiple physiological conditions. The average lifespan of red blood cells (RBCs) is 105–120 days in humans. Approximately 2.5 million RBCs are produced in human bone marrow every second and around 1% are replaced every day under normal conditions. RBC production can rapidly increase during stresses, such as anemia or hemolysis (Lodish, Flygare, & Chou, 2010).

At each stage of RBC production, from erythroid progenitors to the final erythrocyte, intracellular signal transduction proteins activate downstream molecules. Hu et al. examined the expression of lncRNAs during erythropoiesis and identified an erythroid-specific lncRNA, lncRNA-EPS, that has a role in erythroid differentiation and apoptosis. Blocking lncRNA-EPS expression inhibited erythroid differentiation and induced apoptosis, while overexpression had the reverse effect. Hu et al. revealed that the mechanism through which this lncRNA was able to inhibit apoptosis was by repressing transcription of a proapoptotic gene, Pycard. In addition, Lodish and colleagues comprehensively characterized erythropoiesis-associated lncRNAs using E14.5 mouse fetal liver. One of the lncRNAs they discovered, alncRNA-EC7, appeared to be an eRNA, as it was transcribed from an enhancer region that was marked with H3K4me1 and H3K27ac. Targeting shRNAs against alncRNA-EC7 reduced expression of a neighboring gene *Band3*, which is a major anion transporter of the red cell membrane. The alncRNA-EC7 locus has a CTCF binding site, which may allow for direct chromatin looping to activate the expression of *Band3*. This study thus showed that alncRNA-EC7 acts *in cis* to regulate erythropoiesis by enhancer looping (Alvarez-Dominguez et al., 2014).

Finally, Paralkar et al. performed RNA profiling and uncovered hundreds of unannotated, novel erythroid-specific lncRNAs. They discovered murine erythroid progenitor-, erythroblast-, and megakaryocyte-specific lncRNAs, as well as human erythroblast-specific lncRNAs. Functional mouse erythroid lncRNAs are poorly conserved in humans and not expressed in human erythroblasts. Close analysis of these lncRNAs revealed that 75% of them were transcribed from promoters and 25% from enhancers. Many of them were suggested to be regulated by key erythroid TFs including GATA1 and TAL1. RNAi knockdown assay identified seven candidate lncRNAs as functionally important to terminal erythroid maturation (Paralkar et al., 2014). These lncRNAs will provide important directions for future studies and contribute to our continued understanding of the role of ncRNA in erythroid development.

4. ncRNAs IN HEMATOPOIETIC MALIGNANCIES

Impaired expression of important hematopoietic regulators can result in uncontrolled proliferation of blood cells of all different lineages and at different stages of development, ultimately leading to malignancies. LncRNAs are increasingly recognized as important regulators of hematopoietic cell fates, including differentiation and proliferation; they also may play roles in the pathogenesis of hematopoietic malignancies. Unsurprisingly, many of the same lncRNAs that regulate normal hematopoiesis are implicated in cancer development.

Diaz-Beya et al. investigated the association between HOTAIRM1 expression and acute myeloid leukemia (AML) development in 241 patients. They found that high HOTAIRM1 expression was associated with shorter overall survival and higher rates of relapse in AML patients, including in patients with NPM1-mutated AML. They revealed that HOTAIRM1 expression is associated with a specific signature of expression of 33-micro-RNAs that included miR-196b; this signature was suggested as an indicator of poor prognosis in AML (Diaz-Beya et al., 2015).

Similarly, an lncRNA associated with the well-studied RUNX1 gene has been implicated in AML. RUNX1-intragenic lncRNA (RUNXOR) is an unspliced intragenic lncRNA transcribed upstream of RUNX1 that interacts with the RUNX1 promoter and enhancers. Wang et al. reported that expression of *RUNXOR* is increased in AML samples. In response to treatment with the chemotherapeutic agent Ara-C in vitro, expression levels increased. A 3C interaction assay indicated that RUNXOR interacts with regions previously associated with sites of *RUNX1* translocations, including *EVI1*, *ETO*, and *CBFA2T3*. This study showed that RUNXOR participates in intrachromosomal and long-range interchromosomal interactions, suggesting that RUNXOR is a novel lncRNA with a tumor suppressor role in hematopoietic malignancies (Wang, Li, et al., 2014).

Another example is IRAIN, a novel antisense lncRNA located in the *IGF1R* intronic promoter locus. The IRAIN lncRNA is expressed from the paternally derived allele and repressed from the maternally derived allele, indicating that it is imprinted. IRAN expression is lower in high-risk AML patients and leukemic cell lines. IRAIN is thought to be involved in intrachromosomal interaction where it may interact with the IGF1R promoter as well as enhancer regions. Knockdown of IRAIN abolished intrachromosomal interactions as measured in a 3C assay. These data suggest that IRAIN is a new imprinted lncRNA regulating the IGF pathways in hematopoietic malignancies by long-range DNA interactions (Sun, Li, et al., 2014).

Finally, the lincRNAs MONC and MIR100HG are host genes of miR99a/100–125b, which have central roles in HSC homeostasis and leukemogenesis. MONC and MIR100HG are both dysregulated and may act as oncogenes in acute megakaryoblastic leukemia. MIR100HG shows higher expression in erythroid cells, while MONC is predominantly expressed in HSPCs and megakaryocytic cells. Knockdown of both lincRNAs increases apoptosis and reduces cell proliferation and viability. Enforced expression of MONC promotes expansion of immature erythroid precursors at the expense of myeloid differentiation (Emmrich et al., 2014).

Representative examples of lncRNAs involved in the hematopoietic malignancies are summarized in Table 1. Together, these data support a contribution of lncRNAs in the development of malignancies. However, as for their roles in normal hematopoiesis, a great deal more work is needed to understand their relative contribution to cellular differentiation and cancer. Furthermore, with a few exceptions, the mechanisms are not well established. With the many putative lncRNAs now identified through next-generation sequencing studies, the importance of these lncRNAs can now be explored more fully.

5. ncRNAs AS BIOMARKERS

Identifying novel biomarkers with high tissue specificity and sensitivity is important to improving diagnostics, prognostics, and personalized treatment for cancer patients. ncRNAs have emerged as a potential biomarkers and targets for therapeutics (Majem, Rigau, Reventos, & Wong, 2015; Reis & Verjovski-Almeida, 2012), and several studies have attempted to establish specific ncRNAs, or groups of ncRNAs as biomarkers. For example, Lawrie et al. investigated serum and plasma miRNAs and showed that circulating miRNAs are potential biomarkers in B-cell lymphoma (Lawrie et al., 2008). In addition, Mitchell et al. suggested that some circulating miRNAs may be protected from RNase degradation by binding to argonaute proteins (Mitchell et al., 2008), suggesting their use as biomarkers. A number of lncRNAs including HOTAIR, MALAT1, and H19 were discovered as biomarkers across several cancers. In breast cancer, HOTAIR expression is a potential marker of poor prognosis, and it has also been implicated in metastatic cancers. Similarly, MALAT1 is thought to contribute to metastatic cancers, as its expression is lower in nonmetastasizing adenocarcinomas than in metastasizing ones (Kim & Reitmair, 2013). Thus, we anticipate future studies to focus on the value of these and other ncRNAs as potential biomarkers for disease type and severity.

6. LncRNA DATABASES

Several studies have attempted to catalog many of the lncRNAs identified, establishing several databases of great utility. In Table 2, we highlight

Table 2 LncRNA Databases

Database	Source	Species	Website	References
ChIPBase	ChIP-Seq	6 species	http://deepbase.sysu.edu.cn/chipbase	Yang, Li, Jiang, Zhou, and Qu (2013)
LNCipedia	Literature	Human	http://www.lncipedia.org	Volders et al. (2015)
lncRNAdb	Literature	Around 60 species	http://www.lncrnadb.org	Quek et al. (2015)
LncRNADisease	Literature	Human	http://www.cuilab.cn/lncrnadisease	Chen et al. (2013)
LncRNAMap	Literature, seq data	Human	http://lncRNAMap.mbc.nctu.edu.tw	Chan, Huang, and Chang (2014)
NONCODEv4	Literature	10 species	http://www.noncode.org	Xie et al. (2014)
NRED	Design arrays	Human, mouse	http://nred.matticklab.com	Dinger et al. (2009)
StarBasev2.0	CLIP-seq	Human, mouse, *C. elegans*	http://starbase.sysu.edu.cn	Li, Liu, Zhou, Qu, and Yang (2014)

several of these so that the reader may explore these lncRNAs in more depth.

7. CONCLUSIONS

Over the past decade, advances in sequencing and bioinformatics technologies have shed light on the dynamic state of noncoding regulatory RNAs and provided powerful tools for characterizing and quantifying ncRNAs. Systematic characterization of stage- and tissue-specific expression patterns of ncRNAs in the hematopoietic system has provided insights fundamental to our understanding of the development and regulation of hematopoiesis. ncRNAs are not only important regulators of normal hematopoietic development but are also key players in hematopoietic malignancies. Additional examination of the biochemical and genomic features of ncRNAs, including gain- and loss-of-function studies and detailed functional mechanistic approaches, will open new possibilities for novel diagnostic and therapeutic tools against hematopoietic malignancies.

ACKNOWLEDGMENTS

This work was supported by the NIH (DK092883, CA183252, HG007538, CA125123, P50CA126752), the Edward P Evans Foundation, the Adrienne Helis Malvin Medical Research Foundation, and by CPRIT (RR140053).

REFERENCES

Aguilo, F., Avagyan, S., Labar, A., Sevilla, A., Lee, D. F., Kumar, P., et al. (2011). Prdm16 is a physiologic regulator of hematopoietic stem cells. *Blood*, *117*, 5057–5066.

Akashi, K., Traver, D., Miyamoto, T., & Weissman, I. L. (2000). A clonogenic common myeloid progenitor that gives rise to all myeloid lineages. *Nature*, *404*, 193–197.

Alvarez-Dominguez, J. R., Hu, W., Yuan, B., Shi, J., Park, S. S., Gromatzky, A. A., et al. (2014). Global discovery of erythroid long noncoding RNAs reveals novel regulators of red cell maturation. *Blood*, *123*, 570–581.

Andersson, R., Gebhard, C., Miguel-Escalada, I., Hoof, I., Bornholdt, J., Boyd, M., et al. (2014). An atlas of active enhancers across human cell types and tissues. *Nature*, *507*, 455–461.

Aoki, K., Harashima, A., Sano, M., Yokoi, T., Nakamura, S., Kibata, M., et al. (2010). A thymus-specific noncoding RNA, Thy-ncR1, is a cytoplasmic riboregulator of MFAP4 mRNA in immature T-cell lines. *BMC Molecular Biology*, *11*, 99.

Baldridge, M. T., King, K. Y., Boles, N. C., Weksberg, D. C., & Goodell, M. A. (2010). Quiescent haematopoietic stem cells are activated by IFN-gamma in response to chronic infection. *Nature*, *465*, 793–797.

Batista, P. J., & Chang, H. Y. (2013). Long noncoding RNAs: Cellular address codes in development and disease. *Cell*, *152*, 1298–1307.

Benetatos, L., Hatzimichael, E., Dasoula, A., Dranitsaris, G., Tsiara, S., Syrrou, M., et al. (2010). CpG methylation analysis of the MEG3 and SNRPN imprinted genes in acute myeloid leukemia and myelodysplastic syndromes. *Leukemia Research*, *34*, 148–153.

Berg, J. S., Lin, K. K., Sonnet, C., Boles, N. C., Weksberg, D. C., Nguyen, H., et al. (2011). Imprinted genes that regulate early mammalian growth are coexpressed in somatic stem cells. *PLoS One*, *6*, e26410.

Blume, C. J., Hotz-Wagenblatt, A., Hullein, J., Sellner, L., Jethwa, A., Stolz, T., et al. (2015). p53-dependent non-coding RNA networks in chronic lymphocytic leukemia. *Leukemia*, *29*, 2015–2023.

Bousquet, M., Harris, M. H., Zhou, B., & Lodish, H. F. (2010). MicroRNA miR-125b causes leukemia. *Proceedings of the National Academy of Sciences of the United States of America*, *107*, 21558–21563.

Braconi, C., Valeri, N., Kogure, T., Gasparini, P., Huang, N., Nuovo, G. J., et al. (2011). Expression and functional role of a transcribed noncoding RNA with an ultraconserved element in hepatocellular carcinoma. *Proceedings of the National Academy of Sciences of the United States of America*, *108*, 786–791.

Carninci, P., Kasukawa, T., Katayama, S., Gough, J., Frith, M. C., Maeda, N., et al. (2005). The transcriptional landscape of the mammalian genome. *Science*, *309*, 1559–1563.

Carpenter, S., Aiello, D., Atianand, M. K., Ricci, E. P., Gandhi, P., Hall, L. L., et al. (2013). A long noncoding RNA mediates both activation and repression of immune response genes. *Science*, *341*, 789–792.

Chan, W. L., Huang, H. D., & Chang, J. G. (2014). lncRNAMap: A map of putative regulatory functions in the long non-coding transcriptome. *Computational Biology and Chemistry, 50*, 41–49.

Chen, M. T., Lin, H. S., Shen, C., Ma, Y. N., Wang, F., Zhao, H. L., et al. (2015). PU.1-regulated long noncoding RNA lnc-MC controls human monocyte/macrophage differentiation through interaction with microRNA 199a-5p. *Molecular and Cellular Biology, 35*, 3212–3224.

Chen, G., Wang, Z., Wang, D., Qiu, C., Liu, M., Chen, X., et al. (2013). LncRNADisease: A database for long-non-coding RNA-associated diseases. *Nucleic Acids Research, 41*, D983–D986.

Cheng, J., Kapranov, P., Drenkow, J., Dike, S., Brubaker, S., Patel, S., et al. (2005). Transcriptional maps of 10 human chromosomes at 5-nucleotide resolution. *Science, 308*, 1149–1154.

Cheng, J. H., Pan, D. Z., Tsai, Z. T., & Tsai, H. K. (2015). Genome-wide analysis of enhancer RNA in gene regulation across 12 mouse tissues. *Scientific Reports, 5*, 12648.

Dahl, R., Walsh, J. C., Lancki, D., Laslo, P., Iyer, S. R., Singh, H., et al. (2003). Regulation of macrophage and neutrophil cell fates by the PU.1:C/EBPalpha ratio and granulocyte colony-stimulating factor. *Nature Immunology, 4*, 1029–1036.

Diaz-Beya, M., Brunet, S., Nomdedeu, J., Pratcorona, M., Cordeiro, A., Gallardo, D., et al. (2015). The lincRNA HOTAIRM1, located in the HOXA genomic region, is expressed in acute myeloid leukemia, impacts prognosis in patients in the intermediate-risk cytogenetic category, and is associated with a distinctive microRNA signature. *Oncotarget, 6*, 31613–31627.

Dinger, M. E., Pang, K. C., Mercer, T. R., Crowe, M. L., Grimmond, S. M., & Mattick, J. S. (2009). NRED: A database of long noncoding RNA expression. *Nucleic Acids Research, 37*, D122–D126.

Djebali, S., Davis, C. A., Merkel, A., Dobin, A., Lassmann, T., Mortazavi, A., et al. (2012). Landscape of transcription in human cells. *Nature, 489*, 101–108.

Dore, L. C., Amigo, J. D., Dos Santos, C. O., Zhang, Z., Gai, X., Tobias, J. W., et al. (2008). A GATA-1-regulated microRNA locus essential for erythropoiesis. *Proceedings of the National Academy of Sciences of the United States of America, 105*, 3333–3338.

Ebralidze, A. K., Guibal, F. C., Steidl, U., Zhang, P., Lee, S., Bartholdy, B., et al. (2008). PU.1 expression is modulated by the balance of functional sense and antisense RNAs regulated by a shared cis-regulatory element. *Genes & Development, 22*, 2085–2092.

Ellis, B. C., Molloy, P. L., & Graham, L. D. (2012). CRNDE: A long non-coding RNA involved in CanceR, Neurobiology, and DEvelopment. *Frontiers in Genetics, 3*, 270.

Elton, T. S., Selemon, H., Elton, S. M., & Parinandi, N. L. (2013). Regulation of the MIR155 host gene in physiological and pathological processes. *Gene, 532*, 1–12.

Emmrich, S., Streltsov, A., Schmidt, F., Thangapandi, V. R., Reinhardt, D., & Klusmann, J. H. (2014). LincRNAs MONC and MIR100HG act as oncogenes in acute megakaryoblastic leukemia. *Molecular Cancer, 13*, 171.

Fazi, F., Rosa, A., Fatica, A., Gelmetti, V., De Marchis, M. L., Nervi, C., et al. (2005). A minicircuitry comprised of microRNA-223 and transcription factors NFI-A and C/EBPalpha regulates human granulopoiesis. *Cell, 123*, 819–831.

Garding, A., Bhattacharya, N., Claus, R., Ruppel, M., Tschuch, C., Filarsky, K., et al. (2013). Epigenetic upregulation of lncRNAs at 13q14.3 in leukemia is linked to the In Cis downregulation of a gene cluster that targets NF-kB. *PLoS Genetics, 9*, e1003373.

Garzon, R., Garofalo, M., Martelli, M. P., Briesewitz, R., Wang, L., Fernandez-Cymering, C., et al. (2008). Distinctive microRNA signature of acute myeloid leukemia bearing cytoplasmic mutated nucleophosmin. *Proceedings of the National Academy of Sciences of the United States of America, 105*, 3945–3950.

Ghisi, M., Corradin, A., Basso, K., Frasson, C., Serafin, V., Mukherjee, S., et al. (2011). Modulation of microRNA expression in human T-cell development: Targeting of NOTCH3 by miR-150. *Blood*, *117*, 7053–7062.

Goshen, R., Rachmilewitz, J., Schneider, T., de-Groot, N., Ariel, I., Palti, Z., et al. (1993). The expression of the H-19 and IGF-2 genes during human embryogenesis and placental development. *Molecular Reproduction and Development*, *34*, 374–379.

Guo, S., Lu, J., Schlanger, R., Zhang, H., Wang, J. Y., Fox, M. C., et al. (2010). MicroRNA miR-125a controls hematopoietic stem cell number. *Proceedings of the National Academy of Sciences of the United States of America*, *107*, 14229–14234.

Guttman, M., Amit, I., Garber, M., French, C., Lin, M. F., Feldser, D., et al. (2009). Chromatin signature reveals over a thousand highly conserved large non-coding RNAs in mammals. *Nature*, *458*, 223–227.

Guttman, M., Donaghey, J., Carey, B. W., Garber, M., Grenier, J. K., Munson, G., et al. (2011). lincRNAs act in the circuitry controlling pluripotency and differentiation. *Nature*, *477*, 295–300.

Hsu, A. P., Johnson, K. D., Falcone, E. L., Sanalkumar, R., Sanchez, L., Hickstein, D. D., et al. (2013). GATA2 haploinsufficiency caused by mutations in a conserved intronic element leads to MonoMAC syndrome. *Blood*, *121*, 3830–3837.

Hu, G., Tang, Q., Sharma, S., Yu, F., Escobar, T. M., Muljo, S. A., et al. (2013). Expression and regulation of intergenic long noncoding RNAs during T cell development and differentiation. *Nature Immunology*, *14*, 1190–1198.

Hu, W., Yuan, B., Flygare, J., & Lodish, H. F. (2011). Long noncoding RNA-mediated antiapoptotic activity in murine erythroid terminal differentiation. *Genes & Development*, *25*, 2573–2578.

Hung, T., Wang, Y., Lin, M. F., Koegel, A. K., Kotake, Y., Grant, G. D., et al. (2011). Extensive and coordinated transcription of noncoding RNAs within cell-cycle promoters. *Nature Genetics*, *43*, 621–629.

Imam, H., Bano, A. S., Patel, P., Holla, P., & Jameel, S. (2015). The lncRNA NRON modulates HIV-1 replication in a NFAT-dependent manner and is differentially regulated by early and late viral proteins. *Scientific Reports*, *5*, 8639.

International Human Genome Sequencing Consortium. (2004). Finishing the euchromatic sequence of the human genome. *Nature*, *431*, 931–945.

Ji, M., Rao, E., Ramachandrareddy, H., Shen, Y., Jiang, C., Chen, J., et al. (2011). The miR-17-92 microRNA cluster is regulated by multiple mechanisms in B-cell malignancies. *The American Journal of Pathology*, *179*, 1645–1656.

Katayama, S., Tomaru, Y., Kasukawa, T., Waki, K., Nakanishi, M., Nakamura, M., et al. (2005). Antisense transcription in the mammalian transcriptome. *Science*, *309*, 1564–1566.

Kim, T. K., Hemberg, M., & Gray, J. M. (2015). Enhancer RNAs: A class of long noncoding RNAs synthesized at enhancers. *Cold Spring Harbor Perspectives in Biology*, 7, a018622.

Kim, T. K., Hemberg, M., Gray, J. M., Costa, A. M., Bear, D. M., Wu, J., et al. (2010). Widespread transcription at neuronal activity-regulated enhancers. *Nature*, *465*, 182–187.

Kim, T., & Reitmair, A. (2013). Non-coding RNAs: Functional aspects and diagnostic utility in oncology. *International Journal of Molecular Sciences*, *14*, 4934–4968.

Kino, T., Hurt, D. E., Ichijo, T., Nader, N., & Chrousos, G. P. (2010). Noncoding RNA gas5 is a growth arrest- and starvation-associated repressor of the glucocorticoid receptor. *Science Signaling*, *3*, ra8.

Klattenhoff, C. A., Scheuermann, J. C., Surface, L. E., Bradley, R. K., Fields, P. A., Steinhauser, M. L., et al. (2013). Braveheart, a long noncoding RNA required for cardiovascular lineage commitment. *Cell*, *152*, 570–583.

Kondo, M., Scherer, D. C., Miyamoto, T., King, A. G., Akashi, K., Sugamura, K., et al. (2000). Cell-fate conversion of lymphoid-committed progenitors by instructive actions of cytokines. *Nature*, *407*, 383–386.

Kunkeaw, N., Jeon, S. H., Lee, K., Johnson, B. H., Tanasanvimon, S., Javle, M., et al. (2013). Cell death/proliferation roles for nc886, a non-coding RNA, in the protein kinase R pathway in cholangiocarcinoma. *Oncogene*, *32*, 3722–3731.

Lawrie, C. H., Gal, S., Dunlop, H. M., Pushkaran, B., Liggins, A. P., Pulford, K., et al. (2008). Detection of elevated levels of tumour-associated microRNAs in serum of patients with diffuse large B-cell lymphoma. *British Journal of Haematology*, *141*, 672–675.

Lee, J. T. (2009). Lessons from X-chromosome inactivation: Long ncRNA as guides and tethers to the epigenome. *Genes & Development*, *23*, 1831–1842.

Li, Z., Chao, T. C., Chang, K. Y., Lin, N., Patil, V. S., Shimizu, C., et al. (2014). The long noncoding RNA THRIL regulates TNFalpha expression through its interaction with hnRNPL. *Proceedings of the National Academy of Sciences of the United States of America*, *111*, 1002–1007.

Li, J. H., Liu, S., Zhou, H., Qu, L. H., & Yang, J. H. (2014). starBase v2.0: Decoding miRNA-ceRNA, miRNA-ncRNA and protein-RNA interaction networks from large-scale CLIP-Seq data. *Nucleic Acids Research*, *42*, D92–D97.

Ling, K. W., Ottersbach, K., van Hamburg, J. P., Oziemlak, A., Tsai, F. Y., Orkin, S. H., et al. (2004). GATA-2 plays two functionally distinct roles during the ontogeny of hematopoietic stem cells. *The Journal of Experimental Medicine*, *200*, 871–882.

Lodish, H., Flygare, J., & Chou, S. (2010). From stem cell to erythroblast: Regulation of red cell production at multiple levels by multiple hormones. *IUBMB Life*, *62*, 492–496.

Luo, M., Jeong, M., Sun, D., Park, H. J., Rodriguez, B. A., Xia, Z., et al. (2015). Long noncoding RNAs control hematopoietic stem cell function. *Cell Stem Cell*, *16*, 426–438.

Majem, B., Rigau, M., Reventos, J., & Wong, D. T. (2015). Non-coding RNAs in saliva: Emerging biomarkers for molecular diagnostics. *International Journal of Molecular Sciences*, *16*, 8676–8698.

Mattick, J. S. (2007). A new paradigm for developmental biology. *The Journal of Experimental Biology*, *210*, 1526–1547.

McHugh, C. A., Chen, C. K., Chow, A., Surka, C. F., Tran, C., McDonel, P., et al. (2015). The Xist lncRNA interacts directly with SHARP to silence transcription through HDAC3. *Nature*, *521*, 232–236.

Mehta, A., Zhao, J. L., Sinha, N., Marinov, G. K., Mann, M., Kowalczyk, M. S., et al. (2015). The microRNA-132 and microRNA-212 cluster regulates hematopoietic stem cell maintenance and survival with age by buffering FOXO3 expression. *Immunity*, *42*, 1021–1032.

Mitchell, P. S., Parkin, R. K., Kroh, E. M., Fritz, B. R., Wyman, S. K., Pogosova-Agadjanyan, E. L., et al. (2008). Circulating microRNAs as stable blood-based markers for cancer detection. *Proceedings of the National Academy of Sciences of the United States of America*, *105*, 10513–10518.

Mohammad, F., Mondal, T., & Kanduri, C. (2009). Epigenetics of imprinted long noncoding RNAs. *Epigenetics*, *4*, 277–286.

Morceau, F., Chateauvieux, S., Gaigneaux, A., Dicato, M., & Diederich, M. (2013). Long and short non-coding RNAs as regulators of hematopoietic differentiation. *International Journal of Molecular Sciences*, *14*, 14744–14770.

Mossadegh-Keller, N., Sarrazin, S., Kandalla, P. K., Espinosa, L., Stanley, E. R., Nutt, S. L., et al. (2013). M-CSF instructs myeloid lineage fate in single haematopoietic stem cells. *Nature*, *497*, 239–243.

Mourtada-Maarabouni, M., Hasan, A. M., Farzaneh, F., & Williams, G. T. (2010). Inhibition of human T-cell proliferation by mammalian target of rapamycin (mTOR) antagonists requires noncoding RNA growth-arrest-specific transcript 5 (GAS5). *Molecular Pharmacology*, *78*, 19–28.

Mourtada-Maarabouni, M., Hedge, V. L., Kirkham, L., Farzaneh, F., & Williams, G. T. (2008). Growth arrest in human T-cells is controlled by the non-coding RNA growth-arrest-specific transcript 5 (GAS5). *Journal of Cell Science*, *121*, 939–946.

Nakamura, Y., Takahashi, N., Kakegawa, E., Yoshida, K., Ito, Y., Kayano, H., et al. (2008). The GAS5 (growth arrest-specific transcript 5) gene fuses to BCL6 as a result of t(1;3)(q25; q27) in a patient with B-cell lymphoma. *Cancer Genetics and Cytogenetics*, *182*, 144–149.

Natoli, G., & Andrau, J. C. (2012). Noncoding transcription at enhancers: General principles and functional models. *Annual Review of Genetics*, *46*, 1–19.

Nunez, C., Bashein, A. M., Brunet, C. L., Hoyland, J. A., Freemont, A. J., Buckle, A. M., et al. (2000). Expression of the imprinted tumour-suppressor gene H19 is tightly regulated during normal haematopoiesis and is reduced in haematopoietic precursors of patients with the myeloproliferative disease polycythaemia vera. *The Journal of Pathology*, *190*, 61–68.

O'Connell, R. M., Chaudhuri, A. A., Rao, D. S., Gibson, W. S., Balazs, A. B., & Baltimore, D. (2010). MicroRNAs enriched in hematopoietic stem cells differentially regulate long-term hematopoietic output. *Proceedings of the National Academy of Sciences of the United States of America*, *107*, 14235–14240.

Ooi, A. G., Sahoo, D., Adorno, M., Wang, Y., Weissman, I. L., & Park, C. Y. (2010). MicroRNA-125b expands hematopoietic stem cells and enriches for the lymphoid-balanced and lymphoid-biased subsets. *Proceedings of the National Academy of Sciences of the United States of America*, *107*, 21505–21510.

Orkin, S. H., & Zon, L. I. (2008). Hematopoiesis: An evolving paradigm for stem cell biology. *Cell*, *132*, 631–644.

Orom, U. A., Derrien, T., Beringer, M., Gumireddy, K., Gardini, A., Bussotti, G., et al. (2010). Long noncoding RNAs with enhancer-like function in human cells. *Cell*, *143*, 46–58.

Ozsolak, F., Kapranov, P., Foissac, S., Kim, S. W., Fishilevich, E., Monaghan, A. P., et al. (2010). Comprehensive polyadenylation site maps in yeast and human reveal pervasive alternative polyadenylation. *Cell*, *143*, 1018–1029.

Papapetrou, E. P., Korkola, J. E., & Sadelain, M. (2010). A genetic strategy for single and combinatorial analysis of miRNA function in mammalian hematopoietic stem cells. *Stem Cells*, *28*, 287–296.

Paralkar, V. R., Mishra, T., Luan, J., Yao, Y., Kossenkov, A. V., Anderson, S. M., et al. (2014). Lineage and species-specific long noncoding RNAs during erythro-megakaryocytic development. *Blood*, *123*, 1927–1937.

Pelechano, V., & Steinmetz, L. M. (2013). Gene regulation by antisense transcription. *Nature Reviews Genetics*, *14*, 880–893.

Pontier, D. B., & Gribnau, J. (2011). Xist regulation and function explored. *Human Genetics*, *130*, 223–236.

Prensner, J. R., Iyer, M. K., Balbin, O. A., Dhanasekaran, S. M., Cao, Q., Brenner, J. C., et al. (2011). Transcriptome sequencing across a prostate cancer cohort identifies PCAT-1, an unannotated lincRNA implicated in disease progression. *Nature Biotechnology*, *29*, 742–749.

Quek, X. C., Thomson, D. W., Maag, J. L., Bartonicek, N., Signal, B., Clark, M. B., et al. (2015). lncRNAdb v2.0: Expanding the reference database for functional long noncoding RNAs. *Nucleic Acids Research*, *43*, D168–D173.

Ranzani, V., Rossetti, G., Panzeri, I., Arrigoni, A., Bonnal, R. J., Curti, S., et al. (2015). The long intergenic noncoding RNA landscape of human lymphocytes highlights the regulation of T cell differentiation by linc-MAF-4. *Nature Immunology*, *16*, 318–325.

Reis, E. M., & Verjovski-Almeida, S. (2012). Perspectives of long non-coding RNAs in cancer diagnostics. *Frontiers in Genetics*, *3*, 32.

Rodrigues, N. P., Janzen, V., Forkert, R., Dombkowski, D. M., Boyd, A. S., Orkin, S. H., et al. (2005). Haploinsufficiency of GATA-2 perturbs adult hematopoietic stem-cell homeostasis. *Blood, 106*, 477–484.

Rodriguez, A., Vigorito, E., Clare, S., Warren, M. V., Couttet, P., Soond, D. R., et al. (2007). Requirement of bic/microRNA-155 for normal immune function. *Science, 316*, 608–611.

Rossi, L., Lin, K. K., Boles, N. C., Yang, L., King, K. Y., Jeong, M., et al. (2012). Less is more: Unveiling the functional core of hematopoietic stem cells through knockout mice. *Cell Stem Cell, 11*, 302–317.

Saitou, M., Sugimoto, J., Hatakeyama, T., Russo, G., & Isobe, M. (2000). Identification of the TCL6 genes within the breakpoint cluster region on chromosome 14q32 in T-cell leukemia. *Oncogene, 19*, 2796–2802.

Sanyal, A., Lajoie, B. R., Jain, G., & Dekker, J. (2012). The long-range interaction landscape of gene promoters. *Nature, 489*, 109–113.

Sirin, O., Lukov, G. L., Mao, R., Conneely, O. M., & Goodell, M. A. (2010). The orphan nuclear receptor Nurr1 restricts the proliferation of haematopoietic stem cells. *Nature Cell Biology, 12*, 1213–1219.

St Laurent, G., Wahlestedt, C., & Kapranov, P. (2015). The landscape of long noncoding RNA classification. *Trends in Genetics, 31*, 239–251.

Sun, J., Li, W., Sun, Y., Yu, D., Wen, X., Wang, H., et al. (2014). A novel antisense long noncoding RNA within the IGF1R gene locus is imprinted in hematopoietic malignancies. *Nucleic Acids Research, 42*, 9588–9601.

Sun, D., Luo, M., Jeong, M., Rodriguez, B., Xia, Z., Hannah, R., et al. (2014). Epigenomic profiling of young and aged HSCs reveals concerted changes during aging that reinforce self-renewal. *Cell Stem Cell, 14*, 673–688.

Takeuchi, S., Hofmann, W. K., Tsukasaki, K., Takeuchi, N., Ikezoe, T., Matsushita, M., et al. (2007). Loss of H19 imprinting in adult T-cell leukaemia/lymphoma. *British Journal of Haematology, 137*, 380–381.

Taskinen, M., Ranki, A., Pukkala, E., Jeskanen, L., Kaitila, I., & Makitie, O. (2008). Extended follow-up of the Finnish cartilage-hair hypoplasia cohort confirms high incidence of non-Hodgkin lymphoma and basal cell carcinoma. *American Journal of Medical Genetics Part A, 146A*, 2370–2375.

Tsao, H. W., Tai, T. S., Tseng, W., Chang, H. H., Grenningloh, R., Miaw, S. C., et al. (2013). Ets-1 facilitates nuclear entry of NFAT proteins and their recruitment to the IL-2 promoter. *Proceedings of the National Academy of Sciences of the United States of America, 110*, 15776–15781.

van den Berg, A., Kroesen, B. J., Kooistra, K., de Jong, D., Briggs, J., Blokzijl, T., et al. (2003). High expression of B-cell receptor inducible gene BIC in all subtypes of Hodgkin lymphoma. *Genes, Chromosomes & Cancer, 37*, 20–28.

Vargova, K., Curik, N., Burda, P., Basova, P., Kulvait, V., Pospisil, V., et al. (2011). MYB transcriptionally regulates the miR-155 host gene in chronic lymphocytic leukemia. *Blood, 117*, 3816–3825.

Venkatraman, A., He, X. C., Thorvaldsen, J. L., Sugimura, R., Perry, J. M., Tao, F., et al. (2013). Maternal imprinting at the H19-Igf2 locus maintains adult haematopoietic stem cell quiescence. *Nature, 500*, 345–349.

Volders, P. J., Verheggen, K., Menschaert, G., Vandepoele, K., Martens, L., Vandesompele, J., et al. (2015). An update on LNCipedia: A database for annotated human lncRNA sequences. *Nucleic Acids Research, 43*, D174–D180.

Wagner, L. A., Christensen, C. J., Dunn, D. M., Spangrude, G. J., Georgelas, A., Kelley, L., et al. (2007). EGO, a novel, noncoding RNA gene, regulates eosinophil granule protein transcript expression. *Blood, 109*, 5191–5198.

Wang, H., Li, W., Guo, R., Sun, J., Cui, J., Wang, G., et al. (2014). An intragenic long noncoding RNA interacts epigenetically with the RUNX1 promoter and enhancer chromatin DNA in hematopoietic malignancies. *International Journal of Cancer. Journal International du Cancer*, *135*, 2783–2794.

Wang, P., Xue, Y., Han, Y., Lin, L., Wu, C., Xu, S., et al. (2014). The STAT3-binding long noncoding RNA lnc-DC controls human dendritic cell differentiation. *Science*, *344*, 310–313.

Wang, K. C., Yang, Y. W., Liu, B., Sanyal, A., Corces-Zimmerman, R., Chen, Y., et al. (2011). A long noncoding RNA maintains active chromatin to coordinate homeotic gene expression. *Nature*, *472*, 120–124.

Weissman, I. L. (2000). Stem cells: Units of development, units of regeneration, and units in evolution. *Cell*, *100*, 157–168.

Wightman, B., Ha, I., & Ruvkun, G. (1993). Posttranscriptional regulation of the heterochronic gene lin-14 by lin-4 mediates temporal pattern formation in *C. elegans*. *Cell*, *75*, 855–862.

Williams, G. T., Mourtada-Maarabouni, M., & Farzaneh, F. (2011). A critical role for non-coding RNA GAS5 in growth arrest and rapamycin inhibition in human T-lymphocytes. *Biochemical Society Transactions*, *39*, 482–486.

Willingham, A. T., Orth, A. P., Batalov, S., Peters, E. C., Wen, B. G., Aza-Blanc, P., et al. (2005). A strategy for probing the function of noncoding RNAs finds a repressor of NFAT. *Science*, *309*, 1570–1573.

Xiao, C., Calado, D. P., Galler, G., Thai, T. H., Patterson, H. C., Wang, J., et al. (2007). MiR-150 controls B cell differentiation by targeting the transcription factor c-Myb. *Cell*, *131*, 146–159.

Xie, C., Yuan, J., Li, H., Li, M., Zhao, G., Bu, D., et al. (2014). NONCODEv4: Exploring the world of long non-coding RNA genes. *Nucleic Acids Research*, *42*, D98–D103.

Yang, R., Frank, B., Hemminki, K., Bartram, C. R., Wappenschmidt, B., Sutter, C., et al. (2008). SNPs in ultraconserved elements and familial breast cancer risk. *Carcinogenesis*, *29*, 351–355.

Yang, J. H., Li, J. H., Jiang, S., Zhou, H., & Qu, L. H. (2013). ChIPBase: A database for decoding the transcriptional regulation of long non-coding RNA and microRNA genes from ChIP-Seq data. *Nucleic Acids Research*, *41*, D177–D187.

Yildirim, E., Kirby, J. E., Brown, D. E., Mercier, F. E., Sadreyev, R. I., Scadden, D. T., et al. (2013). Xist RNA is a potent suppressor of hematologic cancer in mice. *Cell*, *152*, 727–742.

Yin, Q., Wang, X., McBride, J., Fewell, C., & Flemington, E. (2008). B-cell receptor activation induces BIC/miR-155 expression through a conserved AP-1 element. *The Journal of Biological Chemistry*, *283*, 2654–2662.

Yu, W., Gius, D., Onyango, P., Muldoon-Jacobs, K., Karp, J., Feinberg, A. P., et al. (2008). Epigenetic silencing of tumour suppressor gene p15 by its antisense RNA. *Nature*, *451*, 202–206.

Zhang, X., Lian, Z., Padden, C., Gerstein, M. B., Rozowsky, J., Snyder, M., et al. (2009). A myelopoiesis-associated regulatory intergenic noncoding RNA transcript within the human HOXA cluster. *Blood*, *113*, 2526–2534.

Zhao, J., Sun, B. K., Erwin, J. A., Song, J. J., & Lee, J. T. (2008). Polycomb proteins targeted by a short repeat RNA to the mouse X chromosome. *Science*, *322*, 750–756.

INDEX

Note: Page numbers followed by "*f*" indicate figures, and "*t*" indicate tables.

T

V

X

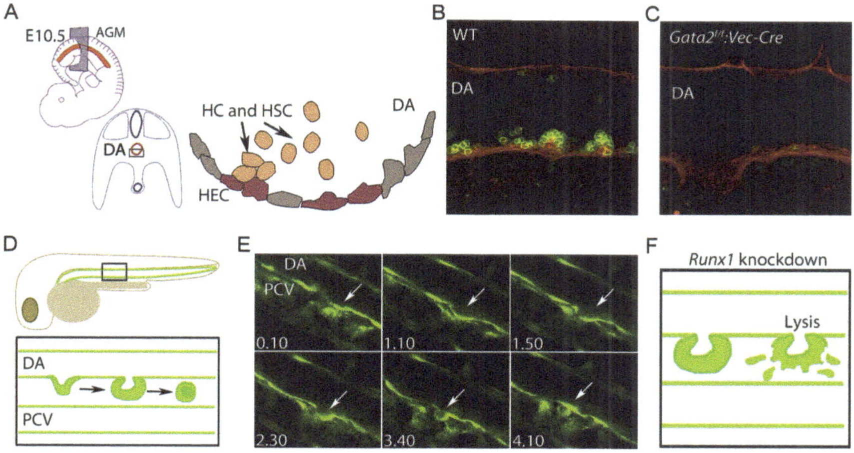

E. Dzierzak and E. de Pater, Fig. 1 The AGM, hematopoietic clusters and the process of endothelial-to-hematopoietic transition (EHT). (A) Graphical representation of EHT in the mouse. An E10.5 mouse embryo is depicted on the *left* where the *square* indicates the level of the cross-section shown in the *middle panel*. The *boxed* area indicates the ventral side of the dorsal aorta. The *right panel* shows where hemogenic endothelial cells in *purple* give rise to clusters of hematopoietic cells and HSCs indicated in *pink*. These shed into the lumen of the dorsal aorta and eventually populate the bone marrow. (B) and (C) *z* projection of a confocal image of (B) a wild-type (WT) mouse embryo and (C) a *Gata2*$^{f/f}$*:Vec-Cre* embryo. Endothelial cells are stained for CD31 (*red*) and hematopoietic cluster cells are stained for c-Kit (*green*). Note the size of WT clusters in (B) and the flat c-Kit positive cluster cells in (C), indicating that not only the number of clusters is severely reduced upon loss of Gata2 but also the process of EHT is affected. (D) Graphical representation of EHT in zebrafish. A schematic diagram of a Tg(*Fli:GFP*) zebrafish embryo (*top panel*) where the dorsal aorta (DA) and the posterior cardinal vein (PCV) are marked in *green* in a embryo. The *bottom panel* shows the *boxed area* where an endothelial cell bulges outward, rounds up to close the gap in the DA and sheds as a hematopoietic cell into the abluminal space between the DA and the PCV. (E) Single *z* scans of time-lapse confocal imaging of the process of EHT *in vivo* using a Tg(*Fli:GFP*) embryo. The DA and PCV are indicated in the first frame and the *arrow* indicates the cell that is undergoing EHT in each frame. The timing is indicated in the *bottom left corner* of each frame and the entire process of EHT is completed after 4 h. (F) Schematic representation of a cell undergoing EHT after knockdown of *Runx1* in the zebrafish embryo. EHT is greatly reduced, but the occasional cell undergoing EHT lyses halfway through the process. *E*, embryonic day; *DA*, dorsal aorta; *AGM*, aorta-gonad-mesonephros region; *HC*, hematopoietic cell; *HEC*, hemogenic endothelium; *HSC*, hematopoietic stem cell; *PCV*, posterior cardinal vein.

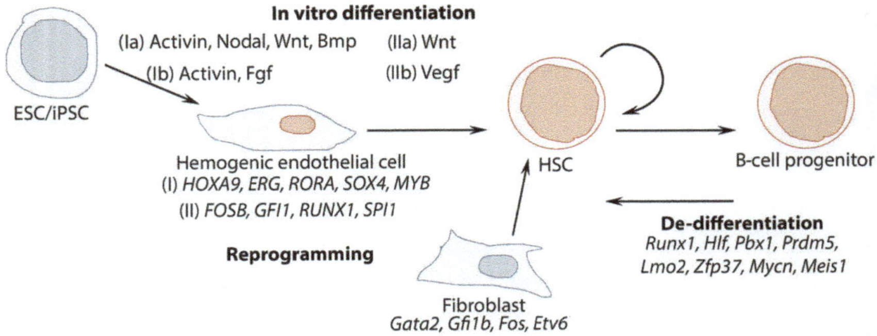

E. Dzierzak and E. de Pater, Fig. 2 Different reprogramming strategies in an effort to obtain HSCs. Two strategies are used. The first approach differentiates ES cells or iPS cells toward a hematopoietic progenitor and/or HSC fate via a hemogenic endothelial intermediate step through the use of a combination of growth and developmental factors. The second approach uses a reprogramming strategy by overexpressing combinations of transcription factors in starting nonhematopoietic cells such as fibroblasts or hematopoietic cells such as B-cell progenitors.

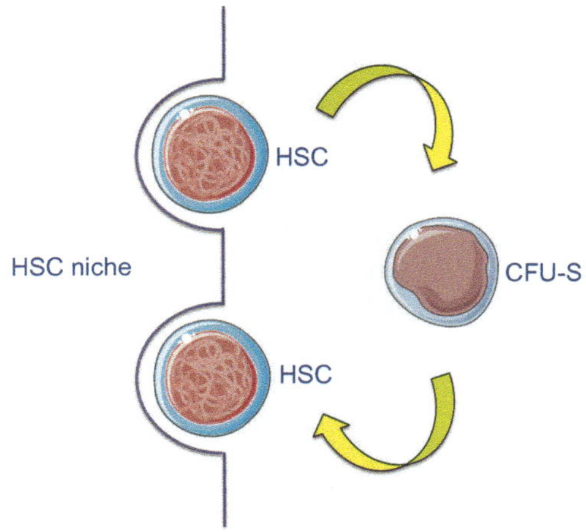

V.W.C. Yu and D.T. Scadden, Fig. 1 The hypothetical view of the stem cell niche from R. Schofield. The stem cell can become the daughter cell, CFU-S, once the stem cell leaves its niche. But if the stem cell finds and reoccupies the niche, it will itself return to its stem cell state. *Adopted from Schofield, R. (1978). The relationship between the spleen colony-forming cell and the haemopoietic stem cell.* Blood Cells, 4, *7–25.*

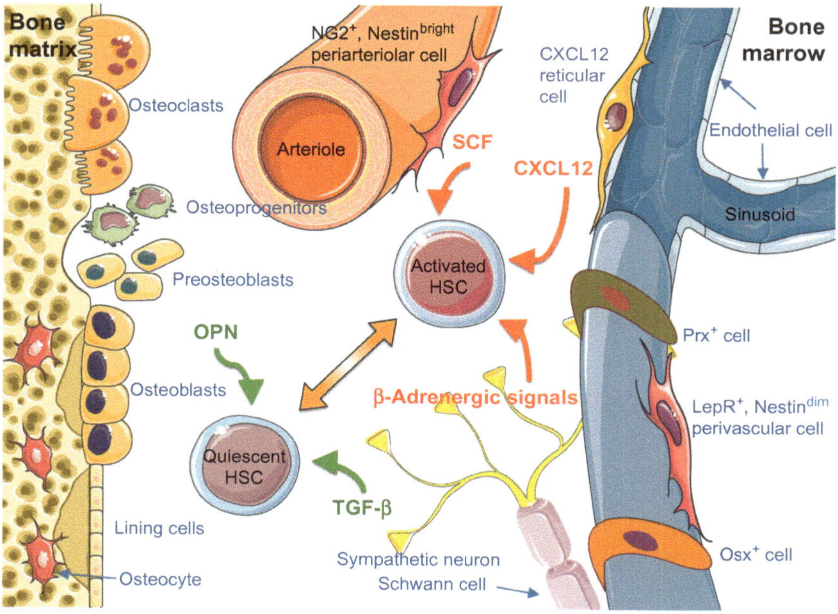

V.W.C. Yu and D.T. Scadden, Fig. 2 The hematopoietic stem cell niche. Extrinsic signals including growth factors, cytokines, morphogens, extracellular matrix proteins, and adhesion molecules regulate the behavior of hematopoietic stem cells (HSCs) and protect them from exhaustion, although few have been pinpointed to derive from a specific cell source. Recent advances in flow cytometry, transgenic mouse models, and intravital microscopy have revolutionized our understanding of the BM niche and how it controls various stem cell behaviors including self-renewal, proliferation, differentiation, lineage commitment, and mobilization. Enlisted in the diagram are cellular and molecular participants that have been experimentally shown to impact different HSC activities through promoter-specific Cre deletion or overexpression. But the diagram is far from complete and represents an active area of ongoing investigation.

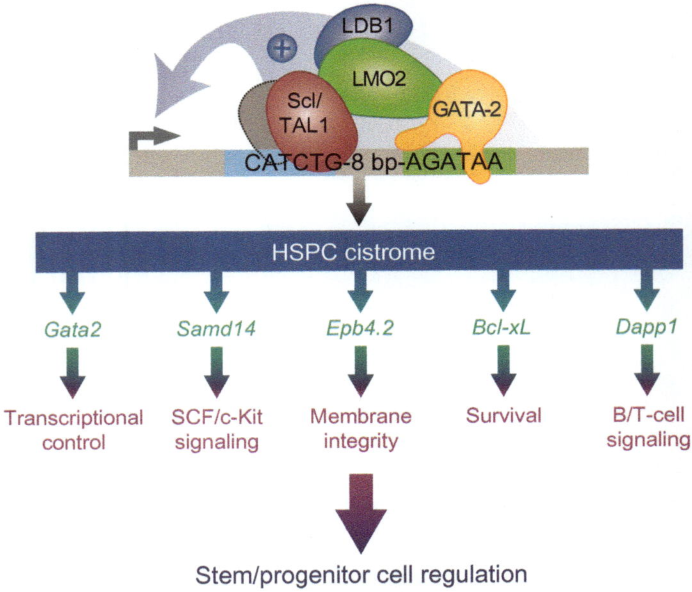

K.J. Hewitt *et al.*, Fig. 1 Composite E-box–GATA elements regulate diverse aspects of hematopoietic stem/progenitor cell development and function. A multimeric complex, consisting of GATA-2, Scl/TAL-1, E2A, LMO2, and LDB1, and almost certainly additional components, forms at a highly restricted subset of E-box–GATA elements in the genome. The GATA-2-activated E-box–GATA elements are important constituents of the HSPC cistrome, which orchestrates vital steps in the stem and progenitor cell transitions that give rise to the full complement of blood cells. Examples of genes controlled by GATA-2 through these elements include *Gata2*, *Samd14*, *Epb4.2*, *Bcl2l1*, and *Dapp1*.

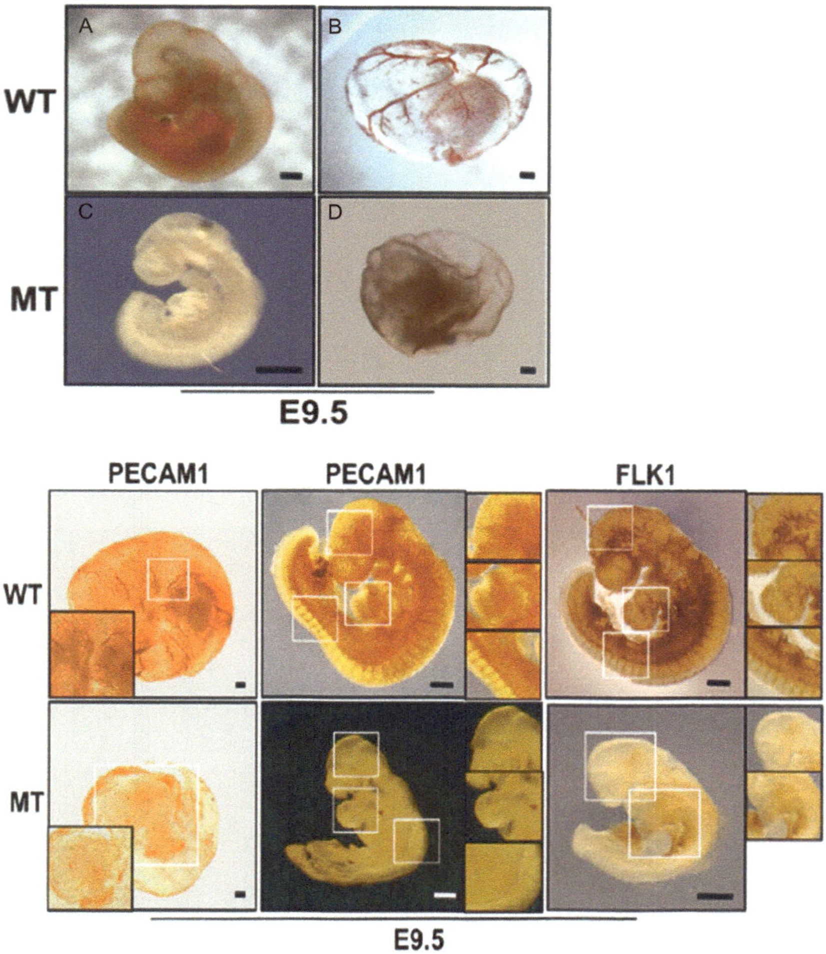

S. Sumanas and K. Choi, Fig. 1 Etv2 (aka Er71) is required for hematopoietic and endothelial cell development. Gross morphology of *Er71*$^{-/-}$ embryos at E9.5 is shown on the *left* (A and C, embryo proper; B and D, yolk sac), Scale bars, 200 μm. Whole-mount PECAM1 or FLK1 staining of E9.5 embryos is shown on the *right*. The *boxed area* is the vasculature of the yolk sac, brain, heart, and intersomitic regions at higher magnification. Scale bars, 200 μm. *Reprinted from Lee, D. J., Park, C., Lee, H., Lugus, J. J., Kim, S. H., Arentson, E., et al. (2008). ER71 acts downstream of BMP, Notch, and Wnt signaling in blood and vessel progenitor specification.* Cell Stem Cell, 2(5), *497–507, Copyright (2008), with permission from Elsevier.*

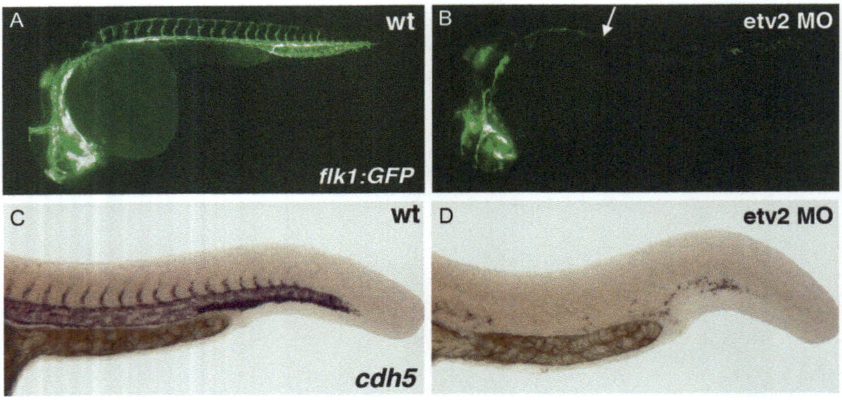

S. Sumanas and K. Choi, Fig. 2 Etv2 morpholino knockdown results in near complete absence of zebrafish transgenic reporter *flk1:GFP* (A and B) and vascular marker *cdh5/ VE-cadherin* (C and D) expression at 24–26 hpf. *Reprinted from Sumanas, S., & Lin, S. (2006). Ets1-related protein is a key regulator of vasculogenesis in zebrafish.* PLoS Biology, 4(1), *e10.*

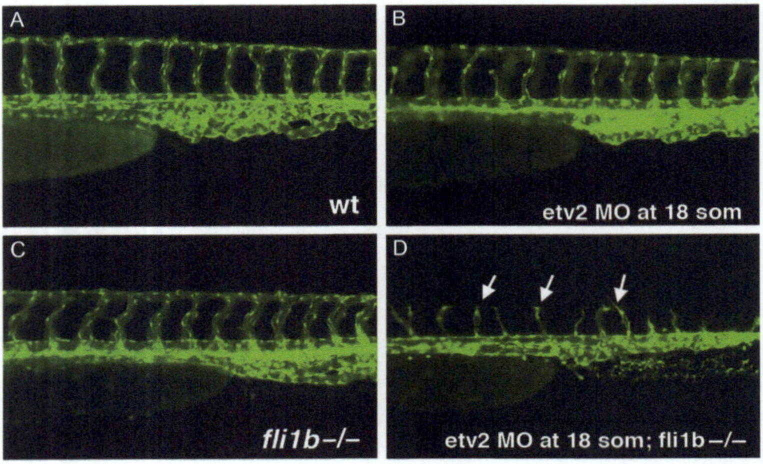

S. Sumanas and K. Choi, Fig. 3 Etv2 functions redundantly with Fli1b in angiogenesis independent from Etv2 early requirement in vasculogenesis. Inducible Etv2 inhibition was performed at the 18-somite stage using photoactivatable Etv2 morpholino in either wild-type (wt) *fli1a:GFP* transgenic reporter embryos (A and B) or *fli1b−/−* insertional mutants (C and D). Embryos were analyzed at 42 hpf. Note that *fli1b−/−* mutants (C) or Etv2-inhibited embryos in wild-type background (B) are phenotypically normal, while inducible Etv2 inhibition in *fli1b−/−* background (D) results in the inhibition of intersegmental vessel angiogenesis (*arrows*), while vasculogenesis and axial vasculature development are not affected. *Reprinted from Craig, M. P., Grajevskaja, V., Liao, H. K., Balciuniene, J., Ekker, S. C., Park, J. S., et al. (2015). Etv2 and fli1b function together as key regulators of vasculogenesis and angiogenesis.* Arteriosclerosis, Thrombosis, and Vascular Biology, 35(4), *865–876.*

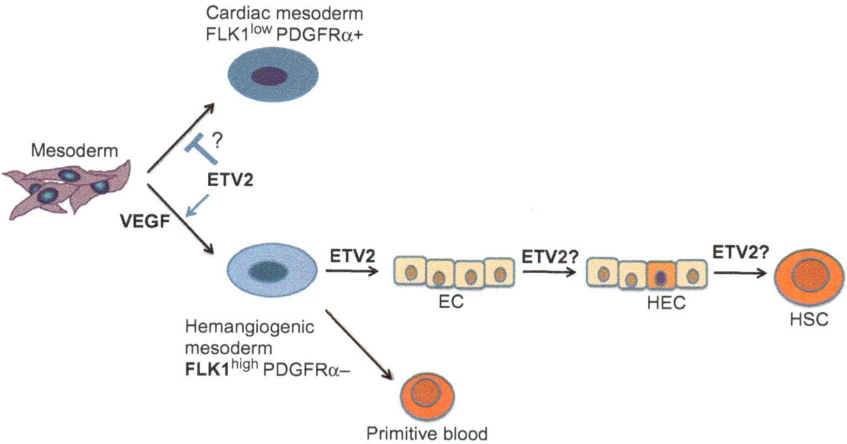

S. Sumanas and K. Choi, Fig. 4 A model depicting ETV2 function in hematopoietic and endothelial cell differentiation. EC, endothelial cell; HEC, hemogenic endothelial cell; HSC, hematopoietic stem cell.

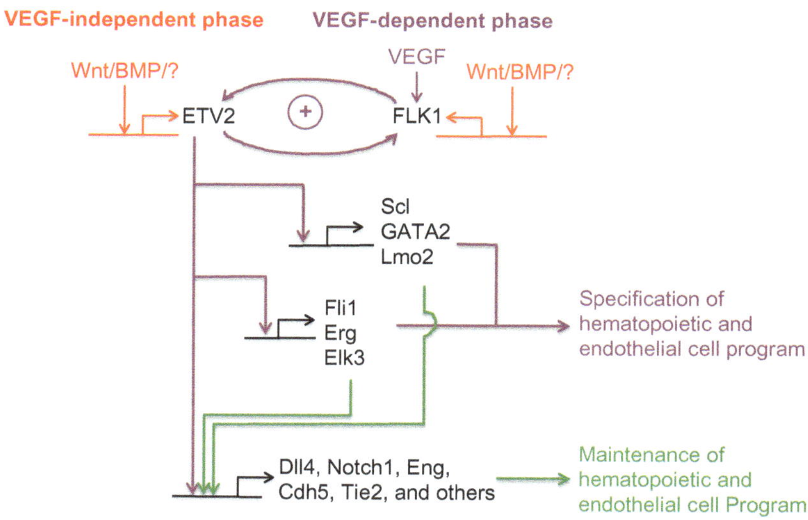

S. Sumanas and K. Choi, Fig. 5 A model depicting ETV2-mediated gene regulatory network controlling the hematopoietic and endothelial cell program.

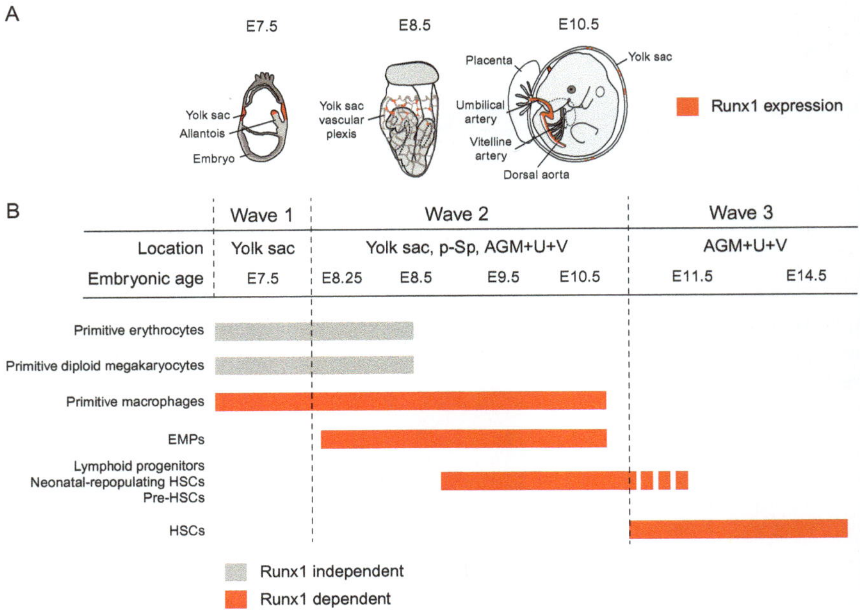

J. Tober *et al.*, Fig. 1 Waves of hematopoietic cell emergence, sites of Runx1 expression, and progenitors/HSCs dependent on Runx1 activity. (A) Schematic diagrams of (*left* to *right*) E7.5 (neural plate), E8.5 (1–4 sp stage), and E10.5 mouse embryos. Sites of Runx1 expression relevant to hematopoietic progenitor/HSC emergence are colored in red. These include the yolk sac blood islands and the distal end of the allantois in the E7.5 embryo; in hemogenic endothelial cells and small clusters of EMPs in the yolk sac vascular plexus of the E8.5 embryo (not shown is Runx1 expression at the juncture of the dorsal aorta, vitelline and umbilical vasculature); and in the ventral aspect of the dorsal aorta, the umbilical artery, vitelline artery, and in small hematopoietic clusters in the yolk sac of an E10.5 embryo. (B) The three waves of blood formation as defined by Yoder (2014). The progenitors that emerge are indicated on the *left*. Progenitors/HSCs that require Runx1 for emergence are indicated with red bars. Lymphoid progenitors were shown to be dependent on functional CBFβ (Chen et al., 2011; Kobayashi et al., 2014). Only the times and sites of progenitor/HSC emergence are shown, and the persistence of the cells in the adult mouse is not illustrated. AGM+U+V, AGM region, umbilical, and vitelline arteries.

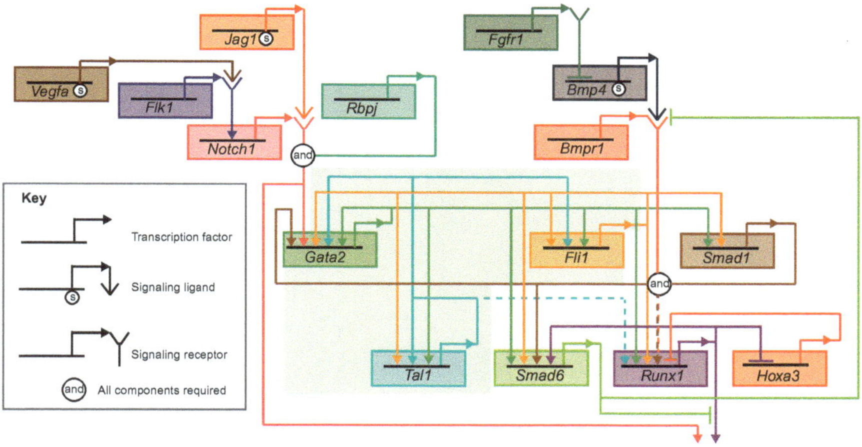

J. Tober *et al.*, Fig. 2 The gene regulatory network in the hemogenic endothelium. References for each of the interactions illustrated are indicated in the main text. The key for the GRN symbols is indicated in the *box*. Shaded in pale green is the interconnected GRN kernel consisting of *Gata1*, *Fli1*, and *Tal1* (Pimanda, Ottersbach, et al., 2007). *Dotted lines* indicate interactions assumed but not yet definitively demonstrated.

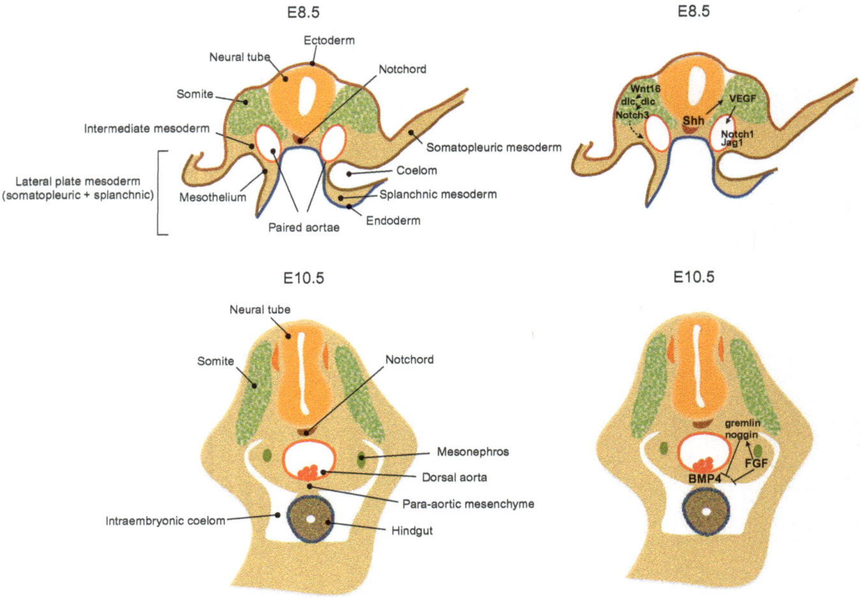

J. Tober *et al.*, Fig. 3 Signaling at the level of the P-Sp/AGM region. On top are schematic diagrams of a transverse section through an E8.5 mouse embryo with the tissues mentioned in the text indicated with labels (*left*), and the relevant signaling pathways involved in migration of the paired aortae to the midline and activation of Notch signaling in aortic endothelial cells on the *right*. Also shown is the Notch3 pathway downstream of Wnt16, which is a separate, earlier requirement for Notch signaling (Clements et al., 2011). Below are schematic diagrams of E10.5 embryos to illustrate the location of the ventral para-aortic mesenchyme (*left*), which is a source of BMP4 (*right*). In zebrafish BMP4 is negatively regulated by FGF signaling (Pouget et al., 2014).

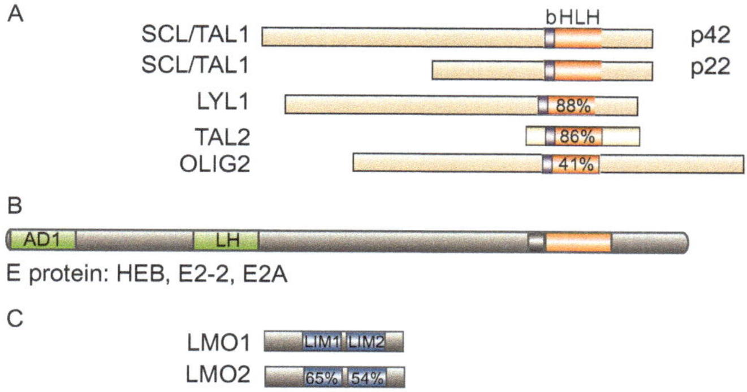

T. Hoang *et al.*, Fig. 1 *Schematic representation of hematopoietic transcription factors of the basic helix–loop–helix (bHLH) family and LIM-only (LMO) proteins.* (A) bHLH transcription factors encoded by genes that are activated by recurring chromosomal translocations in T-ALL. Two major protein isoforms are produced by translation regulation of *SCL*. Shown are the degrees of homology in the HLH domains of the various factors compared to SCL. (B) E proteins (HEB/TCF12, E2-2/TCF4 and E2A/TCF3) are relatively more ubiquitously expressed than hematopoietic bHLH factors. The *E2A* gene is involved in chromosomal translocations in B-ALL. (C) LMO1 and LMO2 proteins are encoded by genes that are activated in T-ALL, either by recurring chromosomal translocations or by retroviral integration. Shown are the homology within LIM1 or LIM2 between LMO1 and LMO2.

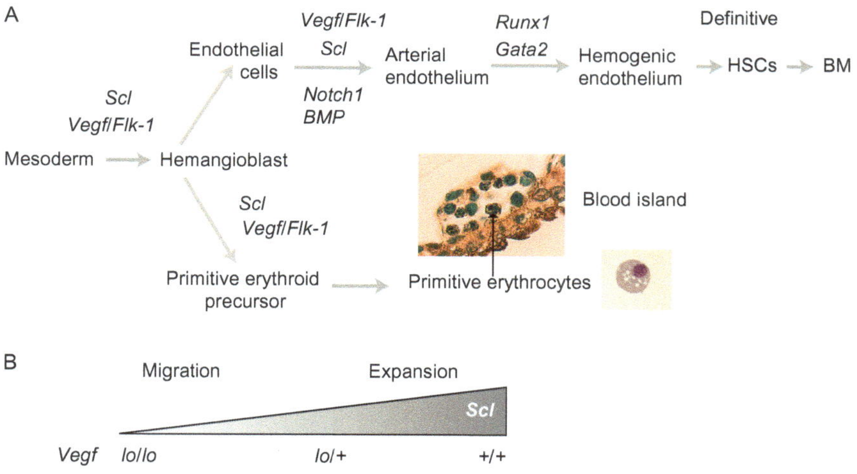

T. Hoang *et al.*, Fig. 2 *Role of Scl and Vegf during the ontogeny of the hematopoietic system.* (A) Spatial and temporal dynamics during hematopoietic development in the mouse. The lateral plate mesoderm gives rise $Flk1^+$ hemangioblasts from which originate both primitive and definitive lineages. The onset of hematopoiesis is characterized by the emergence of primitive erythrocytes within blood islands in the yolk sac (YS; primitive hematopoiesis). Hematopoietic stem cells (HSC) of the definitive myeloid and lymphoid lineages (definitive hematopoiesis) originate from the embryo proper via an endothelial to hematopoietic transition (EHT) to finally seed the bone marrow, the major site of hematopoiesis in the adult. Vegf: vascular endothelial growth factor. (B) Different *Vegf* thresholds control the migration of $Flk1^+$ hematopoietic and endothelial precursors to reach the yolk sac and their expansion during primitive erythropoiesis. *Scl* is also upregulated by VEGF signaling, in vitro and in vivo. $Vegf^{+/+}$: wild-type *Vegf* allele. $Vegf^{lo}$: hypomorphic allele.

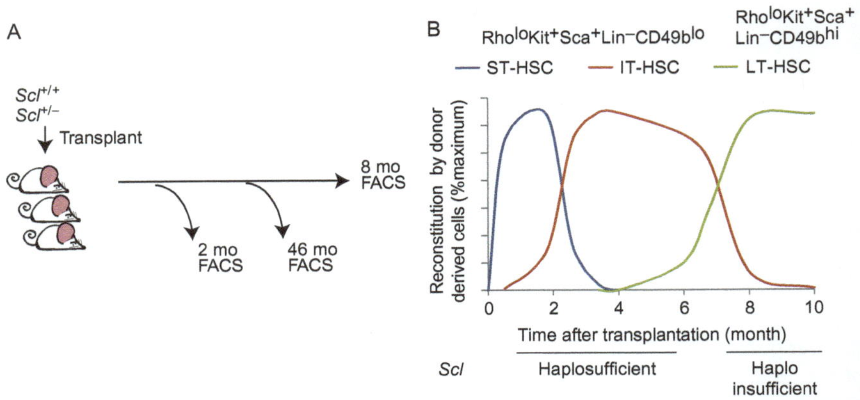

T. Hoang *et al.*, Fig. 3 *Time-line of hematopoietic reconstitution by transplanted hematopoietic stem cells.* (A) Mice are transplanted with adult bone marrow cells of different genotypes. Illustrated is a competitive assay in which HSCs of two different genotypes are transplanted in competition, typically at 1:1 ratio. Mice are sacrificed at the indicated times and analyzed for reconstitution in all hematopoietic organs, blood, bone marrow, thymus, and spleen. Test cells and competitor cells can be distinguished on the basis of CD45 alleles. (B) HSCs can be resolved into different populations based on the duration of reconstitution posttransplant. Short term HSCs (ST-HSCs) provide a transient wave of hematopoietic reconstitution (1–2 month) while long term HSCs (LT-HSCs, >4 month) can confer more durable reconstitution. These two cell types can be separated based on surface markers. Phenotypically, HSCs are enriched in the $Kit^+Sca^+Lin^-$ (KSL). ST-HSCs and LT-HSCs are found within the $CD48^+CD150^-$ and $CD48^-CD150^+$ fractions, respectively (Kiel et al., 2005). More recent evidence indicates that the population of LT-HSCs is heterogenous, containing $CD49b^{lo}$ cells that indeed provide sustained reconstitution (>8 month), but mostly $CD49b^{hi}$ cells providing intermediate reconstitution (IT-HSCs, <6 month) (Benveniste et al., 2010). *Scl* is haploinsufficient in these very long term HSCs (>8 month) while haplosufficient in IT-HSCs (Lacombe et al., 2010).

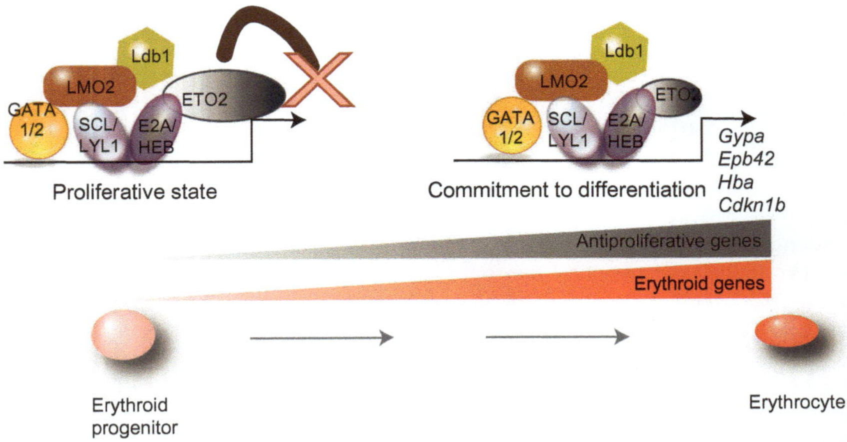

T. Hoang *et al.*, Fig. 4 *Dynamic variations in the stoichiometry of the ETO2/MTGR1 corepressor and SCL/E2A* within the SCL complex determine whether erythroid progenitors proliferate or commit to terminal differentiation. *Cdkn1b*: p27Kip; *Gypa*: glycophorin A; *Epb42*: p4.2; *Hba*: hemoglobin α (Goardon et al., 2006).

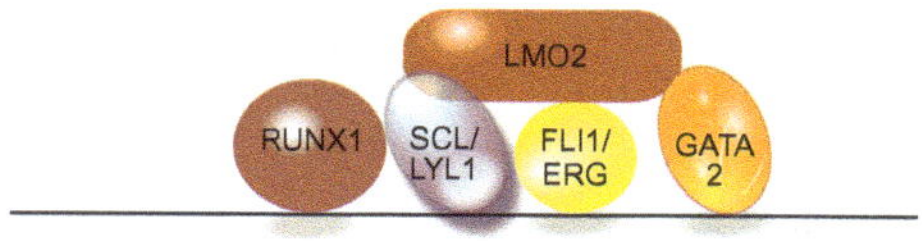

T. Hoang *et al.*, Fig. 5 *Transcription factor heptad*, as revealed by the proximity of DNA-bound peaks in multipotent HPC7 cells (Wilson et al., 2010).

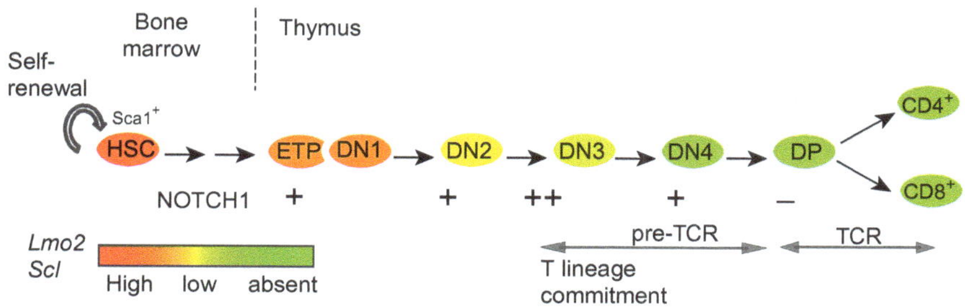

T. Hoang *et al.*, Fig. 6 *Schematic diagram of thymocyte differentiation*. Upon entering the thymus, multipotent hematopoietic progenitors that are produced in the bone marrow are guided toward a T-lineage commitment point controlled by the pre-TCR in concert with elevated Notch1 signaling at the DN3 stage (Thy1^{+}CD4^{-}CD8^{-}CD44^{-}CD25^{+}).

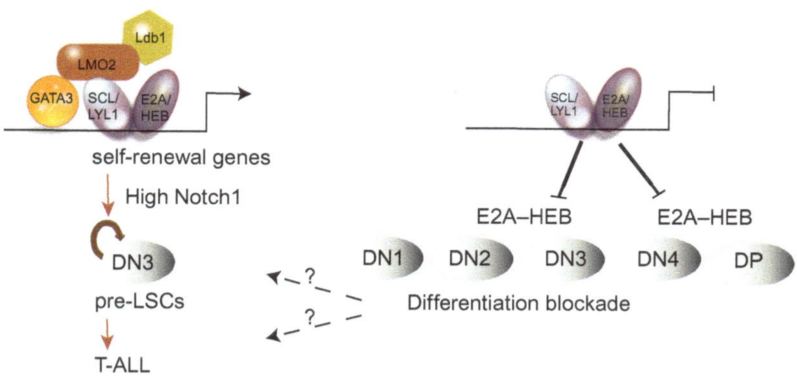

T. Hoang *et al.*, Fig. 7 *SCL acts in two parallel pathways in DN thymocytes*. SCL activates the transcription of self-renewal genes if and only if all members of the transcription complex are present. The SCL complex collaborates with high levels of Notch1 in DN3 thymocytes to reprogram these cells into self-renewing preleukemic stem cells (pre-LSCs). DN3 cells are the only targets of transformation by the SCL complex because Notch1 levels in other DN subpopulations are insufficient for collaboration. These long lasting cells acquire collaborating mutations leading to T-ALL. When bound to E2A or HEB in the absence of other transcription partners, SCL inhibits the activity of E2A/HEB heterodimers or homodimers, thereby causing differentiation blockade via decreased expression levels of E protein target genes that are required for thymocyte differentiation.

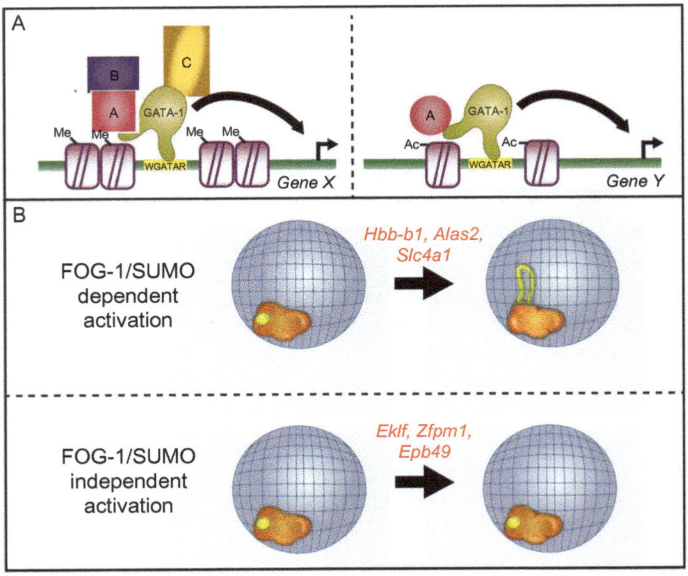

A.W. DeVilbiss *et al.*, Fig. 1 Unique coregulator requirements at distinct GATA factor target genes. (A) GATA-1 requires different combinations of coregulators to activate and repress target genes. The molecular attributes of loci that mandate these requirements are predicted to include local chromatin environment, higher order chromatin structure, and subnuclear localization. (B) GATA-1-activated genes exhibit differential coregulator requirements for localization in specific nuclear neighborhoods. FOG-1/SUMO-sensitive, GATA-1-activated genes are expelled from the nuclear periphery upon activation; whereas FOG-1/SUMO-insensitive genes remain at the periphery (Lee et al., 2009; Lee, Johnson, Boyer, & Bresnick, 2011).

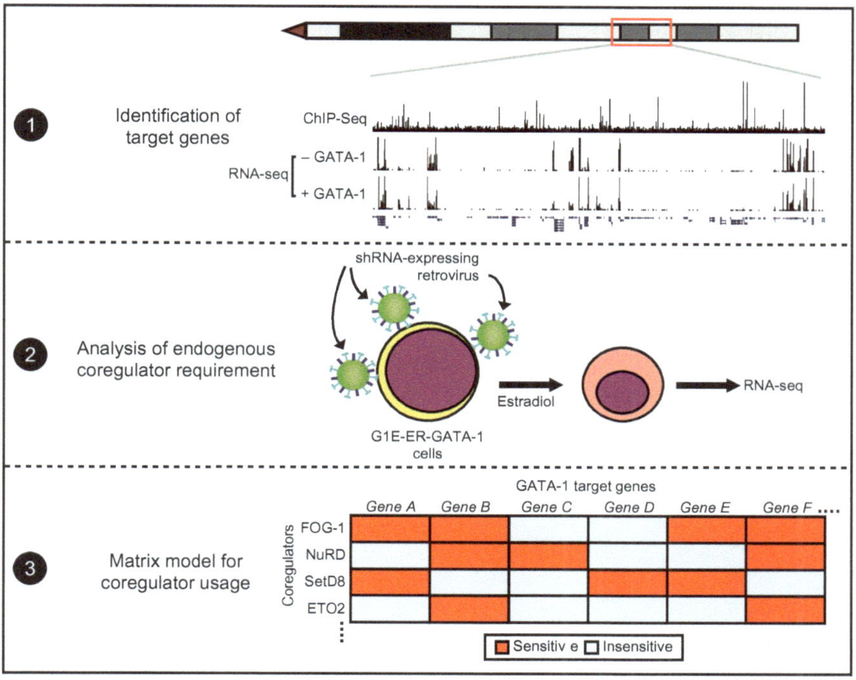

A.W. DeVilbiss *et al.*, Fig. 2 Coregulator matrix model of selective coregulator usage. Developing a coregulator matrix involves: (1) Identifying the target gene ensemble using ChIP-seq combined with genome-wide RNA profiling; (2) Using RNA interference or genome editing, coupled with expression profiling, to determine the requirement of endogenous coregulators for expression of each target gene; (3) Combine regulatory information for each coregulator at each gene into a matrix. This type of matrix constitutes an invaluable resource for identifying locus attributes linked to regulation by a particular cohort of coregulators.

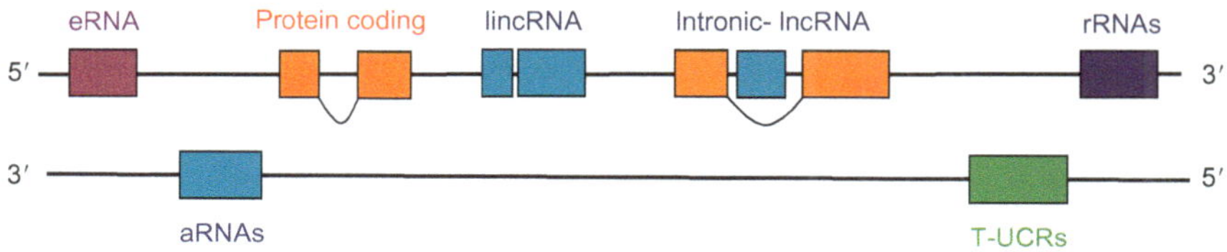

M. Jeong and M.A. Goodell, Fig. 1 Genome-wide landscape of regulatory lncRNAs. The various types of regulatory noncoding RNAs are transcribed from enhancers, promoters, introns of protein-coding or pseudogenes, antisense, or small RNAs within their transcription units. *aRNAs,* antisense RNAs; *eRNAs,* enhancer RNAs; *T-UCRs,* transcribed ultraconserved regions; *lincRNAs,* large intergenic ncRNAs; *rRNAs,* ribosomal RNAs.

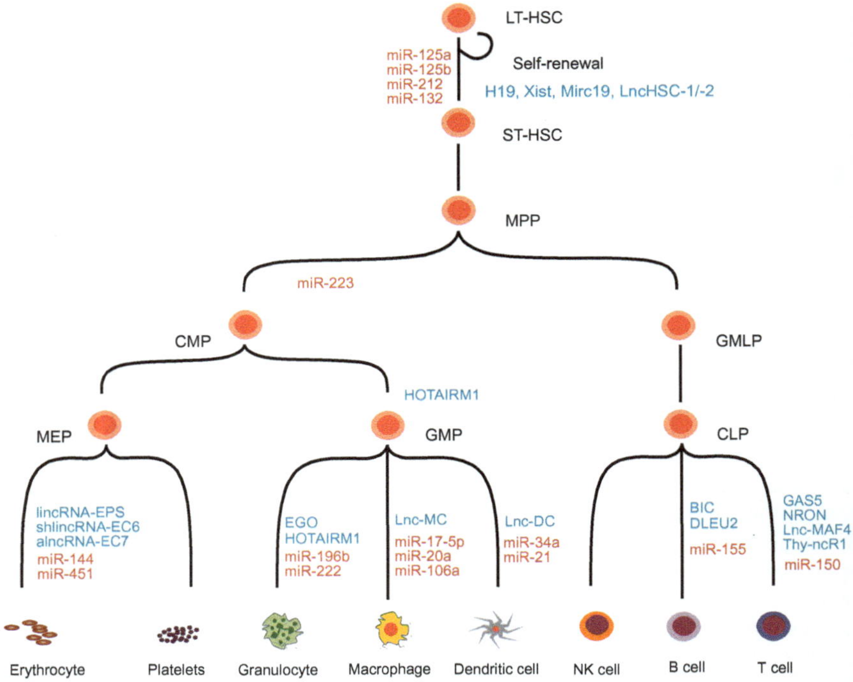

M. Jeong and M.A. Goodell, Fig. 2 The regulation of lncRNAs in hematopoietic development. LncRNAs are involved in regulating the different steps in hematopoiesis, and immune system development and activation. For instance, H19, XIST, and lncHSC-1/2 maintain long-term HSC quiescence and self-renewal. Lnc-DC controls DC differentiation by binding STAT3 (signal transducer and activator of transcription 3). NRON (noncoding RNA, repressor of NFAT) regulates the adaptive immune responses by restricting inappropriate activation of CD4 + T cells in the cytoplasm. LincRNA-Cox2 acts in the innate immune response by regulating TLR activation in macrophages and DCs. *LT-HSC,* long-term hematopoietic stem cell; *ST-HSC,* short-term hematopoietic stem cell; *MPP,* multipotent progenitors; *CMP,* common myeloid progenitor; *CMLP,* common myelo-lymphoid progenitor; *MEP,* megakaryocyte-erythroid progenitor; *GMP,* granulocyte–macrophage progenitor; *CLP,* common lymphoid progenitor.

FSC
www.fsc.org
MIX
Papier | Fördert gute Waldnutzung
FSC® C083411

Kolibri 360 GmbH
Ferdinand-Jühlke-Straße 7
99095 Erfurt, Deutschland
produktsicherheit@kolibri360.de